THE MINERALS INDUSTRIES OF ASEAN AND AUSTRALIA

THE MINERALS INDUSTRIES OF ASEAN AND AUSTRALIA: Problems and Prospects

Edited by

BRUCE MCKERN
AND PRAIPOL KOOMSUP

Sydney
ALLEN & UNWIN
Wellington London Boston

First published in 1988
Allen & Unwin Australia Pty Ltd
An Unwin Hyman company
8 Napier Street, North Sydney NSW 2060 Australia

Allen & Unwin New Zealand Ltd
60 Cambridge Terrace, Wellington, New Zealand

Unwin Hyman Limited
37–39 Queen Elizabeth Street, London SE1 2QB England

Allen & Unwin Inc.
8 Winchester Place, Winchester, Mass 01890 USA

National Library of Australia
Cataloguing-in-Publication entry:

Minerals processing in the industrialisation of ASEAN and Australia.

Bibliography.
Includes index.
ISBN 0 04 330384 6.

1. Mineral industries—Australia. 2. Mineral industries—Asia, Southeastern. 3. Manufacturing processes—Australia. 4. Manufacturing processes—Asia, Southeastern. I. McKern, R.B. (Robert Bruce). II. Praipol Koomsup.

338.2′0994

Set in 10/11pt Times by Graphicraft Typesetters Ltd, Hong Kong
Printed by Stamford Press, Singapore.

Contents

Appendixes

Tables

Appendix tables

Figures

Abbreviations

AAJRP	ASEAN–Australia Joint Research Project
AIEC	Association of Iron-ore Exporting Countries
ASEAN	Association of South-East Asian Nations
ATPC	Association of Tin Producing Countries
bpd	barrels per day
CIPEC	Conseil Intergouvernemental des Pays Exportateurs de Cuivre (Intergovernmental Committee of Copper Exporting Countries)
COMEX	New York Commodities Exchange
EC (or EEC)	European (Economic) Community
IBA	International Bauxite Association
ITA	International Tin Agreement
ITC	International Tin Council
LME	London Metal Exchange
LNG	Liquefied Natural Gas
LPG	Liquid Petroleum Gas
mmscfd	million standard cubic feet per day
mboe	million barrels oil equivalent
OPEC	Organization of Petroleum Exporting Countries
PSV	Present Social Value
RRT	Resource Rent Tax
RV	Retained Value
STOIIP	Stock tank oil initially in place
TPA	Tungsten Producers' Association
UNCTAD	United Nations Committee on Trade & Development
UNIDO	United Nations Industrial Development Organisation

Contributors

Dr Praipol Koomsup, Faculty of Economics, Thammasat University, Bangkok, Thailand

Professor Bruce McKern, Graduate School of Management, Macquarie University, Sydney, Australia

Professor Don Barnett, School of Earth Sciences, Macquarie University, Sydney, Australia

Dr Craig Emerson, Office of Prime Minister, Canberra, Australia

Dr Hank Lim, Department of Economics, National University of Singapore, Singapore

Dr Sahathavan Meyanathan, Faculty of Economics & Administration, University of Malaya, Kuala Lumpur, Malaysia

Dr Teodoro M. Santos, Department of Geology & Geography, University of the Philippines, Quezon City, Philippines

Dr M. Hadi Soesastro, Centre for Strategic & International Studies, Jakarta, Indonesia

Dr Budi S. Sudarsono, Baden Tenaga Atom Nascional, Jakarta Selatan, Indonesia

Dr John Wong, Department of Economics, National University of Singapore, Singapore

Preface

Minerals play an important role in the economies of the six countries of the ASEAN group, and also in Australia, so that a study of the economic relationships between the seven countries must consider trade and investment relationships in the minerals sector. This book is one of two volumes on the minerals industries of the region, prepared under the auspices of the ASEAN–Australia Joint Research Project (AAJRP). The AAJRP arose out of a meeting of heads of government of Australia and the original five members of ASEAN (Indonesia, Malaysia, the Philippines, Singapore and Thailand) in Kuala Lumpur in 1977. (Brunei was not a member of ASEAN until close to the end of the research phase, and was therefore not included in the research project.) Since 1981, research conducted in the six countries originally involved in the project has covered a number of areas concerned with the economic relationships between them, and one of the major sub-projects has been the study of the minerals extraction and processing industries.

The AAJRP was funded by the Australian government and administered by a Joint Steering Committee representative of all six countries. The minerals sub-project was undertaken by researchers from each country, coordinated jointly by the two editors of this volume.

This book concerns the scope of the minerals industries, trade relationships and government policies in the six countries. The major part of the book is devoted to a detailed description of the industry in each country, together with a discussion of its government's policies. These chapters constitute a reasonably complete summary of the state of each country's industry as of 1985. (Statistical data vary in date, the bulk dating from 1981 or 1982.) Three other chapters take a regional viewpoint, examining trade relationships and participation in minerals commodity agreements and contrasting government policies, while one chapter (chapter 2) sets out an economic framework for the evaluation of minerals projects, discussing *inter alia* the relevance of various government policies to the balance of costs and benefits.

The editors hope that this book, and a companion volume on minerals processing in the region, will be of interest to academic and business economists whose work is concerned with the ASEAN countries and Australia and the relationships between them, as well as to mining industry executives and government policymakers in developing countries.

The editors acknowledge with gratitude the assistance of the members of the Joint Steering Committee of the AAJRP in encouraging and directing the efforts of the many contributors to this book. We also acknowledge the support of the six governments in providing a helpful environment for research, and the assistance of officials in the Australian Development Assistance Bureau and the governments of Thailand and Malaysia in smoothing the flow of funds.

We are especially indebted to Professor Heinz Arndt for critical comment and firmness of view, both at seminars in which drafts were exposed and in his reading of chapters. To Hal Hill and Ross Garnaut we extend thanks for managing the administration of the AAJRP in an efficient and enthusiastic manner, and for their criticism and encouragement at every stage. We also thank Florian Pantazi for early editorial assistance and Kate Brown for her thorough editorial work in bringing the manuscripts to their present state. Prue Phillips provided valuable statistical output from the Project's International Economic Data Bank (hereafter referred to as IEDB), and 'All Type Secretarial Services' typed the final drafts with diligence.

These important contributions notwithstanding, the authors of each chapter, abetted by the editors, are alone responsible for any deficiencies the work may have.

R.B. McKern, Sydney
Praipol Koomsup, Bangkok
March 1986

1 Minerals in the economies of the ASEAN countries and Australia

BRUCE MCKERN

THE region which encompasses the ASEAN countries and Australia is rich in minerals, accounting for a quarter of the world's nickel, bauxite and uranium reserves, a major part of its tin and iron reserves and substantial resources of petroleum, natural gas and coal. There are also important deposits of copper, gold, silver, lead and zinc, as well as titanium, zirconium, tantalum, manganese, antimony, cobalt, tungsten and phosphate.

However, the endowment of minerals is unevenly distributed amongst the countries of the region, providing a basis for comparative advantage in trade between them. Indonesia has by far the largest reserves of petroleum and shares with Thailand and Malaysia a large fraction of the world's tin (some 58 per cent). Australia has major reserves of iron ore, coal, bauxite, copper, nickel, uranium, lead and zinc, but lacks self-sufficiency in petroleum.

The minerals industry is of varying importance to the countries of the region. For Indonesia it is crucial, accounting for a quarter of GDP (Table 1.1) and 84 per cent of exports (Table 1.2). The industry is also significant for Singapore as a processor of imported minerals, despite

Table 1.1 Relative importance of minerals sectors, 1980

	Value of output (US$ million)	Share of GDP (per cent)
Indonesia	17 000	26.0
Malaysia	1 170	4.6
Philippines	1 660	4.7
Thailand	660	2.1
Singapore	1 200	13.0
Australia	7 500	5.2

Source: IEDB, AAJRP

Table 1.2 Percentage shares of minerals in trade

	Exports		Imports
	1960	1981	1981
Indonesia	33.0	83.6	30.5
Malaysia	20.0	36.2	30.6
Philippines	10.0	16.9	39.1
Singapore	1.0	37.3	42.2
Thailand	7.0	9.8	43.6
Australia	25.0[a]	36.9	20.8

Note: [a] 1970
Source: IEDB, AAJRP

its total lack of indigenous minerals, accounting for 13 per cent of GDP and 37 per cent of export trade.[1]

For Australia, Malaysia and the Philippines, minerals are of approximately equal relative importance to national product (about 5 per cent of GDP), while in Thailand they are relatively minor, contributing only 2 per cent. Even for Thailand, however, minerals, mainly tin (Table 1.3) are important for the trade balance, contributing 10 per cent of exports. For the others, mineral exports are vital, ranging from 17 per cent of total exports in the Philippines to around 37 per cent for Malaysia and Australia. Brunei is almost totally dependent on exports of petroleum and its derivatives.

Characteristics of the minerals sector

Before going further, it is helpful to review the key characteristics of the minerals industry which greatly influence trade and development possibilities and national policies towards the sector.

The most obvious characteristic is that minerals differ considerably from each other in the ways they are deposited by nature, discovered by man, mined, processed, refined and used. Some minerals are mined predominantly by underground methods (lead, zinc), some mainly by open-cut surface mining (iron ore, bauxite), some by either method (coal, copper, nickel), some mainly by alluvial methods (tin), petroleum (and often sulphur) by sinking wells, and salt by underground mining or surface evaporation. Some minerals require little processing before their ultimate use (coal, phosphate, rich grades of iron ore), while most require several processing steps. The metals usually require concentration (lead, zinc, tin, copper, nickel) or refining (bauxite, nickel) and smelting (lead, zinc, copper, aluminium, nickel, tin), which are steps required to produce the elements in a concentrated and usable form.

A second feature of the sector is the fortuitous distribution of

Table 1.3 Principal exports of minerals and mineral products, 1981

	Mineral	Percentage of total minerals exports
Indonesia	Petroleum	77
	LNG	18
	Tin	2
	Nonferrous ores	2
Malaysia	Petroleum	73
	Tin	22
Philippines	Nonferrous ores, including	
	copper	55
	nickel	11
Singapore	Petroleum products	72
	Petrochemicals	12
	Tin	3
Thailand	Tin	63
	Nonferrous ores	8
	Precious metals, gems	6
Australia (1980)	Coal	25
	Iron ore, iron and steel	19
	Alumina, aluminium	16
	Lead	8
	Nickel	6
	LPG	5
	Zinc	4
	Copper	4

Source: IEDB, AAJRP

mineral deposits around the surface of the earth. The major markets are in the northern hemisphere while deposits are widely distributed, leading to south–north trade of mineral materials. Since for most minerals the grade of metal in the ore is low compared to its desired concentration, there is a strong incentive to undertake some processing close to the mine to reduce weight and transport costs. Thus lead, zinc, nickel, tin and copper are concentrated close to the mine, and often smelted there as well.

A further characteristic is the capital intensity and energy intensity of minerals exploitation. Minerals projects are not large generators of employment, although there is some scope for labour–capital substitution (in the case of tin, for example). Value-added per worker is typically high; approximately US$0.5 million per worker in the case of alumina and aluminium.[2] Energy requirements are high and the form of energy required specific to each mineral—coking coal in the case of iron, electricity for smelting aluminium. The requirements of infrastructure are also usually high and costly, since mining projects are often located in undeveloped areas remote from existing facilities.

Another characteristic of the sector is that it is strongly influenced by technological change, a feature which tends to increase capital intensity and reduce energy requirements. Technological change is also at work on the demand side, seeking cheaper substitutes for existing minerals and, at the same time, new applications. Substitution is a particularly important factor in the relative demand for aluminium, copper and steel, and also in demand for petroleum as compared with other energy forms.

Mineral operations tend to exhibit considerable economies of scale (although recent technological change has offset this feature of the steel industry). For reasons of operational economies and (more importantly) control and risk reduction, there is also a high degree of vertical integration, varying with the mineral. Partly as a consequence, many segments of the minerals industry have been highly concentrated in the past, usually amongst developed economy firms, and it is only in the past decade that major shifts of ownership have occurred. The diffusion of technology and greater availability of capital have enabled new enterprises to enter many segments of the industry. Thus there has been some fragmentation of hitherto oligopolistic structures in copper, oil, nickel and iron ore. Often, state ownership has replaced private ownership, with moderate change in industry concentration. In the case of iron ore, the world structure resembles one of bilateral oligopoly, dominated on the demand side by Japan (which imports 38 per cent of world traded iron ore) and on the supply side by Australia, Brazil and Canada (Vernon and Levy, 1982).

Foreign-based ownership remains an important feature of the minerals industry, with relevance for a country's fiscal revenues, the marketing of its output and the discovery of new deposits. Deposits of minerals are of no value until discovered, and are then of value only if they are exploitable at a cost sufficient to provide an adequate return on capital at prevailing prices. Thus the definition of what mineral resources can be classed as 'reserves' is entirely dependent on extraction costs relative to prices. As both of these variables change, a country's economic 'reserves' of any mineral will also change and, in consequence, the degree of exploitation and the intensity of exploration for new deposits.

One consequence of the characteristics described above has been a relatively low degree of downstream processing in developing countries where minerals are mined. This has also been the case in Australia. In the past decade, downstream processing has been favoured by many countries as a means of increasing 'retained value', that is, the domestic value added arising from mineral production. This issue is addressed in chapter 3 in more depth; here, it is sufficient to say that there is no a priori justification for downstream processing—each case needs to be considered on its cost–benefit merits. The theoretical arguments concerning minerals processing are canvassed fully in a chapter prepared for the AAJRP (Wall, 1987) which, together with others describing the minerals processing industry of each country in detail, will be published as a companion volume to this book (McKern and Praipol, 1987).

Minerals in the developing countries

Minerals arouse special interest amongst government policymakers, particularly in developing countries, and later in this section we will touch on the reasons for this interest, which will be elaborated in later chapters. One of the chief reasons for this interest is their relative importance in the output and export trade of many developing countries, as is the case for the six ASEAN members. Fuel minerals (mainly petroleum) are, of course, the most important by far in value terms for the developing economies as a whole, although the resources are most unevenly endowed among the Third World countries. Amongst non-fuel minerals, 85 per cent of the value of world output is attributable to nine commodities: iron ore, bauxite, copper, manganese, lead, nickel, zinc, tin and phosphate (Bosson and Varon, 1977). All of these figure prominently in the minerals industry of the ASEAN–Australia region.

Developing countries control important fractions of the world's mineral resources. Petroleum's importance to the countries of OPEC is well known, permitting them to effect a massive shift of financial resources in their favour during the 1970s. Although OPEC's strength was greatly diminished in 1984 and 1985, largely as a result of the worldwide recession induced by its pricing policies, there is every possibility of a renewal of its power in the 1990s in a world seduced by lower short-run energy prices to abandon its search for alternative energy sources. In tin, developing countries control some 80 per cent of reserves (the majority in ASEAN), and in several other minerals they control large proportions: bauxite (55 per cent), copper (50 per cent), nickel (48 per cent), phosphate (53 per cent), manganese (31 per cent), and iron ore (30 per cent) (Table 1.4). While it has not been possible for developing countries to avoid the impact of recession on their income from minerals, their importance as owners of large resources and their concern to maximise the benefits of minerals exploitation combine to ensure the importance of minerals in their development programs. Chapter 4 considers ASEAN and Australian participation in commodity agreements in detail.

Minerals trade in the ASEAN–Australia region

In Australia, the development of the minerals industry in the 1960s and early 1970s made a strong contribution to economic growth and the balance of payments, stimulating revaluation of the Australian dollar.

Table 1.4 Mineral reserves of developing countries

Mineral	Percentage of world reserves
Iron ore	30
Bauxite	55
Copper	50
Manganese	31
Lead	21
Nickel	48
Zinc	16
Tin	80
Phosphate	53

Note: Based on estimated reserves recoverable at prices considered to be applicable over the long term.
Source: Bosson and Varon (1977)

Table 1.5 Minerals trade, total, 1971 and 1979 (crude and processed minerals, and metal manufactures; US$ million)

	Exports		Imports	
	1971	1979	1971	1979
Indonesia	538	18 610	218	3 976
Malaysia	492	4 245	370	3 520
Philippines	249	967	328	3 311
Singapore	534	7 832	650	11 640
Thailand	125	669	342	4 384
ASEAN group	1 940	32 324	1 907	26 832
Australia	1 568	7 909	628	4 890

Source: IEDB, AAJRP

More recently Australia's mineral exports have been heavily affected by recession among its major customers, and by Japan's success in diversifying its sources of supply. Despite the discovery and exploitation of off-shore petroleum reserves, Australia remains dependent on imported petroleum for part of its needs (mainly heavier petroleum fractions) and, as a consequence, minerals imports account for some 20 per cent of its total imports.

The other ASEAN economies are even more heavily dependent on imports of minerals or mineral products (except for Brunei).[3] Minerals constitute 30 per cent of the total imports of Indonesia and Malaysia, and around 40 per cent of the imports of the Philippines, Singapore and Thailand (see Table 1.2). In each case (including Indonesia) the predominant mineral imported is crude petroleum or petroleum products (Table 1.6). The second most important mineral import in every case (including Australia) is steel and steel products. (The trade data employed in the research includes not only iron ore, iron and steel, but steel shapes, structural items and fabricated items such as tanks.[4]) Not

Table 1.6 Principal imports of minerals and mineral products, 1981

	Mineral	Percentage of total minerals imports, all sources
Indonesia	Petroleum, crude	17
	Petroleum products	26
	Steel products	33
	Fertiliser (manufactured)	7
Malaysia	Petroleum, crude	25
	Petroleum products	30
	Nonferrous ores	4
	Fertiliser (manufactured)	4
	Steel products	19
Philippines	Petroleum, crude	65
	Petroleum products	11
	Steel products	12
	Fertiliser (manufactured)	4
Singapore	Petroleum, crude	71
	Petroleum products	9
	Steel products	11
Thailand	Petroleum, crude	49
	Petroleum products	17
	Steel products	15
	Fertiliser (manufactured)	4
Australia	Petroleum, crude	36
	Petroleum products	29
	Steel products	13
	Fertilisers (crude and manufactured)	3

Source: IEDB, AAJRP

surprisingly, Indonesia is a major supplier of crude petroleum to Australia, Singapore and the Philippines, although its share in each country's crude imports is not as great as its proximity implies—20 per cent in the case of Australia, 14 per cent for Singapore, and 9 per cent for the Philippines. Singapore's large refining industry makes it an important source of refined petroleum products for Australia (30 per cent of refined imports), Indonesia (88 per cent) and Malaysia (90 per cent). It is clear that Indonesia's rich petroleum resources and relative proximity to Australia and the other ASEAN members have not been sufficient for it to dominate their energy supplies. The same position is evident in the steel sector. Despite its rich reserves of iron ore and coking coal, Australia supplies very little of the steel and steel products imported by ASEAN. Japan's efficient steel industry is able to process iron ore shipped from Australia and Brazil into steel products and gain the major share of imports of these products in ASEAN, and also in Australia. Both South Korea and the EEC are more important suppliers to ASEAN than is Australia. The reasons for these curious patterns of trade are explored in later chapters, particularly chapter 2.

The preceding discussion suggests that minerals trade amongst the ASEAN countries, and between ASEAN and Australia, is not of primary importance in the total minerals trade of each country. This is not quite the case. Minerals trade between ASEAN members represents 17 per cent of total ASEAN imports and a slightly smaller percentage of total ASEAN exports. Minerals trade in ASEAN is somewhat more intense than overall trade amongst its members, which constitutes 13 per cent of their total imports and 17 per cent of total exports. The trade between ASEAN members is, however, dominated by crude refined petroleum, and if these commodities are excluded intra-regional minerals trade is quite small.

Likewise, although ASEAN exports of minerals to Australia were US$843 million in 1981, and its mineral imports from Australia US$674 million, these figures are only 3 per cent and 2.5 per cent respectively of total trade between ASEAN and its world trade partners (Table 1.7). Australia is thus comparatively insignificant for the ASEAN grouping as a trading partner in minerals, but of some

Table 1.7 ASEAN–Australia minerals trade, 1981 (US$ million and percentage)

Principal commodities	ASEAN exports to Australia[a]	ASEAN imports from Australia[b]
Petroleum, crude	374	–
Petroleum products	432	101
Iron ore, iron and steel	13	178
Nonferrous ores and concentrates	1	106
Copper	–	25
Aluminium	–	39
Lead	–	75
Zinc	–	41
Tin	2	2
	843[c]	674[c]
Total ASEAN mineral exports to world		32 324
Percentage to Australia		2.6
Total ASEAN mineral imports from world		26 832
Percentage from Australia		2.5
Total Australian mineral exports to world		7 909
Percentage to ASEAN		8.3
Total Australian mineral imports from world		4 890
Percentage from ASEAN		17.2

Notes: [a]Based on imports reported by Australia
[b]Based on imports reported by ASEAN countries
[c]Totals include small quantities of minerals not listed, such as minor metal manufactures, gold, silverware and jewellery.
Source: IEDB, AAJRP

importance as an export market for Indonesia's crude and Singapore's refined petroleum. On the other hand, ASEAN is a rather more important export market for Australian mineral products (taking 8.3 per cent of Australia's minerals exports) and a very important source of its minerals imports (17.2 per cent of such imports; overwhelmingly in the form of petroleum).

The preceding discussion understates, however, the great importance of minerals in ASEAN's overall trade with Australia. ASEAN's total trade with Australia is quite small by comparison with its world trade (2.2 per cent of its exports of all commodities and 7.5 per cent of its imports), but minerals account for 57 per cent of its exports to Australia and 33.6 per cent of its imports (Table 1.8). Thus, while integration of the ASEAN and Australian economies has not yet progressed very far, a few important minerals (petroleum, steel, steel products and tin) hold the major share of the bilateral trade between the two. This is one of the reasons for the research undertaken on this sector under the AAJRP.

Policies towards the minerals industry

Minerals are regarded by most countries as national assets and their exploitation is controlled in various ways. Chapter 3 sets out a framework for analysis of minerals projects which integrates most of the elements of public policy in a cost–benefit framework. At this point, it is sufficient simply to note that government policy has in the last decade focused on maximising revenue earned from minerals exploitation, but often by seeing existing minerals operations as wealth to be redistributed, rather than to be enlarged. Thus there has been great emphasis on taxation measures, including the novel resource rent tax, domestic ownership, further processing, and export pricing. These preoccupations are understandable. The development of the Australian industry in the 1960s, however, demonstrates the relevance of a favourable policy climate to the discovery and development of mineral deposits, a necessary precursor to redistribution of mineral wealth (McKern, 1976). Chapter 3 canvasses a number of policies adopted by

Table 1.8 Minerals trade: importance in total ASEAN–Australia trade, 1981 (US$ million and percentage)

	ASEAN exports	ASEAN imports
Total trade with world	67 523	26 832
Total trade with Australia	1 475	2 008
Australian trade as percentage ASEAN total	2.2	7.5
Minerals trade with Australia	843	674
Australian minerals trade as percentage of total trade	57.1	33.6

Source: IEDB, AAJRP

governments in the attempt to maximise benefits. It is important to remember that investors are risk-averse, and to compare the policy regimes of different countries in assessing the desirability of investing (when, as is often the case, there are alternative sites for mining available). What may seem an optimal distributional policy for one country, viewed statically and independently, may not optimise the discovery and development of deposits in the longer term, when viewed over time and by comparisons with other countries (Ball, 1984). Government policy towards the mining industry has undergone some evolution during the last two decades. In the Australian case, the comparatively laissez-faire policy of the 1960s (with specific fiscal incentives) gave way to a chauvinistic and distribution-oriented policy in the 1970s, to be followed more recently by a more rationally based, yet interventionist approach. Indonesia had experimented with several approaches, in three broad phases, with more recent emphasis on flexibility, while Singapore has maintained a free-market orientation towards minerals processing from the outset. These issues are discussed in more depth in the chapters dealing with the industry of each country (chapters 5–10) under the headings of fiscal policy (including pricing of minerals), policies towards foreign investment, downstream integration, domestic linkages, localisation of workforce and management, and environmental effects. Our findings on these and other issues are summarised in chapter 11.

Appendix 1.1

Mineral commodities of ASEAN and Australia included in statistical tables, by SITC Code

SITC Code	*Commodity*
271	Fertilisers, crude
276	Other crude minerals
281	Iron ore, concentrates
282	Iron and steel scrap
283	Nonfer base mtl ore, conc
284	Nonferrous metal scrap
285	Silver and platinum ores
286	Uranium, thorium ore, conc
321	Coal, coke, briquettes
331	Crude petroleum, etc
32	Petroleum products
341	Gas, natural and manufactured
521	Coal, petroleum etc, chems
561	Fertilisers, manufactured
663	Other non-metal mineral, mfs
671	Pig iron, etc
672	Iron, steel, primary forms
673	Iron and steel shapes
674	Iron, steel univ., plate, sheet
675	Iron, steel hoop, strip
676	Railway rails etc, iron, steel
677	Iron, steel wire, excluding wire rod
678	Iron, steel tubes, pipes etc
679	Iron, steel castings, unworked
681	Silver, platinum etc
682	Copper
683	Nickel
684	Aluminium
685	Lead
686	Zinc
687	Tin
688	Uranium, thorium, alloys
689	Nonfer base metals, nes
691	Structures and parts, nes
692	Metal tanks, boxes, etc
693	Wire products, nonelectr
694	Steel, copper nails, nuts, etc
695	Tools
696	Cutlery
697	Base metal, household equipment

698	Metal manufactures, nes
897	Gold, silverware, jewellery

nes—not elsewhere specified

Notes

1 This figure is a little misleading in that minerals account for 42 per cent of Singapore's imports, there being no domestic extraction of minerals (other than minor building materials).
2 Based on data for 1982 in Brown and McKern (1986).
3 Brunei's petroleum wealth and small size tend to discourage local manufacture and encourage importation of manufactured goods, rather than mineral raw materials.
4 Where statistics are quoted for trade amongst the six countries, they are usually based on the AAJRP's International Economic Data Bank (IEDB), Australian National University, Canberra. The commodities included in the statistics quoted are listed by SITC code in Appendix 1.1

References

Ball, R. (1984) *The Resource Rent Tax: A Penalty on Risk-Taking* Sydney: Centre for Independent Studies

Bosson, R. and B. Varon (1977) *The Mining Industry and the Developing Countries* New York: World Bank and Oxford University Press

Brown, M. and R.B. McKern (1986) *Minerals Processing, Technology and Economics for Developing Countries* Paris: OECD

McKern, R.B. and Praipol Koomsup eds (1987) *Minerals Processing in the Industrialisation of ASEAN and Australia* Sydney: Allen & Unwin, for the AAJRP (in preparation)

McKern, R.B. (1976) *Multinational Enterprise and Natural Resources* Sydney: McGraw–Hill

Vernon, R. and B. Levy (1982) 'State-Owned Enterprises in the World Economy: The Case of Iron Ore' in L.P. Jones (ed.) *Public Enterprise in Less Developed Countries: Multidisciplinary Perspectives* New York: Cambridge University Press

Wall, D. (1987) 'Comparative Advantage in Minerals Processing' in McKern and Praipol *op. cit.*

2 The ASEAN–Australian trade pattern in fuels and minerals

PRAIPOL KOOMSUP

THE objective of this chapter is to examine the trade pattern of ASEAN and Australia in fuels and minerals during the 1970s. The next section examines the shares of fuels and minerals in international trade conducted by ASEAN and Australia, as well as the fuels and minerals trade balances of these countries. This is followed by an analysis of the direction, dependence, concentration and intensity of fuels and minerals trade by ASEAN and Australia. The export performance of ASEAN–Australian trade in fuels and minerals is then evaluated by the constant market share analysis method. Finally, the future prospects of fuels and minerals trade among the ASEAN countries and Australia are discussed.

Trade share and trade balance

During the 1970s international trade in fuels and minerals formed a significant portion of total trade flows for the world, as well as for ASEAN and Australia. For the whole decade, the share of fuels and minerals in world trade was 28 per cent (Table 2.1). Their shares in total exports ranged from 10 per cent for Thailand to a very high level of 70 per cent for Indonesia. The corresponding figures for the ASEAN group and Australia were 39 per cent and 32 per cent respectively. On the import side, fuels and minerals constituted only 12 per cent of Australia's total imports, compared with 28 per cent for ASEAN. For Thailand, the Philippines and Singapore, the share of fuels and minerals in imports was around 30 per cent. Between 1970 and 1979, the share of fuels and minerals in total imports rose significantly for every ASEAN country and for Australia. However, the only country in the group to experience significant increases in the share of fuels and minerals in total exports was Indonesia. For other countries, this share either remained constant or declined over the period.

Table 2.1 Trade in fuels and minerals by ASEAN and Australia, 1970–79

	Fuels and minerals exports				Fuels and minerals imports			
	1970	1979	1970–79	Annual growth rate (%)	1970	1979	1970–79	Annual growth rate (%)
Value (US$ million)								
-Indonesia	467	10 558	48 091	35	11	1 471	8 045	54
-Malaysia	509	3 301	13 980	21	372	2 474	9 102	21
-Philippines	251	858	4 076	14	340	2 133	9 566	20
-Singapore	407	3 955	19 254	25	721	5 823	24 463	23
-Thailand	131	640	2 428	18	313	2 317	10 660	22
-ASEAN	1 765	19 312	88 549	27	1 757	12 218	59 836	23
-Australia	1 597	5 629	34 558	13	489	2 633	12 933	19
-World	65 331	457 030	2 198 000	22	65 331	457 030	2 198 000	22
Share in total exports or imports of each country (%)								
-Indonesia	42	68	70	–	1	20	20	–
-Malaysia	30	30	30	–	26	32	25	–
-Philippines	24	19	17	–	27	32	29	–
-Singapore	26	28	32	–	29	33	30	–
-Thailand	18	12	10	–	24	32	32	–
-ASEAN	29	38	39	–	24	31	28	–
-Australia	33	28	32	–	10	14	12	–
-World	23	30	28	–	23	30	28	–

Source: The source for all the tables in this chapter is the International Economic Data Bank (IEDB), Australian National University.

It should be noted at this point that, among all minerals traded internationally by ASEAN and Australia, fuel products, mainly crude oil and petroleum products, have played a dominant role. As seen in Table 2.2, the weights of fuel products to all minerals in intra-ASEAN trade in ASEAN's total exports and imports, in Australia's total imports, in ASEAN's exports to Australia, and in total world trade were at least 60 per cent. The only exceptions were in the cases of Australia's exports to the rest of the world and ASEAN's imports from Australia, where fuel minerals accounted for about one-third of all mineral products. Crude oil and petroleum products were consistently two of the three largest mineral import items for all ASEAN countries and for Australia (Table 2.3). It is interesting to observe that, even for net oil exporters like Indonesia and Malaysia, these two products were quite significant imports (in a refined form). On the export side, crude oil and petroleum products were very important foreign exchange earners for Indonesia and Malaysia. As a regional oil refining centre, Singapore derived a large part of its export earnings from petroleum products, produced from imported crude oil. The monetary importance of crude oil and petroleum products in world trade as well as in trade by ASEAN and Australia was brought about by dramatic increases in oil prices during the 1970s. The increasing value of these fuel products in international trade was the main factor which increased the share of fuels and minerals in the imports of ASEAN and Australia; it was also the main contributor to their increased share in the exports of Indonesia (the only OPEC country in the group).

Table 2.4 shows the trade-balance positions of ASEAN and Australia in fuels and minerals during the 1970s compared with the rest of the world. It is clear that both ASEAN and Australia were net exporters of these products, with a combined export surplus of over US$50 billion in that decade. For ASEAN the surplus was a result of Indonesia's large net exports—particularly of crude oil, which amounted to over US$40 billion. Malaysia was another net exporter, with sizeable surpluses in nonferrous metals (mainly tin) and crude oil. The Philippines, Singapore and Thailand were clearly net importers of the products, with a combined trade deficit of US$15 billion in the 1970s.

Once fuels trade is excluded, ASEAN is seen to be a net *importer* of non-fuel minerals, while Australia remains a net exporter of US$18 billion worth of non-fuel minerals. In these minerals all the ASEAN countries, except Malaysia and the Philippines, experienced a trade deficit with the rest of the world. These two ASEAN countries managed to gain export surpluses in non-fuel minerals because of large net exports, of tin in the case of Malaysia, and copper in the case of the Philippines.

Australia had export surpluses for most of the six categories of fuels and minerals, the largest belonging to metallic ores and concentrates, particularly iron ore and concentrate. ASEAN as a whole was a net

Table 2.2 Proportions of individual mineral groups to total minerals trade, 1970–79 (per cent)

	World trade	Intra-ASEAN trade	ASEAN's exports to world	ASEAN's imports from world	Australian exports to world	Australian imports from world	ASEAN's exports to Australia	ASEAN's imports from Australia
Crude minerals	1.62	1.05	0.42	1.52	1.24	4.60	0.26	1.62
Ores, concentrates and scrap of metals	6.11	2.83	6.82	2.95	38.53	1.08	0.34	11.61
Fuel minerals	63.60	88.48	80.76	62.12	33.34	66.32	96.94	32.30
Semiprocessed iron and steel products	16.93	4.01	1.27	21.53	10.66	21.42	1.57	20.67
Nonferrous metals	8.87	1.22	10.01	5.73	15.85	3.72	0.03	32.46
Some mineral manufactures	2.87	2.41	0.72	6.15	0.38	2.86	0.86	1.34
Total mineral products	100.00	100.00	100.00	100.00	100.00	100.00	100.00	100.00

Source: See Table 2.1

Table 2.3 The three most important export and import items in fuels and minerals trade, ASEAN and Australia, 1970–79

	Rank	Indonesia	Malaysia	Philippines	Singapore	Thailand	Australia
Exports	1	Crude oil (84)	Crude oil (51)	Copper ore and concentrate (73)	Petroleum products (87)	Tin metal (67)	Coal (27)
	2	Petroleum products (7)	Tin metal (41)	Iron ore and concentrate (9)	Iron and steel tubes, pipes and fittings (2)	Tungsten ore and concentrate (14)	Iron ore and concentrate (23)
	3	Natural gas (4)	Petroleum products (5)	Nickel metal (7)	Nickel metal (7)	Fluorite (8)	Ores and concentrates of nonferrous base metals: copper, nickel, bauxite lead, and zinc (13)
Imports	1	Petroleum products (22)	Crude oil (29)	Crude oil (50)	Crude oil (61)	Crude oil (34)	Crude oil (38)
	2	Iron and steel universals, plates and sheets (15)	Petroleum products (24)	Iron and steel in ingots and primary forms (9)	Petroleum products (15)	Petroleum products (17)	Petroleum products (28)
	3	Crude oil (14)	Iron and steel universals, plates and sheets (9)	Iron and steel universals, plates and sheets (8)	Iron and steel tubes, pipes and fittings (5)	Iron and steel universals, plates and sheets (15)	Iron and steel universals, plates and sheets (9)

Notes: Items are classified by 3-digit SITC (see also note 3 to Table 2.4). The relevant SITC codes are listed in Appendix 1.1.

Figures in parentheses are the shares of the items in exports and imports of all fuels and minerals by each country.

Three most important items in world trade of fuels and minerals are crude oil, petroleum products, and iron and steel universals, plates and sheets. The shares of these items are 44 per cent, 14 per cent and 5 per cent, respectively.

Source: See Table 2.1

Table 2.4 Net exports and imports of fuels and minerals by ASEAN and Australia, 1970–79, and importance in world trade (US$ million and per cent)

	Indonesia	Malaysia	Philippines	Singapore	Thailand	ASEAN	Australia
Crude minerals	−134 (0.39)	−263 (0.77)	−144 (0.43)	−23 (0.45)	18 (0.57)	−546 (2.55)	−165 (1.67)
Ores, concentrates and scrap of metals	1314 (1.01)	−311 (0.53)	3209 (2.58)	199 (0.32)	−132 (0.37)	4279 (4.50)	13174 (9.91)
Fuels	42451 (3.25)	2754 (0.55)	−3596 (0.35)	−2265 (1.37)	−4997 (0.36)	34347 (5.12)	2946 (0.82)
Semiprocessed iron and steel products	−3179 (0.87)	−1850 (0.51)	−2000 (0.56)	−2413 (0.87)	−2321 (0.65)	−11763 (3.46)	914 (0.99)
Nonferrous metals	465 (0.59)	5169 (2.93)	−227 (0.28)	−613 (0.54)	645 (0.87)	5438 (4.55)	4996 (2.81)
Some mineral manufactures	−871 (1.54)	−621 (1.03)	−564 (0.93)	−94 (0.95)	−893 (1.43)	−3043 (5.84)	−240 (0.59)
Non-fuel minerals	−2405 (0.63)	2124 (0.78)	274 (0.49)	−2944 (0.66)	−2684 (0.57)	−5635 (2.83)	18679 (2.88)
All fuels and minerals	40046 (2.19)	4878 (0.64)	−3322 (0.39)	−5209 (1.11)	−7681 (0.44)	28712 (4.03)	21625 (1.57)

Notes: Net imports indicated by minus sign (−)

Figures in parentheses are the percentage shares of exports (in the case of net exports) or imports (in the case of net imports) in total world trade for each group of commodities.

Crude minerals include commodities under SITC 271 and 276 (eg crude fertilisers and refractory minerals). Ores, concentrates and scrap of metals include commodities under SITC 28 (eg ores and concentrates of copper, nickel, lead, and iron). Fuels include commodities under SITC 321, 331, 332 and 341 (eg coal, crude oil, petroleum products, and natural gas). Semiprocessed iron and steel products include commodities under SITC 67 (eg pig iron, and some primary forms of iron and steel products). Nonferrous metals include commodities under SITC 68 (eg metals of tin, copper, lead, zinc and aluminium). Some mineral manufactures include commodities under SITC 521, 561 and 663 (eg manufactured fertilisers, ceramic and cement products).

Source: See Table 2.1

exporter of fuels, metallic ores and concentrates, and nonferrous metals. However, Indonesia and Malaysia were the only ASEAN net exporters of fuels (mainly crude oil), while the rest of the group were net fuel importers. The most important export item among the metallic ores and concentrates from the ASEAN group was copper from the Philippines, while tin metal from Malaysia, Thailand and Indonesia was ASEAN's main nonferrous metal export item. All ASEAN countries were net importers of semiprocessed iron and steel products, and some mineral manufactures. In fact, semiprocessed iron and steel products have always been one of the most important groups of mineral imports for the ASEAN countries, reflecting their early stage of development, which requires imports of most iron and steel products, and their lack of indigenous iron ore.

For most fuels and minerals items, ASEAN and Australia were, by world standards, fairly small exporters and importers, accounting for less than 5 per cent of world trade. The ASEAN export items which were significant in terms of world market share were tin metal from Malaysia (38 per cent of world exports), tin metal from Thailand (11 per cent), tin metal from Indonesia (6 per cent), copper ore from the Philippines (6 per cent), petroleum products from Singapore (6 per cent), and crude oil from Indonesia (4 per cent). From Australia the significant minerals export items were iron ore and concentrate (19 per cent of world exports), coal (12 per cent), uranium (12 per cent), zinc metal (9 per cent), and alumina–bauxite (9 per cent). On the import side, there were hardly any fuels or minerals products in which ASEAN and Australia were important buyers in the world market. The most significant imports in terms of world market share were manufactured fertilisers and zinc metal, in which imports by ASEAN as a group accounted for about 6–7 per cent of world trade in the 1970s.

Trade direction and trade intensity

Table 2.5 shows the percentages of trade in fuels and minerals exported from ASEAN and Australia during the 1970s, broken down by destination. These percentages are used here as a measure of export dependence by these countries. It is clear that, for fuels and minerals exports from each ASEAN country and Australia, at least 70 per cent were destined for countries outside the region. The ASEAN–Australia region therefore had quite a high dependence on the rest of the world as far as fuels and minerals exports were concerned. However, this export dependence on the rest of the world was lower in 1979 than in 1970, for ASEAN as a whole and also for Australia. Conversely, the intra-ASEAN export dependence increased from 11 per cent in 1970 to 15 per cent in 1979.

For each ASEAN country, the export dependence on the rest of ASEAN varied. The Philippines had the lowest dependence, at only

Table 2.5 Breakdown by destination of all fuels and minerals exports from ASEAN and Australia, 1970–79 (percentage of total)

		Exported by						
		Indonesia	Malaysia	Philippines	Singapore	Thailand	ASEAN	Australia
Imported by								
ASEAN	1970	7.07	19.84	2.79	8.35	11.45	11.05	3.88
	1979	12.54	17.06	1.52	23.06	2.50	14.65	8.45
	1970–79	8.74	15.74	1.68	14.56	5.20	10.89	5.91
Australia	1970	6.64	2.75	na	5.16	na	3.68	–
	1979	1.39	0.58	0.23	7.31	0.47	2.37	–
	1970–79	0.58	0.44	0.19	7.18	0.66	1.97	–
Rest of world	1970	86.30	77.41	97.21	86.49	88.55	85.27	96.12
	1979	86.07	82.37	98.25	69.63	97.03	82.98	91.55
	1970–79	90.67	83.81	98.13	78.26	94.14	87.14	94.09
Total % for each period		100.00	100.00	100.00	100.00	100.00	100.00	100.00
Total exports (US$ million)	1970	467	509	251	407	131	1 765	1 597
	1979	10 558	3 301	858	3 955	640	19 312	5 269
	1970–79	48 091	13 980	5 244	19 254	1 980	88 549	34 558

Note: na—not available
Source: See Table 2.1

2 per cent, while Malaysia and Singapore showed the highest level, 15–16 per cent, during the decade. Singapore in particular showed an increasing trend of export dependence on the rest of ASEAN, the proportion increasing from 8 per cent in 1970 to 23 per cent in 1979. Conversely, Singapore's dependence on the rest of the world declined from 86 per cent to 70 per cent. ASEAN's export dependence on Australia averaged only 2 per cent during the decade and seemed to decline, whereas Australia's export dependence on ASEAN averaged 6 per cent during the decade, doubling from 4 per cent in 1970 to over 8 per cent in 1979. Among the ASEAN countries, Singapore had the highest degree of export dependence on Australia, at 7 per cent. Therefore, as far as exports of fuels and minerals are concerned, Singapore's dependence on the rest of ASEAN and Australia seemed to be highest, and rising. In summary, there is growing intra-ASEAN export dependence in minerals trade, but little dependence on Australia, whereas for Australia ASEAN is an increasingly important market for its minerals products.

In general, other countries of the world constituted the main market for the minerals exports of ASEAN and Australia (Tables 2.5 and 2.6), and also the main source—although they were rather more important as a market than as a source. Conversely, trade between the ASEAN members, and also between ASEAN and Australia, showed the opposite pattern: the ASEAN nations and Australia were more important in each other's imports of minerals than in each other's export trade. Moreover, their mutual import dependence increased more rapidly than their export dependence during the 1970s. However, their mutual trade was much less important than their trade with the rest of the world: in 1979, 73 per cent of ASEAN's minerals imports and 83 per cent of Australia's were still supplied by other countries.

Import interdependence among the ASEAN countries increased significantly, from 11 per cent in 1970 to 23 per cent in 1979, averaging 16 per cent during the decade. All the ASEAN countries except Indonesia increased their import dependence on the rest of ASEAN during the period, and their degree of dependence was above 23 per cent in 1979. ASEAN's import dependence on Australia during the decade was quite stable at just over 3 per cent, with the highest ratio (6 per cent) for the Philippines and Malaysia, and the lowest (2 per cent) in Singapore. On the other hand, Australia's import dependence on ASEAN increased from 13 per cent in 1970 to 17 per cent in 1979, and averaged 14 per cent during the decade.

Concentration ratios for exports and imports

Table 2.7 shows the concentration ratio, which is defined as exports or imports of a region destined for, or supplied by, countries in the same region as a percentage of that region's total exports or imports. Thus the ratio is the degree of the intra-regional export or import

Table 2.6 Breakdown by source of all fuels and minerals imports to ASEAN and Australia 1970–79
(percentage of total)

		Imported by						
		Indonesia	Malaysia	Philippines	Singapore	Thailand	ASEAN	Australia
Exported by								
ASEAN	1970	45.45	4.57	15.00	10.68	14.38	11.10	13.29
	1979	0.88	23.20	28.95	26.41	28.55	23.15	17.39
	1970–79	2.65	18.72	16.48	19.76	15.27	16.11	13.51
Australia	1970	54.55	1.88	6.47	2.91	2.24	3.53	–
	1979	6.32	5.30	6.88	1.56	4.02	3.64	–
	1970–79	4.75	5.58	5.59	1.61	2.71	3.41	–
Rest of world	1970	–	93.55	78.53	86.41	83.39	85.37	86.71
	1979	92.29	71.50	64.17	72.03	67.43	73.20	82.61
	1970–79	92.60	75.70	77.73	78.62	82.02	80.48	86.49
Total % for each period		100.00	100.00	100.00	100.00	100.00	100.00	100.00
Total imports (US$ million)	1970	11	372	340	721	313	1 757	489
	1979	1 471	2 474	1 133	5 823	1 317	12 218	2 633
	1970–79	8 045	9 102	8 566	24 463	9 660	59 836	12 933

Source: See Table 2.1

dependence. Two groups are considered here: ASEAN, and ASEAN plus Australia. For all minerals products combined the intra-ASEAN trade as a percentage of exports and imports increased during the decade. For most product groups the concentration ratios tended to increase over time, particularly for crude minerals and some minerals manufactures. Trade concentration was quite high in fuel minerals (22 per cent of total imports), crude minerals (27 per cent of total exports), semiprocessed iron and steel products (34 per cent of total exports), and some mineral manufactures (36 per cent of total exports). The lowest trade concentration was in nonferrous metals (1 per cent of total exports and 3 per cent of total imports), semiprocessed iron and steel products (3 per cent of total imports), and metal ores, concentrates and scrap (4 per cent of total exports).

It should be noted that for all groups of commodities, if ASEAN was a net exporter, the concentration ratio on the import side was higher than that on the export side. The opposite is true in the case of a net importer. The explanation for this is that a region which is a net exporter must have more exports to the rest of the world than imports; thus there is a tendency for the region's exports to be more geographically diversified than its imports. On the other hand, a net importer, having a trade deficit with the rest of the world, tends to depend on countries outside the region as import sources relatively more than as export destinations. Therefore, for such commodities as fuel minerals, metallic ores and concentrates, and nonferrous metals, in which ASEAN was a net exporter, import trade was more concentrated within the region than export trade. The opposite was true in the cases of such net imports as semiprocessed iron and steel products, and crude minerals.

If the region's share in total world trade of those commodities can be interpreted as its *expected* concentration ratio, then, when its *actual* concentration ratio is larger than its share in world trade, the trade can be described as being more concentrated than expected.[1] The figures in Table 2.7 clearly show that this is the case for most fuels and minerals traded by ASEAN. The only two exceptions were its exports of nonferrous metals, where the trade concentration was less than expected, and its exports of metal ores, concentrates and scrap, in which the trade concentration was comparable to its world trade share.

When Australia is grouped with ASEAN the concentration ratio in most cases declines. This implies that, as far as minerals are concerned, the trade tie between Australia and ASEAN in the 1970s was still relatively less significant than the intra-ASEAN trade. This conclusion is reinforced by the figures quoted earlier. The only exception was exports of nonferrous metals, in which the concentration ratio was 1.30 per cent for intra-ASEAN trade, whereas it was 7.03 per cent for ASEAN-Australia trade. This was because Australia's exports of nonferrous metals, particularly lead and zinc, to ASEAN, formed a

Table 2.7 Concentration ratios of ASEAN and Australia in fuels and minerals, 1970–79 (US$ million)

		ASEAN				ASEAN and Australia				World trade
		Exports	Imports	Concentration ratio		Exports	Imports	Concentration ratio		
				As % of exports	As % of imports			As % of exports	As % of imports	
All fuels and minerals	1970	1 764	1 757	11.05	11.10	3 362	1 917	7.45	13.18	65 331
	1979	19 312	12 218	14.65	23.15	25 750	17 159	12.18	18.35	457 030
	1970–79	88 549	59 836	10.89	16.11	123 107	72 769	10.05	15.33	2 198 000
		(4.03)	(2.72)			(5.60)	(3.31)			
Crude minerals	1970	16	29	6.21	3.49	29	78	5.56	1.39	1 549
	1979	80	183	37.94	16.68	172	196	23.13	16.04	5 860
	1970–79	364	911	27.24	10.88	794	1 506	18.30	6.87	35 671
		(1.02)	(2.55)			(2.23)	(4.22)			
Ores, concentrates and scrap of metals	1970	423	82	1.21	6.27	1 088	91	0.56	6.07	7 638
	1979	1 199	433	1.94	5.37	2 869	445	4.11	5.51	23 198
	1970–79	6 044	1 765	4.44	15.20	19 358	1 896	3.09	14.39	134 350
		(4.50)	(1.31)			(14.41)	(1.41)			
Fuel minerals	1970	850	758	18.91	21.18	1 310	1 003	13.48	22.10	27 117
	1979	16 864	9 746	14.08	24.36	18 762	11 486	13.09	24.58	312 580
	1970–79	71 515	37 169	11.70	22.50	83 038	45 746	11.34	21.89	1 398 000
		(5.11)	(2.66)			(5.94)	(3.27)			
Semiprocessed iron and steel products	1970	43	384	43.60	4.91	172	519	24.81	3.70	15 433
	1979	254	2 530	55.83	5.61	834	3 278	29.30	4.53	67 142
	1970–79	1 122	12 844	33.75	2.94	4 806	13 614	20.13	2.59	372 020
		(0.30)	(3.45)			(1.29)	(4.20)			

Nonferrous metals	1970	421	104	0.26	1.05	749	132	2.86	0.90	11 558
	1979	1 984	1 035	1.91	3.67	2 998	1 114	6.30	3.41	35 462
	1970–79	8 867	3 429	1.30	3.35	14 344	3 910	7.03	2.95	194 990
		(4.55)	(1.76)			(7.36)	(2.01)			
Some minerals manufactures	1970	11	69	11.38	1.79	14	94	15.37	4.23	2 036
	1979	100	599	81.75	13.63	115	639	77.52	12.81	12 786
	1970–79	637	3 680	35.85	6.20	767	4 050	34.75	6.00	62 970
		(1.01)	(5.84)			(1.22)	(6.43)			

Notes: For definitions of categories, see explanatory note to Table 2.4.

Figures in parentheses are the regions' exports and imports as percentages of world trade during the 1970s in each product group.

The concentration ratio is the ratio of exports (or imports) of the region destined for (or supplied by) countries in the same region, as a percentage of the region's total exports (or imports) of these commodities.

Source: See Table 2.1

substantial part of its total exports of these mineral products. As a result, the export concentration ratio of nonferrous metals for the ASEAN–Australia grouping was closer to the expected value, as reflected by the group's share in world trade. On the other hand, the inclusion of Australia in the group reduced the export concentration ratio of metal ores, concentrates and scrap to a level below the expected value because of Australia's heavy reliance on non-ASEAN countries in exports of these products.

A further analysis can be made of the ASEAN and Australian trade pattern in fuels and minerals using the trade intensity index.[2] The index is defined as the share of one country's (or region's) exports to another country (or region), divided by the latter's import share of world trade. It is thus an indicator of one country's importance in the import trade of another, normalised by the latter's relative dependence on imports. The index is influenced by two factors. The first, called the 'complementarity index', measures the degree to which the commodity pattern of the former's exports matches the latter's import pattern. The second, called the 'bias index', indicates the extent to which bilateral trade is influenced by such biases as distance, transport and communications costs, and other trade impediments. The estimates of these three trade indexes for fuels and minerals traded by ASEAN and Australia are shown in Table 2.8.

Table 2.8 Intensity, complementarity and bias indexes in fuels and minerals trade among the ASEAN countries and Australia, 1970–79

		Exported by					
		Rest of ASEAN			Australia		
		I	C	B	I	C	B
Imported by							
Indonesia	1970	22.24	0.72	31.07	22.54	0.61	36.98
	1979	0.38	0.79	0.48	5.36	0.81	6.61
	1970–79	1.40	0.92	1.53	2.95	0.61	4.84
Malaysia	1970	2.40	1.23	1.95	0.81	1.04	0.78
	1979	6.12	0.92	6.66	4.55	0.98	4.65
	1970–79	5.48	1.13	4.85	3.53	0.86	4.10
Philippines	1970	6.49	0.83	7.82	2.66	0.59	4.52
	1979	6.72	0.59	11.36	5.98	1.24	4.82
	1970–79	4.34	0.90	4.80	3.67	0.84	4.35
Singapore	1970	5.13	0.77	6.62	1.21	0.66	1.83
	1979	7.24	1.04	6.97	1.34	0.78	1.71
	1970–79	6.21	1.27	4.91	1.01	0.29	3.44
Thailand	1970	5.70	0.82	6.99	0.91	0.60	1.53
	1979	6.57	0.75	8.81	3.49	1.30	2.68
	1970–79	3.87	0.80	4.84	1.72	0.59	2.94

ASEAN	1970	–	–	–	1.74	0.74	2.35
	1979	–	–	–	2.54	0.83	3.06
	1970–79	–	–	–	2.91	0.53	5.48

		Imported by					
		Rest of ASEAN			Australia		
		I	C	B	I	C	B
Exported by							
Indonesia	1970	3.22	1.08	2.99	8.73	0.94	9.31
	1979	4.37	1.00	4.37	2.40	0.94	2.55
	1970–79	3.70	1.07	3.45	0.98	0.88	1.12
Malaysia	1970	12.23	0.65	18.79	3.54	0.52	6.73
	1979	6.43	0.96	6.70	1.00	0.75	1.33
	1970–79	6.79	0.99	6.83	0.75	0.61	1.23
Philippines	1970	1.71	0.69	2.48	0.15	0.29	0.52
	1979	0.51	0.57	0.90	0.46	0.18	2.53
	1970–79	0.72	0.70	1.03	0.33	0.54	0.61
Singapore	1970	7.72	0.73	10.54	6.69	1.15	5.80
	1979	11.96	0.91	13.08	12.51	0.93	13.39
	1970–79	8.95	1.29	6.94	12.06	1.88	6.41
Thailand	1970	6.86	0.59	11.69	0.05	0.80	0.06
	1979	0.87	0.86	1.01	0.69	0.41	1.70
	1970–79	2.26	0.96	2.35	1.13	0.44	2.56
ASEAN	1970	–	–	–	4.78	0.75	6.33
	1979	–	–	–	3.76	0.84	4.48
	1970–79	–	–	–	3.18	1.01	3.16

Notes: I—Intensity index
C—Complementarity index
B—Bias index
I—C × B
Source: See Table 2.1

A number of observations can be made on these figures. The first is that in most cases the three indexes for fuels and minerals trade among ASEAN countries were higher than those for trade between ASEAN and Australia in the 1970s. That is, the intra-ASEAN trade in fuels and minerals during the period was more intense, more complementary, and suffered less resistance than the corresponding ASEAN–Australia trade. This conclusion supports the above finding that intra-ASEAN trade was more concentrated than ASEAN–Australia trade.

Second, as far as fuels and minerals trade among ASEAN countries is concerned, Singapore's trade with other ASEAN countries had the highest degree of intensity and complementarity, and the fewest resistances. On the other hand, the smallest degree of intensity and complementarity was found in trade between the Philippines and the rest of ASEAN. In most cases the trade intensity index is rather high and significantly above unity. And, while the complementarity index is very close to unity, as might be expected, the bias index has a very high value in almost all cases. With a few exceptions, the intensity of intra-ASEAN trade in fuels and minerals tended to increase over the 1970–79 period. Since the bias factor was always more significant than the complementarity factor, the changes in trade intensity were a result of changes in the trade bias rather than in the complementarity of commodity mixtures, no doubt reflecting the geographic proximity between the ASEAN nations.

Third, for fuels and minerals trade between Australia and individual ASEAN countries the intensity index varies but is generally above unity. The export intensity index for Australia tended to increase during the period, while the import index fell. The degree of intensity is particularly high (about 12) in the case of exports from Singapore to Australia, but it is quite low (about 0.33) for exports from the Philippines to Australia. Trade intensity between Australia and ASEAN as a group has been rather steady, 3 being the average index. As in intra-ASEAN trade, ASEAN-Australia trade exhibits a bias index which is much stronger than the complementarity index.[3]

Export performance

Asean's exports to Australia

Table 2.9 clearly shows that during the 1970s, over 90 per cent of total fuels and minerals exports from ASEAN to Australia were petroleum products and crude oil. While most of Singapore's trade was in refined petroleum products, Indonesia and Malaysia exported mainly crude oil. These petroleum items accounted for almost all of the total fuels and minerals exports from these three countries to Australia in the whole decade. From the Philippines, fuels and minerals exports to Australia between 1970 and 1973 consisted mainly of copper ore and concentrate, but in 1979 iron and steel products became equally important, although the total trade remained small. Australian fuels and minerals imports from Thailand were largely iron and steel products, which, as in the case of the Philippines, mostly comprised tube and pipe products.

The share of ASEAN in Australian fuels and minerals imports, as shown in Table 2.10, declined from 13 per cent in 1970 to 8 per cent in 1973, and rose to 17 per cent in 1979. The increase was the result of the sharp oil price rises after 1973, which raised ASEAN's share of Australian fuel imports from 17 per cent in 1973 to 26 per cent in 1979. On the other hand, Australia's share in ASEAN's fuels and minerals exports exhibited a similar trend, declining between 1970 and 1973 and increasing to about 2 per cent in the second half of the decade (Table 2.11). The same changes can also be observed for the share of intra-

ASEAN exports in the 1970s. These movements were mainly influenced by the prices of petroleum products, which formed the large bulk of ASEAN–Australia and intra-ASEAN trade.

To analyse ASEAN's performance in exporting fuels and minerals to Australia, the constant market share analysis is applied during three intervals: 1970–73, 1973–76, and 1976–79.[4] ASEAN's exports to Australia are evaluated against the world's exports to Australia and, separately, against the exports of all other LDCs to Australia, and the rate of growth is decomposed into three components. These components are:

1 standard growth, that is, the growth of Australian imports from the rest of the world;
2 a commodity composition effect reflecting shifts in the growth of particular commodities, and
3 a 'competitiveness' residual.

The analysis was performed separately in terms of the rest of world export growth and LDC exports. The results presented in Table 2.12 show that during 1970 to 1973 ASEAN's fuels and minerals exports to Australia declined, mainly because of the negative competitiveness effect relative to both the rest of the world and to all other LDCs. The same was true for the individual cases of Indonesia, Malaysia and the Philippines. On the other hand, exports from Singapore and Thailand to Australia grew substantially through large gains in competitiveness relative to non-ASEAN exporters. During the period 1973–76, which saw the first oil crisis, ASEAN's exports to Australia more than tripled and its share was increasing relative to the world and also relative to other LDCs because it was more competitive, particularly in crude oil from Indonesia and petroleum products from Singapore. Both countries, whose fuels exports accounted for most of ASEAN's minerals exports to Australia, clearly gained their shares against other exporters due primarily to the competitiveness effect. As for the other ASEAN countries, their fuels and minerals exports to Australia declined or remained stagnant during that time interval, either because of an unfavourable commodity composition (as in Malaysia), or because of declining competitiveness (as in Thailand), or because of a combination of both (as in the Philippines). Between 1976 and 1979, ASEAN continued to increase its share relative to other exporters. However, the gain was due more to its commodity composition than to improving ASEAN competitiveness. This was true in the case of exports from Indonesia and Malaysia, where crude oil was the main item. The gain in Thailand's mineral exports to Australia stemmed from its increasing competitiveness, while the Philippines' gain was explained by a very large commodity composition effect (relative to all LDCs) and by a

Table 2.9 Australian imports of fuels and minerals from ASEAN and rest of world: commodity composition

From		Percentage of each group in total fuels and minerals imports						
		Crude minerals	Ores and concentrates	Fuels	Iron and steel products	Nonferrous metals	Mineral manufactures	Total fuels and minerals
World	1970	9.9	1.9	50.0	27.5	5.6	5.1	100.0
	1973	10.1	2.6	43.2	33.9	5.7	4.5	100.0
	1976	6.0	1.0	72.5	14.6	3.0	2.9	100.0
	1979	0.5	0.5	66.1	28.4	3.0	1.5	100.0
ASEAN	1970	0.1	0.7	93.8	0.4	0.8	4.2	100.0
	1973	1.5	0.7	92.4	2.6	0.5	2.3	100.0
	1976	0.0	0.1	97.6	2.1	0.1	0.1	100.0
	1979	0.2	0.3	97.7	1.4	0.3	0.1	100.0
Indonesia	1970	0.0	0.0	100.0	0.0	0.0	0.0	100.0
	1973	0.0	0.0	100.0	0.0	0.0	0.0	100.0
	1976	0.0	0.0	100.0	0.0	0.0	0.0	100.0
	1979	0.0	0.0	100.0	0.0	0.0	0.0	100.0

Malaysia	1970	0.0	0.5	96.0	0.0	3.5	0.0	100.0
	1973	0.0	4.2	81.7	1.6	12.5	0.0	100.0
	1976	0.0	0.0	67.2	21.9	9.9	1.0	100.0
	1979	0.0	0.0	93.0	0.4	6.3	0.3	100.0
Philippines	1970	0.0	100.0	0.0	0.0	0.0	0.0	100.0
	1973	0.0	87.3	11.3	1.4	0.0	0.0	100.0
	1976	0.0	7.0	30.5	60.5	1.2	0.8	100.0
	1979	0.0	49.5	0.0	50.2	0.0	0.3	100.0
Singapore	1970	0.3	0.4	84.8	1.1	0.1	13.3	100.0
	1973	1.6	0.0	94.9	1.0	0.0	2.5	100.0
	1976	0.0	0.1	99.6	0.3	0.0	0.0	100.0
	1979	0.1	0.1	98.5	1.2	0.0	0.1	100.0
Thailand	1970	2.0	2.0	0.0	91.8	0.0	4.1	100.0
	1973	0.1	0.0	19.9	80.1	0.0	0.0	100.0
	1976	0.0	0.0	0.0	99.0	0.8	0.2	100.0
	1979	18.8	0.4	0.0	77.8	1.6	1.4	100.0
Rest of world	1970	11.5	2.0	43.4	31.6	6.3	5.2	100.0
	1973	10.9	2.8	39.0	36.6	6.1	4.6	100.0
	1976	7.0	1.1	68.3	16.7	3.5	3.4	100.0
	1979	0.5	0.5	59.4	34.1	3.6	1.9	100.0

Note: For definitions of categories, see explanatory note to Table 2.4
Source: See Table 2.1

Table 2.10 Shares of ASEAN and rest of world in Australian imports of fuels and minerals (per cent)

		Crude minerals	Ores and concentrates	Fuels	Iron and steel products	Nonferrous metals	Mineral manufactures	Total fuels and minerals
Australian imports from:								
ASEAN	1970	0.1	4.9	24.9	0.2	1.8	11.1	13.3
	1973	1.2	2.1	16.7	0.6	0.7	4.1	7.8
	1976	0.0	1.6	19.0	2.1	0.6	0.1	14.1
	1979	6.8	10.5	25.8	0.9	1.6	0.7	17.4
Rest of world	1970	99.9	95.1	75.1	99.8	98.2	88.9	86.7
	1973	98.8	97.9	83.3	99.4	99.3	95.9	92.2
	1976	100.0	98.4	81.0	97.9	99.4	99.9	85.9
	1979	93.2	89.5	74.2	99.1	98.4	99.3	82.6

Note: For definitions of categories, see explanatory note to Table 2.4
Source: See Table 2.1

Table 2.11 Shares of Australia, rest of world and ASEAN in ASEAN's exports of fuels and minerals

		Percentage of ASEAN's exports destined for the country groups listed						
		Crude minerals	Ores and concentrates	Fuels	Iron and steel products	Nonferrous metals	Mineral manufactures	Total fuels and minerals
ASEAN's exports to:								
Australia	1970	0.4	0.1	7.2	0.7	0.1	25.2	3.7
	1973	2.9	0.1	1.8	2.7	0.0	2.1	1.4
	1976	0.1	0.0	2.3	4.6	0.0	0.2	2.0
	1979	1.1	0.1	2.7	2.6	0.1	0.3	2.3
Rest of world	1970	93.4	98.7	73.9	55.7	99.6	63.4	85.6
	1973	86.0	92.6	89.1	86.4	99.0	84.1	90.9
	1976	30.9	97.1	87.3	45.2	98.5	6.3	88.2
	1979	60.9	97.9	83.3	41.5	97.4	18.0	84.3
ASEAN	1970	6.2	1.2	18.9	43.6	0.3	11.4	10.7
	1973	11.0	7.3	9.1	10.9	1.0	13.8	7.7
	1976	69.0	2.9	10.4	50.2	1.5	93.5	9.8
	1979	37.9	1.9	14.1	55.8	2.5	81.8	13.4

Note: For definitions of categories, see explanatory note to Table 2.4
Source: See Table 2.1

Table 2.12 ASEAN fuels and minerals exports to Australia: constant market share analysis

		Decomposition of growth in ASEAN exports to Australia						
	Period	Growth in Australian imports from ASEAN	Growth in Australian imports from: Rest of world	All other LDCs[i]	Commodity composition effect relative to: Rest of world	All other LDCs	Competitiveness residual relative to: Rest of world	All other LDCs
ASEAN	1970–73	−19	28	34	−5	−3	−43	−51
	1973–76	345	177	225	30	−21	138	140
	1976–79	137	53	47	61	102	23	−12
Indonesia	1970–73	−97	34	42	−24	−22	−107	−117
	1973–76	1 038	190	237	23	−24	824	824
	1976–79	1 455	54	48	−43	−37	1 445	1 444
Malaysia	1970–73	−81	23	25	93	43	−197	−149
	1973–76	−52	192	240	−223	−320	−21	28
	1976–79	11 458	63	59	48	−141	11 347	11 540
Philippines	1970–73	−19	22	24	86	−66	−127	23
	1973–76	−7	192	240	109	−340	−308	93
	1976–79	474	64	61	−30	24 711	440	−24 298
Singapore	1970–73	186	16	16	25	37	145	133
	1973–76	338	178	227	28	−20	131	130
	1976–79	44	67	64	60	61	−84	−81
Thailand	1970–73	27 600	22	24	−14	−18	27 592	27 593
	1973–76	1	192	240	25	−18	−215	−229
	1976–79	82	64	61	−68	6	86	15

Note: See Table 2.4, note 3
Source: See Table 2.1

significant competitiveness effect (relative to the whole world). Singapore became less competitive in its fuels and minerals exports to Australia, and in fact slightly lost its share in the late 1970s.

Australian exports to ASEAN

Table 2.13 shows the commodity composition of ASEAN's fuels and minerals imports in the 1970s from Australia, as well as from within ASEAN and from the rest of the world. Unlike imports from other sources, which were predominantly petroleum items, those from Australia consisted mainly of nonferrous metals, iron and steel products, and metallic ores and concentrates, which together accounted for at least 70 per cent of ASEAN's fuels and minerals imports from Australia. The main components of Australian nonferrous metals were zinc, copper, lead and aluminium. Australia's metallic ores exported to ASEAN were mostly tin destined for refineries in Malaysia, and some iron ore, and iron and steel scrap to the five ASEAN countries. Most of the iron and steel semiprocessed products imported from Australia were items in primary forms, and iron and steel plates and sheets. Fuel products, which accounted for 8–18 per cent of Australia's total fuels and minerals exports to ASEAN, were mostly petroleum products shipped to Singapore (for entrepôt trade) and Indonesia. Exports of Australian coal (which increased significantly in 1979) went to the Philippines and Thailand.

Australia's exports of metallic ores and nonferrous metals amounted to about 20 per cent of ASEAN's total imports of these commodity groups, while the corresponding shares of other fuels and minerals products were generally less than 10 per cent (Table 2.14). In total, Australia's market share in ASEAN's fuels and minerals imports remained relatively stable at 3–4 per cent during the decade. However, the share of ASEAN in Australian fuels and minerals exports increased from about 4 per cent in 1970 to 8.5 per cent in 1979; the increase was quite significant in some minerals manufactures, iron and steel products, and nonferrous metals (Table 2.15). While it is obvious that Australia cannot be a major supplier of fuels (except coal), it is also clear that its share of trade in iron and steel products is very low (at 4 per cent of ASEAN imports—see Table 2.14).

A constant market share analysis was undertaken for Australia's exports to ASEAN relative to the rest of the world. The decomposition also included a 'market distribution' effect, reflecting shifts in the emphasis of Australian exports amongst the members of ASEAN.

The results of the constant market share analysis (Table 2.16) indicate that, while Australia made some gains in the ASEAN import market between 1970 and 1973, and between 1976 and 1979, its market share declined during the 1973–76 period, when ASEAN's oil import bill grew rapidly. The gains during the two intervals were due to Australia's favourable commodity composition and to the increased

Table 2.13 ASEAN fuels and minerals imports from Australia, ASEAN and rest of world: commodity composition (per cent)

Imports from:		Crude minerals	Ores and concentrates	Fuels	Iron and steel products	Nonferrous metals	Mineral manufactures	Total fuels and minerals
World	1970	1.7	3.8	63.4	21.9	5.3	3.9	100.0
	1973	2.0	3.9	47.1	31.9	7.4	7.7	100.0
	1976	1.6	1.8	70.1	18.5	4.7	3.3	100.0
	1979	1.5	3.8	62.4	19.7	7.6	4.9	100.0
Australia	1970	0.8	5.1	18.3	41.5	32.3	2.0	100.0
	1973	0.4	17.6	7.8	27.3	43.0	3.9	100.0
	1976	6.2	18.7	8.8	32.2	33.4	0.8	100.0
	1979	2.1	21.2	18.1	23.0	33.9	1.7	100.0
ASEAN	1970	0.5	2.7	85.4	10.1	0.6	0.7	100.0
	1973	1.0	11.3	81.1	1.9	2.1	2.7	100.0
	1976	1.3	1.8	88.3	4.7	1.3	2.6	100.0
	1979	1.1	0.9	88.3	5.3	1.4	3.0	100.0
Rest of world	1970	1.8	3.9	62.4	22.6	4.8	4.4	100.0
	1973	2.3	2.3	45.2	35.6	6.2	8.4	100.0
	1976	1.5	1.2	69.2	20.3	4.3	3.5	100.0
	1979	1.6	3.9	56.9	23.8	8.2	5.6	100.0

Note: For definitions of categories, see explanatory note to Table 2.4
Source: See Table 2.1

Table 2.14 Shares of Australia, ASEAN and rest of world in ASEAN's fuels and minerals imports (per cent)

		Crude minerals	Ores and concentrates	Fuels	Iron and steel products	Nonferrous metals	Mineral manufactures	Total fuels and minerals
ASEAN's imports from:								
Australia	1970	1.6	4.4	1.0	6.3	20.5	1.7	3.3
	1973	0.8	20.1	0.7	3.8	25.6	2.3	4.4
	1976	11.0	29.6	0.4	4.9	20.0	0.7	2.8
	1979	5.1	20.2	1.1	4.3	16.2	1.3	3.7
ASEAN	1970	3.5	7.6	14.4	4.9	1.2	1.8	10.7
	1973	5.0	29.1	17.3	0.6	2.8	3.5	10.0
	1976	11.5	14.1	17.3	3.5	3.9	10.9	13.8
	1979	16.7	4.9	31.1	5.9	4.1	13.6	22.0
Rest of world	1970	94.9	88.0	84.6	88.8	78.3	96.5	86.0
	1973	94.2	50.7	82.0	95.6	71.6	94.2	85.5
	1976	77.4	56.3	82.3	91.6	76.1	88.4	83.4
	1979	78.2	74.9	67.8	89.8	79.7	85.1	74.3

Note: For definitions of categories, see explanatory note to Table 2.4
Source: See Table 2.1

Table 2.15 Shares of ASEAN and rest of world in Australian fuels and minerals exports (per cent)

		Crude minerals	Ores and concentrates	Fuels	Iron and steel products	Nonferrous metals	Mineral manufactures	Total fuels and minerals
Australian exports to:								
ASEAN	1970	3.9	0.4	2.3	19.0	5.8	32.9	3.7
	1973	1.9	2.0	1.6	12.9	10.6	47.5	5.0
	1976	22.1	2.1	0.8	14.9	11.0	7.9	4.0
	1979	10.2	5.7	4.2	17.7	14.9	50.2	8.5
Rest of world	1970	96.1	99.6	97.7	81.0	94.2	67.1	96.3
	1973	98.1	98.0	98.4	87.1	89.4	52.5	95.0
	1976	77.9	97.9	99.2	85.1	89.0	92.1	96.0
	1979	89.8	94.3	95.8	82.3	85.1	49.8	91.5

Note: For definitions of categories, see explanatory note to Table 2.4
Source: See Table 2.1

Table 2.16 Australian fuels and minerals exports to ASEAN: constant market share analysis (per cent)

		Decomposition of growth in Australian exports to ASEAN				
	Period	Growth in ASEAN imports from Australia	Growth in ASEAN imports from rest of world	Commodity composition effect	Market distribution effect	Competitiveness residual
ASEAN	1970–73	95	70	224	−212	12
	1973–76	62	249	−117	−35	−35
	1976–79	185	82	81	−56	78
Indonesia	1970–73	446	193	868	–	−615
	1973–76	117	554	−394	–	−44
	1976–79	116	70	−13	–	58
Malaysia	1970–73	305	−3	39	–	268
	1973–76	53	160	−221	–	114
	1976–79	266	94	90	–	82
Philippines	1970–73	73	53	−40	–	60
	1973–76	39	262	−149	–	−74
	1976–79	223	66	72	–	85
Singapore	1970–73	−16	103	−120	–	0
	1973–76	33	262	−48	–	−181
	1976–79	212	80	−33	–	165
Thailand	1970–73	138	108	−37	–	67
	1973–76	66	196	−78	–	−52
	1976–79	119	100	5	–	13

Source: See Table 2.1

competitiveness of its fuels and minerals exports to ASEAN. Not being a net oil exporter, Australia obviously experienced a negative impact in the commodity composition, market distribution and competitiveness effects of its exports to ASEAN between 1973 and 1976. It is worth noting that the market distribution effect of Australia's exports to ASEAN was unfavourable throughout the 1970s. In general, the performance of Australia's fuels and minerals exports to individual ASEAN countries paralleled that of Australia's exports to ASEAN as a whole; the only exception is Australia's exports to Singapore, where Australia lost its market share between 1970 and 1976, and later started to gain as its fuels and minerals exports became more competitive.

Future trade prospects

The drastic oil price increases in the 1970s, while having increased the trade share of fuels during the decade, also encouraged most oil importing countries to conserve energy and switch away from oil to coal, natural gas, and other indigenous energy sources. As a result, oil prices are expected to be under downward pressure until the late 1980s and to pick up in the 1990s. Meanwhile, trade in coal and LNG is likely to increase and substitute for the declining trade volume of oil. Nevertheless, barring a major collapse in oil prices, oil and other fuel minerals will still play a significant role in world trade, as well as in trade by ASEAN and Australia.

Based on the papers prepared for the ASEAN–Australia Research Project reviewing production, trade and processing of fuels and minerals in ASEAN and Australia,[5] the future prospects for fuels and minerals trade in ASEAN and Australia are reviewed in the following sections.

Coal

While Australia is well endowed with coal and is already prominent as a world exporter of coal, there are plans for increased consumption of coal in most ASEAN countries. Indonesia is expected to expand its coal production to satisfy its growing demand, and will probably have some surplus for export in the 1990s. Malaysia, the Philippines and Thailand will increase the role of coal in their energy mixtures by importing and using it in generating power, and in cement and steel production. It is probable that by sometime in the 1990s these three countries will be importing large amounts of coal to supplement their existing local energy sources. With relatively short distances between Australia and ASEAN, there is strong potential for increasing coal imports from Australia to these three ASEAN countries. There is also a possibility that some coal will be exported from Indonesia to other ASEAN countries in the 1990s.

Oil

The expansion of refining capacity in Indonesia, Malaysia and Thailand is likely to reduce the third-party processing of Malaysia's and Indonesia's crude oil in Singapore, as well as cutting down Singapore's exports of petroleum products to the three countries. Consequently, the role of Singapore as the region's refining centre and supplier of refined products will decline, and less trade in petroleum between Singapore and the other ASEAN countries can be expected. However, the importation of crude oil by Thailand and the Philippines from Indonesia and Malaysia may increase in the future as the demand for light products in the two importing countries expands and ASEAN begins to realise the mutual benefits of oil trade. There is a possibility that Singapore might adopt the role of middleman for the region by processing semifinished products from the Persian Gulf and trading balances in finished products. It is not clear how the oil trade between ASEAN and Australia, which has so far been dominated by Singapore's export of refined products to Australia, will be affected. If Singapore is successful in diversifying its export market, there is a chance that the share of its refined products in Australia's imports will be expanded.

Natural gas

Large reserves of natural gas have been found in Indonesia, Malaysia, Australia and Thailand. Trade in natural gas has been, and will continue to be, in the form of LNG exports from Indonesia, Malaysia and Australia to Japan and South Korea. There is a plan for Singapore to import natural gas through pipelines from Malaysia and Indonesia.

Iron and steel

All the ASEAN countries except Singapore have planned to build their own integrated steel plants in order to reduce imports of iron and steel products. Malaysia seems to have the most definite plan for setting up a plant. If most of these plans can be implemented, it is very likely that ASEAN's imports of iron ore and pellets from Australia, which has vast reserves of the ore, will be quite significant. However, given the currently depressed state of the steel industry worldwide, such developments will need to be implemented with caution. One way to reduce the initial investment outlay is to adopt electric arc-furnace processes on a relatively low scale, using a large proportion of steel scrap as feed (Brown and McKern, 1986). This approach could also offer potential for Australia to export pre-reduced iron agglomerates.

Tin

The trade volume of tin has been small among ASEAN countries, as well as between ASEAN and Australia, because most of the region's exports are destined for industrial countries outside the region. It is expected that the regional trade in tin will decline further as the tin smelting activity in Singapore slows down. Australia's exports of its tin concentrate to Malaysia's smelters will also decrease if the tin smelting and refining capacity of Australia is expanded in the future.

Other nonferrous minerals

Australia is rich in several nonferrous minerals, including, among others, bauxite, copper, lead, zinc, nickel and uranium. In most cases, there is potential for increasing trade between ASEAN and Australia as a result of higher demand in industries and more minerals processing activities in ASEAN. Plans for the construction of an aluminium smelter in Indonesia, and probably also in the Philippines, could increase their imports of Australia's bauxite and alumina. ASEAN is likely to increase further its reliance on Australia's lead because of its increasing industrial demand and its small reserves of the ore. The Philippines will soon import uranium oxide yellowcake from Australia to fuel its first nuclear power plant. In the cases of zinc, copper and nickel, there will be more competition from some ASEAN countries against Australian exports. Production from Thailand's first zinc smelter has already substantially reduced its imports of zinc metal from Australia. It is possible that Thailand's zinc exports to other ASEAN countries will further reduce Australia's share in the ASEAN market. Copper from the Philippines, Indonesia and Malaysia will compete with Australia's copper in the Japanese and ASEAN markets. Competition is also expected between refined nickel from Australia and that from the Philippines.

It is interesting to note that, while there is a tendency for a higher degree of minerals processing in both ASEAN and Australia, its effects on trade vary, depending on whether a country is self-sufficient in mineral ores. Where the country does not possess ore reserves adequate for local processing, this implies more imports of ores, and fewer imports (or more exports) of processed products, as seen in the processing of iron and steel and aluminium in ASEAN. On the other hand, if the country if well endowed with mineral ores, further processing will reduce its ore exports and increase the exports (or reduce the imports) of processed products, as in the cases of the processing of tin, zinc and copper in ASEAN and Australia.

The above analysis of future trade prospects suggests that ASEAN and Australia will remain net exporters of fuels and minerals. The large reserves in Australia of iron ore, coal, natural gas and several nonferrous minerals should enable it to further expand exports of these fuels and minerals, subject to an improvement in worldwide demand. For ASEAN, however, the picture is rather mixed. Singapore, the Philippines and Thailand will not be self-sufficient in energy. However, the Philippines and Thailand are expected to reduce their dependence on imported oil and rely more on indigenous energy sources—natural gas and lignite in the case of Thailand, and geothermal energy and coal in the case of the Philippines. Both will eventually have to import coal unless their petroleum and coal reserves expand significantly. Indonesia and Malaysia will still be net exporters of oil, and will also increase their LNG exports to Japan and South Korea. In the next decade, non-fuel mineral exports of ASEAN will still be concentrated in a few metals. Tin metal is likely to remain the most important mineral export of Malaysia, Indonesia and Thailand, while copper will be the most significant for the Philippines. As a whole, ASEAN will continue to be a net exporter of fuels and a net importer of non-fuel minerals.

With a good potential for coal and iron imports from Australia to ASEAN, the volume of ASEAN–Australia trade in fuels and minerals should expand in the future. It is not clear how the intra-ASEAN trade in these commodities will change. Much depends on how Singapore can adapt its role in oil trading. In any case, the trade-bias factor, which was found to be quite high in the 1970s, is expected to be also high in the future. This factor should therefore contribute to maintaining a high degree of intensity and concentration in fuels and minerals trade among the ASEAN countries. Even though more minerals trade amongst ASEAN and Australia will be regionalised, major portions of their fuels and minerals exports will continue to be sold to industrial countries, notably Japan.

Summary and conclusions

In the 1970s, trade in fuels and minerals was quite a significant part of total trade for the world, as well as for ASEAN and Australia. Since the first oil crisis in 1973–74, oil has played a dominant role in minerals and fuels trade. Even though Singapore, Thailand and the Philippines were net oil importers, the huge oil surpluses exported by Indonesia and Malaysia still enabled ASEAN as a group to become a net exporter of fuels.

ASEAN's most important non-fuel minerals exports were copper from the Philippines and tin from Malaysia, Indonesia and Thailand. ASEAN's net imports in non-fuel minerals were caused mainly by large trade deficits in iron and steel products experienced by all member countries.

With its abundant mineral resources, Australia has always been a net exporter of most fuels and minerals, notably coal, iron ore and concentrates, and some nonferrous metals.

During the 1970s, at least 70 per cent of the fuels and minerals exports from each ASEAN country and Australia were destined for countries outside the region. However, there has been growing intra-ASEAN export dependence, while ASEAN has become an increasingly important market for Australia's minerals products. At the same time, import interdependence among ASEAN and Australia in these products has been more significant and has increased more rapidly than their mutual export dependence. Despite this shift, most of the minerals imported into the two regions were supplied by the rest of the world.

For most minerals products, intra-ASEAN trade as a percentage of ASEAN's total exports and imports (the concentration ratio) increased over the decade, and was larger than ASEAN's share in world trade. However, when Australia is grouped with ASEAN, the concentration ratio in most cases declines, implying that the minerals trade tie between Australia and ASEAN was less significant than intra-ASEAN trade.

Estimates of the trade intensity, complementarity and bias indexes indicate that intra-ASEAN trade in fuels and minerals was more intense, more complementary and suffered less trade resistance than the corresponding ASEAN–Australia trade. In both cases, the bias index was much more pronounced than the complementarity index.

The analysis of ASEAN's fuels and minerals exports to Australia reveals that over 90 per cent of these exports have been petroleum products (from Singapore) and crude oil (from Indonesia and Malaysia) throughout the 1970s. The prices of petroleum products therefore had a great influence on ASEAN's share in Australian fuels and minerals imports, which declined during 1970 and 1973 and increased thereafter. The constant market share analysis shows that from 1970 to 1973 ASEAN's fuels and minerals exports to Australia declined, mainly because of the negative competitiveness effect relative to the rest of the world (and also relative to all other LDCs). During the period 1973–76, ASEAN's exports to Australia substantially increased their share by being more competitive in petroleum products (from Singapore) and crude oil (from Indonesia). Between 1976 and 1979, ASEAN's gain in the Australian market continued through improvements in its commodity composition.

Of Australia's fuels and minerals exports to ASEAN in the 1970s, at least 70 per cent consisted of nonferrous metals (notably zinc, copper, lead and aluminium), iron and steel products, and metallic ores and concentrates. Australia's share in the ASEAN import market fell during the 1973–76 period, when the commodity composition, market distribution and competitiveness of its exports to ASEAN deteriorated. In other periods, it managed to make some gains through better commodity composition and competitiveness effects.

As for the future prospects of trade in minerals and fuels by ASEAN and Australia, it is likely that they will continue to be net exporters of these products, though Singapore, the Philippines and Thailand will still have to import oil. There is a good potential for increases in fuels and minerals trade between ASEAN and Australia. Most of the increases are likely to be in the form of exports from Australia to ASEAN, notably Australian coal to Malaysia, the Philippines and

Thailand, iron ore and pellets or agglomerates to Malaysia, bauxite and alumina to Indonesia, and uranium to the Philippines.

On the other hand, there are developments which will tend to reduce the intra-ASEAN and ASEAN–Australia trade. The expansion of local oil refineries in Indonesia, Malaysia and Thailand is expected to lower Singapore's exports of petroleum products to these three countries. The smelting of zinc in Thailand and the refining of copper in the Philippines will compete with Australian products in ASEAN and non-ASEAN markets.

In the next decade, ASEAN will continue to be a net importer of non-fuel minerals and an important exporter of a small number of metal items: tin from Malaysia, Indonesia and Thailand, and copper from the Philippines. Australia is expected to expand further its exports of a wide variety of minerals products. Although minerals trade by ASEAN and Australia will tend to become more regionalised, most of their minerals exports are still expected to rely on markets in industrial countries outside the region, particularly Japan.

Appendix 2.1

The trade intensity index

The trade intensity index was first used by Kojima (1964), and then by Drysdale (1969). For country i's exports to country j, it is defined as the share of i's exports destined for j relative to the share of j's imports in total world imports net of i's imports. It can be expressed symbolically as follows:

$$I_{ij} = \frac{X_{ij}}{X_i} \bigg/ \frac{M_j}{M_w - M_i} \tag{1}$$

where
I_{ij} = the trade intensity index of country i's exports to country j
X_{ij} = country i's exports to country j
X_i = country i's total exports
M_j = country j's total imports
M_w = total world imports
M_i = country i's total imports

In a world where transport costs and other impediments to trade are negligible, country i's exports of commodity h to country j are expected to be the product of country j's total imports of this commodity and country i's share in total world exports of the commodity. This can be expressed as:

$$\bar{X}_{ij}^h = M_j^h \left(\frac{X_i^h}{X_w^h} \right) \tag{2}$$

Therefore, the expected value of country i's total exports to country j can be defined as the sum of the expected values of exports of all commodities from country,i to country j, or

$$\bar{X}_{ij} = \sum_h \bar{X}_{ij}^h \tag{3}$$

where $\sum_h$ means the sum for all exports h.

When the actual value of exports from country i to country $j(X_{ij})$ is replaced by the expected value $(\bar{X}_{ij})$ in equation (1), we obtain

$$C_{ij} = \frac{\bar{X}_{ij}}{X_i} \bigg/ \frac{M_j}{M_w - M_i} \tag{4}$$

where C_{ij} is an index which measures the degree to which the commodity pattern of country i's exports matches the import pattern of country j. C_{ij} is called the *complementarity* index.

The divergence between the actual and the expected value of exports from country i to country j is defined as

$$B_{ij} = X_{ij}/\bar{X}_{ij} \tag{5}$$

where B_{ij} measures the degree of trade bias caused by various trade impediments, such as transport, communication and other costs associated with trade. B_{ij} is called the *bias* index.

It should be noted that these three indexes are defined in equations (1), (4) and (5) such that the trade intensity index is the product of the complementarity index and the bias index:

$$I_{ij} = C_{ij} \cdot B_{ij} \tag{6}$$

Therefore the trade intensity of country *i*'s exports to country *j*, as measured by I_{ij}, is determined by the following two factors:

1 the degree to which the commodity pattern of country *i*'s exports is complementary to the commodity pattern of country *j*'s imports, as measured by C_{ij};
2 the degree to which this bilateral trade is biased by various impediments, as measured by B_{ij}.

Appendix 2.2

Constant market share analysis

Constant market share analysis is a method of evaluating a country's export performance relative to its competitors. A set of indicators is calculated to show whether the country has maintained its shares of all (or a group of) importers' markets for all (or a group of) commodities during a time interval. The rate of change over time in the country's exports E.. can be broken down into four terms, each with its own meaning:

$$\frac{E'_{..} - E_{..}}{E_{..}} = r \quad \text{which is called 'standard growth'}$$

$$+ \frac{\sum_i (r_i - r) E_{i.}}{E_{..}} \quad \text{which is called 'commodity composition effect'}$$

$$+ \frac{\sum_i \sum_j (r_{ij} - r_i) E_{ij}}{E_{..}} \quad \text{which is called 'market distribution effect'}$$

$$+ \frac{\sum_i \sum_j (E'_{ij} - E_{ij} - r_{ij} E_{ij})}{E_{..}} \quad \text{which is called 'competitiveness residual'}$$

where

$$r = \frac{W'_{..} - W_{..}}{W_{..}}$$

$$r_i = \frac{W'_{i.} - W_{i.}}{W_{i.}}$$

$$r_{ij} = \frac{W'_{ij} - W_{ij}}{W_{ij}}$$

Definition of terms

- E' is the exporting country's (focus country's) exports in the last year of the time period and E is its exports in the first year
- $E_{i.}$ is the total value of its exports of commodity i
- $E_{.j}$ is the total value of its exports to country j
- E_{ij} is the value of its exports of commodity i to country j
- $E_{..}$ is the total value of all its exports worldwide
- $W_{i.}$ is the total value of world exports of commodity i
- $W_{.j}$ is the total value of all world exports to country j
- W_{ij} is the total value of world exports of commodity i to country j
- $W_{..}$ is the total value of all world exports.

Notes

1 While the *actual* concentration ratio is defined as the exports (or imports) of a region destined for (or supplied by) countries in the same region as a percentage of that region's total exports (or imports), the region's share of world trade is the *expected* concentration ratio, in the sense that it is the

exports (or imports) of the region destined for (or supplied by) all countries as a percentage of all regions' total exports (or imports).

2 See the methodological explanation of trade intensity in Appendix 2.1.

3 This finding is similar to that reported by the Pacific Community Task Force on Minerals and Energy (n.d.), where it was found that the intensity index between ASEAN (excluding Singapore) and Australasia (Australia and New Zealand) ranged from 2.0 to 3.5, and that bias was relatively more important than complementarity.

4 See the brief explanation of the constant market share analysis methodology in Appendix 2.2. For a more detailed explanation, see Leamer and Stern (1970); Richardson (1971); and Tyers and Phillips (1984: Appendix 11).

5 See Barnett (1984); Meyanathan (1984; 1985a, b); Praipol (1984); Soesastro and Sudarsono (1984); Wong and Lim (1984); and Wong (1985).

References

Barnett, D.W. (1984) The Minerals Industry in Australia 1970–1980, mimeo prepared for the ASEAN–Australia Joint Research Project, Kuala Lumpur and Canberra

Drysdale, P.D. (1969) 'Japan, Australia and New Zealand: The Prospects for Western Pacific Integration' *Economic Record* 45, iii, pp. 321–42

Brown, M. and R.B. McKern (1986) *Minerals Processing, Technology and Economics for Developing Countries* Paris: OECD (in press)

Kojima, K. (1964) 'The Pattern of International Trade Among Many Countries' *Hitotsubashi Journal of Economics* 5, 1, pp. 16–36

Leamer, E.E. and R.M. Stern (1970) *Quantitative International Economics* Boston: Praeger

Meyanathan, S. (1984) Mineral Trade and Economic Development in Malaysia, mimeo prepared for the ASEAN–Australia Joint Research Project, Kuala Lumpur and Canberra

—— (1985a) ASEAN–Australia Interests in Mineral Commodity Associations and Agreements, mimeo prepared for the ASEAN–Australia Joint Research Project, Kuala Lumpur and Canberra

—— (1985b) Current State of Minerals Processing in Malaysia, mimeo prepared for the ASEAN–Australia Joint Research Project, Kuala Lumpur and Canberra

Pacific Community Task Force on Minerals and Energy (n.d.) 'Pacific Minerals and Energy Trade' (background paper)

Praipol Koomsup (1984) Minerals and Energy in Thailand: Production, Consumption and Trade, mimeo prepared for the ASEAN–Australia Joint Research Project, Kuala Lumpur and Canberra

Richardson, J.D. (1971) 'Constant Market Shares Analysis of Export Growth' *Journal of International Economics* 1, 2

Santos, T. (1984) Philippines Mineral Industry, Development and Trade, mimeo prepared for the ASEAN–Australia Joint Research Project, Kuala Lumpur and Canberra

Soesastro, H. and B. Sudarsono (1984) Mineral and Energy Development in Indonesia, mimeo prepared for the ASEAN–Australia Joint Research Project, Kuala Lumpur and Canberra

Tyers, R. and P. Phillips (1984) *Australia, ASEAN and Pacific Basin Mer-*

chandise Trade: Factor Composition and Performance in the 1970s ASEAN–Australia Economic Papers No. 13, ASEAN–Australia Joint Research Project, Kuala Lumpur and Canberra

Wong, J. and H. Lim (1984) The Changing Role of Singapore in the Trading and Processing of Mineral Resources, mimeo prepared for the ASEAN–Australia Joint Research Project, Kuala Lumpur and Canberra

Wong, J. (1985) Economic Assessment of Government Policies on Mineral Development: The Singapore Case, mimeo prepared for the ASEAN–Australia Joint Research Project, Kuala Lumpur and Canberra

3 Issues in minerals policy

CRAIG EMERSON

In recent years the governments of mineral-rich countries have sought to increase the national benefits from the development of their mineral resources. Frequently they have required higher levels of local equity participation in mining projects, progressive localisation of the mine workforce, the undertaking of local business development programs and the purchase of domestically produced mine inputs. At the same time, governments have attempted to increase their fiscal receipts from mining projects through the application of conventional taxes at higher rates and the introduction of novel forms of taxation such as progressive income taxes and taxes based on resource rents. Governments have also seen advantages in promoting the domestic further-stage processing of locally extracted minerals. Indeed, there seem to have been strong moves by the governments of many mineral-rich countries, and by international agencies such as UNIDO, UNCTAD and the United Nations Centre on Transnational Corporations, in support of increased domestic processing of minerals.

The governments of all minerals producing ASEAN countries place high priority on the domestic processing of minerals. They consider the establishment of large-scale mineral processing facilities an important component of their industrialisation programs. Australian Governments, too, at both the State and Federal levels, have favoured the domestic processing of minerals.

There has been much confusion about the nature of the national benefits that mining and mineral processing bring. In large part, this confusion has arisen out of the use of partial and inappropriate indicators of the worth of a project to the nation. These indicators have included foreign exchange effects, contributions to value added, employment generation and linkage effects. It is not uncommon to hear, for example, that the mining industries should not be taxed heavily because they make large contributions to foreign exchange, or that minerals processing should be encouraged because of the high value added that is associated with the processing of raw minerals domestically.

Apart from these various partial indicators of project worth, a number of formal methodological approaches have been suggested and used in evaluating the effects on national welfare of mining and minerals processing projects. These approaches have included the calculation of retained value and present social value. Retained value (RV) is the proportion of the foreign exchange earnings of a project retained in the host country. Present social value (PSV) is the discounted value of the lifetime stream of net benefits of a project evaluated at shadow prices that reflect the value society places on the inputs and outputs of the project.

This chapter demonstrates that PSV is the only conceptually sound method of assessing the national welfare effects of a project. It goes on to develop a methodology for evaluating the welfare effects of mining and processing projects which recognises the limitations on the availability of information that are usually faced by project analysts. This methodology provides a framework for the subsequent analysis of the various government policies aimed at increasing national welfare from mineral development.

The next section discusses alternative methodological approaches to project evaluation, and in light of that discussion it assesses critically some of the benefits that have been claimed for mining and minerals processing projects. This is followed by sections which outline the suggested practical methodology, utilising PSV, for evaluating mining and processing projects, and make use of this methodology in analysing the welfare effects of a range of government policies related to minerals development. The final section summarises and concludes.

Alternative measures of national welfare effects

Formal methods that have been used in attempting to measure the national welfare effects of projects include Retained Value (RV), the Effective Protection Rate (EPR), Domestic Resource Cost (DRC) and Present Social Value (PSV). Two well known approaches to the calculation of PSV are those developed by UNIDO (1972) and Little and Mirrlees (1969; 1974). The discussion following assesses the suitability of RV and PSV, since these are the main methods that have been used to evaluate mining and processing projects. Critiques of EPR and DRC can be found elsewhere (for example, Emerson, 1983; Warr, 1983). The discussion then proceeds to a critical assessment of the purported benefits of mining and processing projects, most of which rely upon partial indicators of project worth. These partial indicators are compared with the comprehensive measure of PSV.

Present social value

Projects involve the employment of factor and non-factor (that is, intermediate) inputs which, if used in some other employment, are normally capable of yielding output that is valued by society. The true cost of the resources used in a project, therefore, is their opportunity cost—that is, the value of the output forgone from employing the resources in the project instead of in their best alternative use. The resources used in a project produce outputs which are valued by society. These are the benefits of the project. The net benefits of the project (the net welfare gains to society) are simply the benefits minus the costs. Since the prices used in calculating these benefits and costs reflect social valuations, they may differ from actual market prices. These 'social' prices are called shadow prices.

Costs and benefits occur over time, usually producing an uneven flow of net benefits. Typically, net benefits are negative in the early years of a project as construction costs are incurred. Later, they can be expected to become positive as the value of output exceeds the costs of production. If all individuals affected by the project were indifferent to the timing of costs and benefits, the project's net benefits would be obtained simply by adding together the net benefits in each time period calculated in real terms (that is, in terms of constant purchasing power). But individuals generally prefer receiving a unit of purchasing power in the present to receiving the same unit of purchasing power in the future. In having this preference, individuals reveal a positive real rate of time preference for money. Therefore, in order for the net benefits occurring in different periods to be comparable with each other, they must be discounted to the present at an appropriate discount rate. This converts all net benefits into present-value terms. Adding them together gives the net present value (NPV) of the project, which, when inputs are valued at the appropriate shadow prices, is the project's PSV.

Two sets of circumstances may be distinguished in which project appraisal is to be undertaken. The first is where the question is whether or not to undertake a project (the alternative being to do nothing). In this case the project should be undertaken only if its PSV exceeds zero. The second case is where there is more than one project under consideration and the projects are mutually exclusive. For example, the projects may be proposed for the same site, or the number of publicly financed projects that can be undertaken may be limited by a government budgetary constraint. Then the project with the highest PSV should be undertaken, so long as its PSV exceeds zero.

A measure which has been derived from PSV is the internal rate of return (IRR). The IRR is simply the discount rate which makes the PSV equal to zero. It is the discounted cash flow rate of return on the project. The IRR has shortcomings not displayed by PSV. In some circumstances the IRR can rank mutually exclusive projects incorrectly.[1] So long as the use of the IRR is confined to situations in which the projects being evaluated are not mutually exclusive, it is a useful complement to the calculation of PSV. However, it should never be a replacement for PSV.

Retained value

Retained value is simply the share of the export revenue of a project that is retained locally. It may be expressed in percentage terms or in absolute money amounts. The amount of RV may be defined as export revenue minus the value of imported intermediate inputs, minus factor incomes paid abroad (such as wages, salaries, interest and dividends). Often a further deduction is made for the import content of local intermediate inputs.[2] Retained value has been used in a large number of studies of mining industries and projects (for example, Mikesell, 1971; Pearson and Cownie, 1974; Thoburn, 1977, 1981; ESCAP/CTC, 1981).

Extensions to the notion of RV have been made by relating it to the final consumer value of the end product of the export industry. It has been claimed that in many cases RV in countries exporting primary commodities is low in relation to the value of the final product (for example, ESCAP/CTC, 1978: 4).[3] Claims of this sort have been used in support of domestic processing of minerals (for example, UNIDO, 1980).

Advocates of RV claim that it provides useful insights into the nature and level of benefits that a project generates for the host country. They argue that a major advantage of RV is its simplicity and hence the ease with which it may be estimated, especially compared with more sophisticated measures of a project's worth such as PSV.

However, there are many pitfalls associated with the use of RV as an indicator of project worth, even as a 'first glance' method. These pitfalls include the following:

1 Retained value is a gross measure which in no way takes account of the opportunity cost of resources used in the activity. Costs could exceed benefits greatly, and yet RV may be very high. It follows that measures to increase RV may involve additional costs to the nation in return for little or no extra benefit. For example, RV could be raised through the host government's providing loan finance to the project operators at subsidised interest rates, which might negate the benefits which the country could have gained through an inflow of foreign capital. Or it could be raised through the establishment of an inefficient steel mill to supply construction material, when the domestic resources used in steel production might have been better used in some other activity. Since RV does not consider the opportunity cost of resources, one project may have a much higher RV than another, and yet offer smaller net benefits to the nation.

2 The prices used in calculating RV are actual market prices rather than shadow prices. These prices therefore do not necessarily reflect the value society places on the project inputs and outputs. For example, imported intermediate inputs subject to tariffs are priced at their tariff-ridden price (market price) in the calculation of RV, whereas their 'social' price does not include the tariff component (see Little and Mirrlees, 1974).
3 RV provides only a 'snapshot' of a project. Actual cash flows associated with a project are likely to be highly uneven over time. Retained value focuses on only one or a few of the years in the life of a project. It therefore can present a very misleading picture of the value of a project, depending on the years upon which it focuses. Moreover, RV does not provide for the discounting of net benefits over time.
4 There is no reference point or criterion in the RV method that can be used as a basis for decisionmaking. What is an acceptable level of RV: 30 per cent? 60 per cent? 80 per cent?

For these reasons, RV is not a useful measure of project worth. Its simplicity, which is claimed to be its main virtue, is actually its main deficiency. There is no easy, short-cut method of evaluating the welfare effects of mining and processing projects.[4] Estimated PSV is the only valid indicator of project worth.

Purported benefits of mining and minerals processing

Benefits commonly claimed for mining and processing projects include contributions to foreign exchange earnings; contributions to value added; direct and indirect employment generation; and the creation of backward and forward linkages. In addition, a number of other benefits have been suggested. They include technology transfer, skills creation and positive demonstration effects.

Contributions to foreign exchange earnings

Although it is often claimed that mining and minerals processing projects contribute to net foreign exchange earnings, the net foreign exchange effects are often defined incorrectly. Increased exports are a gross measure of foreign exchange effects. The value of exports minus the value of domestic intermediate inputs that otherwise would have been exported is sometimes considered to be the net effect on foreign exchange earnings. For example, the foreign exchange effects of a copper processing plant might be considered to be the value of processed copper exported, minus the value of copper concentrate that would have been exported had the processing plant not been established. But this is incorrect. Such an estimate does not include the foreign exchange content of the domestic intermediate inputs that otherwise would not be exported; for example, the import content of domestically produced chemicals used in copper processing that otherwise would have been sold elsewhere on the domestic market. And it does not include the alternative foreign exchange earnings or savings of domestic factors used in the project; for example, the export-replacing or import-replacing goods that would have been produced by employees of the copper processing plant had they not been employed at the plant. Nor does it include the capital inflows and outflows associated with foreign debt and equity financing (loans and equity capital received, principal repayments, interest payments and dividends).

A true measure of net foreign exchange effects can be obtained by calculating the foreign exchange equivalent of all inputs and outputs. This is nothing but the net value of the project (using foreign exchange as the numeraire) which, when discounted to the present, is the project's net present value to the nation, or its PSV. There is no justification for separately using some partial indicator of the foreign exchange effects of a project as a criterion in assessing its worth to the nation.

Contributions to value added

Very large increases in the value of many minerals can be observed as they pass through successive stages of processing and refinement. It may seem attractive for minerals producing countries to capture the large additions to value added associated with the processing of minerals, rather than exporting the minerals in unprocessed or semi-processed form. But any project is capable of making a gross contribution to value added. The relevant question is whether it makes a net contribution to value added, taking into account the opportunity cost of employing resources in the project. This cost is the value of the benefits forgone from using resources in the project rather than in their best alternative use. Unfortunately this cost is often ignored in speaking of the contribution of a project to value added.

The steps in arriving at the correct definition of value added for the purposes of assessing project worth are as follows:

1 The value of the project's output minus the value of intermediate inputs equals the project's *gross domestic* value added;
2 Gross domestic value added minus net factor incomes payable abroad is the project's *gross national* value added, or its gross contribution to GNP;
3 Gross national value added minus the opportunity cost of factor inputs is the project's *net national* value added, or its net contribution to GNP.

So long as the prices used in calculating the net national value added reflect social valuations, and the annual amounts of net national value added are then discounted to the present at the social discount rate, the project's net discounted national value added is equivalent to its PSV. Any gross measure of value-added or any measure of domestic rather than national value added is incomplete.

Employment generation
Direct and indirect employment generation is also an inappropriate indicator of the national worth of a project. It is often argued that, although few job opportunities are created directly by mining and processing projects, the multiplier effects of these projects can result in substantial indirect employment generation, which would be beneficial to the nation. These arguments do not consider in any systematic way the value of output forgone from attracting labour away from its alternative uses. Employment generation would represent an unambiguous net national welfare gain only under the unrealistic assumption that all labour employed as a result of the project would otherwise produce nothing at all.

Workers who otherwise would have been employed in the normal labour market would have produced marketable goods or services. And workers who otherwise would not have been employed in the formal labour market usually would have produced some goods or services valued by society, such as handmade products or household services. The value of these outputs, although probably small, must be deducted in calculating the net effects of a project on employment generation, which may be positive or negative. Whatever the net welfare gain or loss, it should enter directly into the calculation of PSV; employment generation should not be considered separately as a benefit of mining or mineral processing. Employment generation, therefore, is only a partial and gross measure of the national worth of a project. This is not to disagree with the concern national governments have for employment, but to argue that the net effects should be determined as part of a full PSV analysis.

Backward and forward linkages
Like employment generation, backward and forward linkage effects are not beneficial in themselves, although they are sometimes assumed to be. Mining projects create a demand for intermediate inputs such as electricity, explosives, chemicals, and food for the mine workforce. Minerals processing projects also demand these sorts of goods and services. In addition, it is sometimes suggested that minerals processing projects facilitate the establishment of labour-intensive downstream activities, such as small-scale metal fabricating facilities.

A linkage industry can be established only by devoting scarce resources to it. The opportunity cost of these resources needs to be considered. The possibility of a mining or processing project forging backward and forward linkages is not a valid argument for proceeding with the project without first subjecting it to an economic evaluation. After all, the original mining or processing project may not be socially beneficial in its own right. Moreover, there can be no general presumption that the creation of linkage industries in itself will make a positive net contribution to national welfare. And even if it could be determined that a linkage industry itself probably would be socially beneficial, it might be possible to establish that industry by relying on foreign markets for its inputs and outputs, rather than proceeding with a mining or processing project simply for the sake of encouraging linkage industries.

If the establishment of a mining or processing project were considered to have major advantages for a linkage industry with respect to reductions in uncertainty of supply and in transport costs, that industry would have to be included in the economic evaluation of the original project. Since the resources devoted to the mining or processing project might otherwise have created linkage effects in their alternative uses, it is only the welfare effects from the net increase in linkages that should be taken into account (difficult as this may be to estimate).

Other purported benefits

Purported benefits such as technology transfer, skills creation and the positive demonstration effects of a project on prospective investors are intangible in nature. Consequently their welfare effects, while quite possibly large, are extremely difficult to quantify. This creates the risk of exaggeration of these effects by those who have a vested interest in favouring particular mining and processing projects on noneconomic grounds. On the other hand, their positive effects might be overlooked.

A useful method of dealing with this complication is first to estimate the PSV of a project while ignoring these effects. Then, if the PSV is negative, the value of these intangible benefits must exceed the negative amount of PSV in order for the project to be acceptable. If the PSV is positive without these benefits, the project is socially desirable in any case (so long as there are not intangible costs that are even larger, such as pollution and social disruption). This approach provides a stronger basis for rational judgment on the desirability for the nation of proceeding with a project. Again, in making these judgments, it is important to bear in mind that activities which are forgone as a result of proceeding with a particular mining or processing project would also have produced intangible benefits. It is only the expected net increase in intangible benefits that should be taken into account.

A methodology for evaluating mining and processing projects

Any practical methodology for the economic evaluation of mining and mineral processing projects needs to recognise the constraints on the availability of information that are typically faced by project analysts. Given the limited information and time usually available, it is sensible, in estimating shadow prices, to concentrate on the main inputs and outputs of the project. Less attention can be devoted to the shadow pricing of minor inputs and outputs. The methodology proposed here devotes greatest attention to the shadow pricing of the inputs and outputs that are believed to be most important in typical mining and processing projects. These projects have several characteristic features which suggest areas in which shadow pricing effort might be usefully concentrated. These features are:

1 mining and processing projects are typically capital-intensive, requiring very large investments;
2 the projects often are joint activities of private firms, both foreign and local, and host governments;
3 host governments usually attempt to appropriate a substantial part of the income from mining and processing projects through taxation;
4 the projects, especially processing projects, usually require large inputs of energy, so that the issue of energy pricing becomes important;
5 the lengthy periods needed to establish large-scale mining and processing facilities makes the rate of discount a crucial variable in the analysis, but complicated borrowing arrangements often make the discount rate difficult to estimate; and
6 mining and processing projects are typically subject to a high degree of uncertainty, which limits the usefulness of any single estimate of project profitability.

These features suggest that the main effort in shadow pricing should be concentrated on physical capital, foreign finance, taxation, energy pricing and the discount rate. The proposed methodology considers these, as well as suggesting how the less important inputs and outputs might be priced with relatively less effort. The systematic treatment of uncertainty is also important. Sensitivity analysis is a useful method of treating uncertainty. It involves varying the assumed values of selected variables so as to determine the importance of those variables to the project's social profitability, as well as to gain an appreciation of the degree of uncertainty surrounding the project's viability.

But before proceeding to the shadow pricing issues and sensitivity analysis, three preliminary practical issues need to be addressed. These are: the sources of data; the choice between real and nominal values in project appraisal; and the need for omissions and simplifying assumptions in the benefit–cost analysis.

Sources of data

For projects which are proposed or planned but which have not yet commenced, feasibility studies conducted by or on behalf of the project proponents will usually be the main source of data. Access to feasibility studies, however, is not always easy, since companies are usually reluctant to release information on the expected profitability of their operations. Governments, too, may wish to preserve the confidentiality of information provided by the companies.

For the government analyst or academic researcher, caution should be exercised in interpreting the results of a PSV calculation based on company feasibility data. Feasibility studies very often seem to yield estimates of private profitability which suggest that the project will be just marginal. It is not inconceivable that the data in some feasibility studies have been manipulated so as to produce a projected level of profitability which is not so low as to make the government curious about why the company is proceeding with the project in the first place, and not so high as to entice the government into implementing measures aimed at increasing the national share of benefits from the project. For this reason it is important to conduct independent investigations into the main variables where possible, perhaps relying upon feasibility study data as a baseline. Where there is a great deal of uncertainty about the appropriate value of a major important variable, it is important that a range of alternative values be considered in the sensitivity analysis.

Where the evaluation is of a project that has been in existence for some time, actual historical data will obviously be required. Again, companies and governments are reluctant to release comprehensive data, especially on costs and profitability. Perseverance is required here too, as is usually the cooperation of the government as a source of information.

Real or nominal values?

A practical question that has led to some confusion in the evaluation of future cash flows is whether to use nominal (inflation-included) or real (inflation-excluded) values. If nominal values are used, the appropriate discount rate is a nominal discount rate, based on the nominal costs of capital. There can be some practical advantages in using nominal values. First, feasibility study data are often expressed in nominal terms. Second, if there are any mechanisms in the project, such as a progressive income tax, which can affect real values through their effect on nominal values, it is essential to use nominal values. These nominal values may then be converted into real PSV terms by discounting them at a nominal social discount rate. Nominal and real values should not be mixed.

Omissions and simplifying assumptions

Where there is insufficient information to allow a reliable estimate to be obtained of the value of a variable, or information is too aggregated to allow a more detailed analysis to be conducted, some simplifying assumptions and omissions will need to be made in the evaluation of a project. Where possible, the simplifications and omissions should be made in such a way as to yield a bias in the same direction as the estimated PSV. For example, assumptions can be made which all result in an overstatement of the project's true PSV. Then, if the estimated PSV is marginal or negative, it can be concluded that the project in reality would look even worse if the simplifying assumptions and omissions were not made. Making simplifying assumptions and omissions that result in biases in different directions would greatly increase the likelihood of inconclusive results. While bias in one direction accumulates and yields a possibly too pessimistic view, this is unlikely to be serious if the effects are themselves of minor importance.

Shadow pricing rules

It is not feasible here to deal comprehensively with all the alternative approaches to shadow pricing that have been proposed in the works of Little and Mirrlees (1969, 1974), UNIDO (1972) and in subsequent handbooks and guides. Rather, the discussion attempts to synthesise the existing techniques of project appraisal and to adapt them to the circumstances of evaluating mining and minerals processing projects.

The discussion covers the choice of numeraire; the shadow pricing of intermediate inputs and outputs; the shadow pricing of labour; the treatment of foreign financing; the treatment of taxation; the treatment of externalities; the treatment of distributional issues; and the choice of a discount rate.

THE CHOICE OF NUMERAIRE

It makes no difference to project ranking which numeraire is used. The numeraire chosen in the UNIDO Guidelines is expressed in terms of domestic prices, whereas that used by Little and Mirrlees is expressed in terms of convertible foreign exchange. Since many of the inputs and outputs of mining and processing projects are traded, and since feasibility study data are often in foreign currency terms, it is usually convenient to use foreign exchange as the numeraire.

INTERMEDIATE INPUTS AND OUTPUTS

These may be classified into three groups: fully traded goods, tradable goods which are not traded, and nontraded goods. Fully traded goods are those whose increased usage or output resulting from a project affects only the country's imports or exports and not domestic consumption or production. Tradable goods which are not traded are those that would be fully traded if the government were to pursue optimal trade policies but which, because of nonoptimal trade policies, are not traded at the margin. Non-traded goods usually have high transportation costs, and would not be traded even if the government were to pursue optimal trade policies.

The correct shadow price of a fully traded good is its relevant border price: the cif price if the good is an imported input or an output that replaces imports; and the fob price if the good is an input that would otherwise have been exported, or an output that is exported. The use of border prices rests on the small-country assumption. However, if a project is expected to influence the world price of a good, account has to be taken of the terms-of-trade effects in the shadow pricing of traded goods (see Little and Mirrlees, 1974: 68, 161; Corden, 1974: 401–5). Although the establishment of a mining or processing project will usually require an investment that is large relative to other investments in the country, any one project normally will not materially affect the world price of the output or of the traded inputs.

When trade distortions created by the application of tariffs, quotas and other protective devices are expected to persist, the correct shadow price of a tradable good that is not fully traded is the second-best shadow price; that is, these goods should be treated as non-traded goods (see Joshi, 1972: 14–19; Sen, 1972: 488–90).

The shadow pricing of nontraded goods is far more difficult to estimate than that of fully traded goods. It is usually assumed in benefit–cost analysis that nontraded goods are in less than perfectly elastic supply and demand. An increase in the usage of a nontraded good resulting from a project will therefore affect its supply and demand domestically. The project's demand for the good will raise its price, causing an increase in domestic supply and a reduction in domestic demand elsewhere in the economy. The shadow price of the good is then a weighted average of the marginal social benefit (MSB) which is forgone, as reflected by the market price of the good, and the marginal social cost (MSC) of producing the good. The weights are given by the proportions in which the project's requirements of the good are satisfied by a reduction in domestic consumption elsewhere and an increase in domestic supply (the Harberger weighted-average formula; see Harberger 1969, 1971; Warr 1977, 1982a). This formula is a partial equilibrium formula, since it does not take account of induced price changes in other domestic markets. The general equilibrium formula would be even more complicated (see Boadway, 1975; Warr, 1977, 1982a).

As a short-cut method of estimating the shadow prices of nontraded inputs, Little and Mirrlees (1974) propose the use of conversion factors. These are rough estimates of the shadow prices of major categories of nontraded inputs such as transportation and construction. For less important nontraded inputs Little and Mirrlees propose the use of a standard conversion factor (SCF), which is an estimate of the weighted average of the shadow prices of all goods in the economy divided by their market prices. Thus the SCF measures the average deviation of shadow prices from market prices in the economy. However, it is by no means obvious that the correct shadow price of any particular nontraded good will be closer to the SCF (the economy-wide average shadow price) than to its actual market price (see Warr, 1982a).

Fortunately, the most important inputs and outputs of mining and processing projects, especially in developing countries, are usually fully traded. (A notable exception is electricity, which is discussed separately below.) In the pricing of less important nontraded goods there can be no presumption that the use of such highly aggregated conversion factors as the SCF will provide a better approximation of the correct shadow price than the use of actual domestic market prices. Moreover, the use of domestic market prices (converted into the numeraire at the official exchange rate)[5] has important advantages over the use of highly aggregated conversion factors. Domestic market prices are readily observable and verifiable, and they are objective. When domestic market prices are used in project appraisal there is less scope for those with vested interests, such as politicians and bureaucrats, to manipulate the results of benefit–cost analyses to support the projects they favour on noneconomic grounds.

As for the shadow pricing of electricity, the relevant indicator is the long-run marginal cost (LRMC) of supplying electricity to the project. The main categories of costs will be capital, fuel, labour, materials, overheads, and transmission costs. A detailed discussion of electricity pricing appears in Munasinghe and Warford (1982). The discussion below outlines some of the main issues in estimating the LRMC of supplying electricity.

A public utility supplying electricity to a mining or processing project often will have to install additional generating capacity to meet the needs of the project. A portion of the capital cost of this additional capacity should be allocated to the project, the portion being the share of the additional capacity that will be devoted to serving the project. An estimate will be needed of the interest rate to apply to the provision of this capital. The relevant interest rate is the social opportunity cost of supplying capital to electricity generation.[6] The capital charge, including interest, is then annualised.

Next, the shadow price of the energy used in electricity generation is estimated. The appropriate reference point is the value of the fuel in its best alternative use. For example, coal used in electricity generation could often be exported, so that the fob price provides a benchmark. Adjustments will have to be made for quality differentials and for the costs of transportation to a seaport.

As for the other categories of costs, the actual market prices will often suffice, since these costs usually represent a fairly small proportion of the total cost of electricity supply to a mining or processing project.

Once all costs have been estimated in annual terms, they can be converted to the cost per kilowatt hour (kWh) of electricity supplied to the project and used directly in the project evaluation. Where there is room for doubt about the correct shadow price of some inputs into electricity supply (especially the social opportunity cost of capital and the shadow price of energy), a range of alternative prices would need to be considered in the sensitivity analysis.

Shadow wage rates

In addition to reflecting the social opportunity cost of a worker, shadow wage rates (SWRs) have been adjusted in complex ways in the literature in an attempt to reflect a government's desire to achieve redistribution of income between groups and over time (Little and Mirrlees, 1969, 1974; UNIDO, 1972; Squire and van der Tak, 1975). Serious doubts have been raised about the theoretical basis of these adjustments (see Warr, 1982b).

Deriving estimates of the social opportunity cost of a worker is difficult enough. The procedures involved have been the subject of considerable controversy (for a critical discussion see Lal, 1974: 35–9; Warr, 1982b: 21–8). Since mining and minerals processing projects are not labour-intensive, and since the use of complex SWR formulas is not firmly established in theory, it seems reasonable, in the pricing of unskilled and skilled local labour, to use the market wage converted at the official exchange rate. As for foreign labour, the cost to the nation is the foreign exchange used in employing that labour.

Foreign financing

Foreign capital inflows are a benefit to the host country, while foreign capital outflows are a cost. Capital inflows associated with a project consist of loans and equity contributions; capital outflows comprise interest, dividends, principal repayments and any repatriations of equity capital. For each year in the life of a project the net capital flow can be calculated by adding together the inflows and the outflows. In the early years it is likely that the net foreign capital flow will be positive for the host country, with debt and equity contributions outweighing any remittances. Subsequently, with the repayment of loans and the remittance of dividends, the net foreign capital flow normally will be negative. Assuming the foreign financing is tied to the project (such that otherwise it would not enter the country), it is evaluated by calculating the NPV of the stream of net foreign capital flows using the social discount rate. If the resulting NPV is positive, the present value of inflows exceeds the present value of outflows, and the foreign capital is being supplied on 'concessionary' terms.[7]

The next step is to add the NPV of the foreign financing to the NPV of the flow of goods and services, giving the PSV of the project. This can be done in practice by adding the stream of financial flows to the stream of flows of goods and services, and then discounting the combined stream at the social discount rate.

Where the foreign financing is not tied to the particular project (for example, in the case of untied grants of foreign aid), it is necessary to include an estimate of the social opportunity cost to the nation of the foreign funds. This involves deducting from the NPV of the foreign financial flows an estimate of the PSV of the most socially profitable project forgone as a result of using the foreign funds in the project under evaluation.

Taxation

Governments apply a variety of taxes which affect the social profitability of mining and processing projects. It is not feasible to examine all of them here. Instead, the discussion below lays out some principles for the treatment of taxation revenue in the benefit–cost analysis of mining and processing projects.

Revenue generated by taxes which otherwise would have been lost to the nation (through profits to foreign investors) is a national benefit. Taxes paid by host-country nationals represent pure transfers and therefore do not constitute a national benefit, except when the tax arises from income which would not otherwise have been earned.[8] Each year in the life of a project it is necessary to include as a national benefit the proportion of revenue from various taxes which otherwise would have accrued to foreigners. For example, if foreign equity in a project is 40 per cent and income tax revenue in a given year is $60 million, the national benefit from income tax in that year is 0.4 multiplied by $60 million, or $24 million.

Special taxes on mining are usually levied by governments, and it is not uncommon for governments to use them as a vehicle for encouraging the domestic processing of locally extracted minerals. These are discussed separately in the next section.

Even if some of these effects prove too difficult to estimate, they should not be ignored altogether. A discussion of these effects, together with estimates of possible orders of magnitude of the associated national income gains or losses, should be included in the overall evaluation of the minerals processing project.

Externalities

An argument often used in favour of mining and processing projects is that they generate positive externalities,[9] such as labour training and the construction of roads which are used free of charge by individuals not associated directly with the project. It is not easy to value these externalities in benefit–cost analysis.

In the evaluation of some mining and processing projects the exclusion of positive externalities may not be an important omission. The resources diverted into these projects are likely to have generated positive externalities in their alternative uses if the resources were already available. It is therefore only the *increase* in the social value of positive externalities which is relevant. It is not evident that mining and processing projects generate more positive externalities than other projects. In fact, it is quite possible that they generate more negative externalities. Mining and processing projects can be polluting, and those harmed by the pollution are often not compensated, or not fully compensated, out of project proceeds.

On balance, it seems unlikely that any failure to quantify the positive and negative externalities associated with mining and processing projects will result in an underestimation of their social profitability. A qualitative treatment of these effects is desirable.

Distributional issues

Governments may attempt to use projects to help achieve the desired redistribution of income. Theorists on project appraisal have developed income weighting systems which they claim can be supportive to government distributional objectives. For example, they may weight project income received by the poor more heavily than income received by the wealthy. Major criticisms have been levelled at these weighting systems, relating to their theoretical foundations and to their arbitrariness (for example, Warr, 1982b). Furthermore, the use of (non-uniform) distributional weights in project evaluation is based upon a presumption that the government is constrained in its ability to raise sufficient taxation revenue from other sources to achieve the distribution of income that it desires, so that it has to use projects for redistributive purposes. In the case of mining and processing projects the government presumably can use taxation revenue derived from them for redistributing income, rather than using distributional weights to affect the estimated PSV (and, hence, the perceived social profitability) of the projects. These considerations suggest that the use of distributional weights should be avoided in the economic evaluation of mining and processing projects.

The choice of a discount rate

If a country is a net borrower on international capital markets, the effective international borrowing rate for the country provides a lower bound for the social discount rate.[10] This is the rate at which the country can substitute between present and future consumption. If a loan is being extended to the project on commercial terms by international lenders, the interest rate on that loan is the appropriate international borrowing rate. The *effective* interest rate is the internal rate of return (IRR) on the stream of financial flows associated with the loan. These flows comprise loan drawdowns (positive), and loan repayments, interest payments, financing fees and bank guarantees (all negative).

There are circumstances in which the social discount rate for the project will exceed the international borrowing rate. These circumstances have the common feature that they necessitate some of the finance for the project being obtained from domestic rather than international sources. The first is where a country faces problems of creditworthiness. When a country has already borrowed heavily on international capital markets, foreign bankers may respond, not by increasing the interest rate on the proposed loan but by refusing to extend the loan altogether. Alternatively, the country's central bank might refuse permission to borrow further in an effort to protect the country's reputation for creditworthiness. At least some of the required funds must then be provided by the local capital market. This also occurs when governments require local equity in projects.

When some of the project financing must be obtained locally, the marginal productivity of that capital enters into the calculation of the social discount rate. The social discount rate becomes a weighted average of the international borrowing rate for the project and the marginal productivity of local capital, the weights reflecting the proportions in which the funds are raised from the two sources.

Treatment of uncertainty

Several features of mining and processing projects make them subject to an unusually high degree of uncertainty. These include the following:

1 the projects are typically energy-intensive and future movements in energy prices are difficult to predict;
2 metal prices are often highly volatile;
3 once construction has begun, the very large investments involved make the social profitability of the projects sensitive to even minor delays in the commencement of production;
4 the disposal of byproducts such as sulphuric acid, which would be polluting if unsold or untreated, can pose problems that are not easily anticipated;
5 large and unpredictable movements in international interest rates usually have an important bearing on the social profitability of mining and processing projects, through interest obligations on foreign loans and through the effect of international interest rate movements on the social discount rate.

There are two basic methods of treating this uncertainty: probability analysis and sensitivity analysis. While probability analysis in principle is superior to sensitivity analysis, it is far more demanding on information (see Pouliquen, 1970; Reutlinger, 1970; Anderson, Dillon and Hardaker, 1977). Sensitivity analysis will usually have to suffice. Sensitivity analysis should be conducted for each variable, including those discussed above, that is expected to have a significant bearing upon the project's PSV. It is useful to calculate the *switching value* for each of these variables. This is the value that just changes the sign of the PSV. A variable will have a switching value that is very close to its assumed value, indicating that the project's social profitability is sensitive to even small changes in such variables. Sensitivity analysis is useful in indicating how marginal a project is under changing circumstances. It can also identify variables for which greater effort in estimation is warranted.

Analysis of policy measures

It is widely recognised that mining projects tend to perform as enclaves, especially in developing countries. That is, their effect on aggregate private national income, apart from the indirect effects of the expenditure of government revenue obtained from the projects, is typically small (see Bosson and Varon, 1977: 101; Gillis et al., 1978: 6–8; Zorn, 1978: 11; and Diaz Alejandro, 1979: 289). Governments aware of the weakness of the backward and forward linkage effects of mining projects have applied policies aimed at increasing national welfare from mineral development. These policies have included requirements relating to taxation; lease allocation; public and local private investment; domestic purchases of inputs; manpower training and localisation; and downstream processing, fabricating and marketing activities.

Taxation

Taxation is usually the main source of national benefit from mining projects, especially when the projects perform as enclaves. (For a review of the empirical evidence, see Emerson, 1982: 562–3.) The role of taxation is established in the system of property rights governing mineral development. In most countries mineral resources are community property. On behalf of the community, governments allow private companies exclusive rights to explore for and develop mineral deposits in particular areas. No other companies may operate in allocated lease areas. These restrictions on entry are necessary to allow efficient exploration and mineral development. A 'free-for-all' is in no one's interests, since it results in inefficient exploration and production.

The restrictions on entry into these areas allow for the existence of pockets of resource rent. Resource rent is the profit remaining after *all* costs have been deducted. These costs comprise exploration outlays, project development costs and cash operating costs. They include a return on capital which is just sufficient to attract investment to the area (the minimum required return on investment).

In this respect an economist's notion of costs, and therefore of profit, differs from an accountant's notion. An economist treats the entire minimum required return on investment as a cost and defines any return in excess of that minimum as profit. In the resources area this profit is known as resource rent. An accountant, however, does not regard all of the minimum required return on investment as a cost, but only the actual interest charges on debt. Any return to equity capital, including the minimum required part, is regarded by an accountant as profit. Thus an economist does not differentiate between the source of investible funds, whereas an accountant does, by treating the returns on equity differently from the returns on debt.

Resource rent arises because some mineral deposits are naturally richer in quality or closer to a final market than others. These quality and locational advantages are not of the investor's making; they are given by nature. Some exploration prospects and operating mines therefore will be naturally more attractive than others. Those exploration prospects or mineral developments which offer an investor a return just equal to the investor's minimum required return are regarded as marginal, since they have no expected rent. Exploration prospects or mineral developments offering a return in excess of the minimum required return are better than marginal and have positive expected rent. Other prospects or developments will offer a submarginal return.

So at any time there may be a whole range of prospects and developments—some marginal, some submarginal, others a little better than marginal, and still others that may be much better than marginal. In return for allowing a private investor exclusive access to community property, a government has a legitimate interest in obtaining for the community the highest possible share of the economic profits (resource rent) from resource projects that are better than marginal, while allowing the investor to achieve his or her minimum required return.

A resource rent tax aims to do this. In principle, resource rent can be taxed without deterring investment in exploration or mineral production. Because it avoids this deterrence, a tax on resource rent is consistent with high levels of exploration and mineral development. This in turn provides for the highest possible tax base. Mineral deposits that would be discovered in the absence of taxation are not left undiscovered because of the imposition of a resource rent tax, and known deposits that would have been developed and extracted are still developed and extracted under a resource rent tax.

But other taxes can deter exploration and full mineral development. These include taxes based on the volume or value of output (specific and ad valorem duties) and taxes based on the accountant's notion of profits. Because they have these deterring effects (that is, they are 'non-neutral'), these taxes are not consistent with maximising the taxation revenue (and hence the community's return) from a country's mineral resource endowment.

Specific and ad valorem duties are payable on every mineral unit produced, regardless of the cost of producing that unit. Recall that some mines will be only marginal or a little better than marginal. In addition, some production from deposits within a given mine will be only marginal or a little better than marginal; that is, the costs of extraction will rise as lower quality and less accessible parts of the deposit are encountered. At some point there will be a commercial cut-off grade in the deposit, beyond which the cost of extraction exceeds the price of the mineral. Obviously there are advantages for both government and investor in the extraction of all parts of the deposit above that cut-off grade, since their extraction can yield both profits to the investor and taxation revenue to the government.

By adding to the costs of extraction, specific and ad valorem duties raise the cut-off grade; that is, they prevent the full economic recovery of minerals from operating mines. To illustrate, suppose that the expected price of a mineral is $25 per unit. The mine operator will extract all the mineral in the mine whose expected cost of extraction is below $25 per unit. But now suppose a 10 per cent ad valorem royalty is applied. At the mineral price of $25 this ad valorem royalty is equivalent to a specific royalty of $2.50 per unit. The operator now will extract only those mineral units in the deposit whose expected extraction cost is below $22.50, since the operator now has to pay a royalty of $2.50 per unit in addition to the normal extraction costs. The parts of the deposit whose expected extraction costs lie between $22.50 and $25 per unit will remain unused as a result of the royalty. Everyone loses from this deterrence to full economic recovery from operating mines.

Specific ad valorem duties not only act as a wasteful deterrent to full economic recovery from operating mines; they can completely deter the development of mines that are only marginal or a little better than marginal, and they can have the same effect upon exploration. The extent of this deterrence will be greater the higher the rate of duty.

A specific or ad valorem duty applied on a sliding scale can be an improvement on one applied at a constant rate, insofar as the minimising of deterring effects is concerned. Under a sliding scale system the rate of duty rises with the volume or price of output. The lower rates of duty on lower levels of output or price are in recognition of the tendency for profitability to be lower at low levels of output or price. But because a specific or ad valorem duty does not recognise actual costs at all, any reflection of actual costs through the sliding scale of duty rates can be only approximate at best. There will be some projects from which the sliding scale duty would collect more than the resource rent. So, while a specific or ad valorem duty applied on a sliding scale might be less deterring than one applied at a constant rate which was designed to collect the same amount of revenue, the deterring effects of a sliding scale duty can nevertheless be substantial.

A tax based on the accountant's notion of profit also has deterring effects, since it taxes part of the minimum required return on investment. The ordinary company income tax is the best known example of an accounting profits-based tax. Other examples are royalties which provide the same sorts of deductions as company tax. An accounting profits-based tax differs from a resource rent tax in two important respects. First, it permits the deduction of capital expenditures only over a number of years, whereas a rent tax allows the immediate deduction of these and all other costs. Second, it provides for the deduction of interest charges only, whereas a rent tax recognises the cost of providing both debt and equity capital by permitting the investor to earn a minimum required return on total (debt plus equity) investment before tax becomes payable.

The various non-rent taxes also are often incapable of capturing a large share of the rents from highly successful mineral resource developments. Since they are not based on economic profits, they do not adjust automatically to variations in actual profitability between projects and over time. Governments applying these taxes usually come under pressure to raise the tax rates or to introduce new taxes when mining profits are believed to be large. Investors are aware of this and tend to view taxes that require frequent discretionary changes by the government as being inherently unstable. This instability raises investors' perceptions of political or sovereign risk, which is the risk that a government will change the fiscal rules after investment has been made in a venture. Higher levels of sovereign risk tend to raise the minimum required return on investment. In turn, this lowers the rent value of mineral resources.

Taxes based on the actual rent value of mines increase the risk to the government that it will receive little or no revenue from a given mine. But they also increase the amount of government revenue that can be expected to be obtained from the minerals sector at large, since more mines are likely to be opened under a rent-based taxation system and mineral recovery from operating mines is likely to be greater. Through these effects a rent-based taxation system actually broadens the taxation revenue base compared with the base that would exist under a system of non-rent taxes designed to collect a similar amount of revenue.

A tax which has been developed with the taxation of resource rent explicitly in mind is the resource rent tax (RRT) designed by Garnaut and Clunies Ross (1975). The RRT is payable only if and when an investor has recouped the total investment, together with some minimum return on investment. The RRT may be interpreted as a more practical variant of the tax first proposed by E. Cary Brown (1948), in which the government makes payments to the investor whenever negative cash flows are realised and receives payments from the investor whenever positive cash flows are realised. Governments might be reluctant to enter into or honour commitments relating to the making of potentially large cash payments to mining companies. The RRT attempts to avoid making these payments, by carrying forward any negative net cash flows to be offset against positive cash flows. However, if there are insufficient positive cash flows against which to offset accumulated negative cash flows, the RRT can collect more than the full rent value of a project. The most realistic rent tax might be one which combines the features of the RRT and the Brown Tax, such that government payments are made on high-risk exploration, and in the event of any unrecouped losses at the end of a project, but negative cash flows at the development stage are carried forward to be offset against positive cash flows. These proposals for rent taxes, together with other issues of practical application, are developed more fully elsewhere (Garnaut and Clunies Ross, 1983; Emerson and Garnaut, 1984; Emerson, 1984).

The enclave nature of mining projects is consistent with their making large contributions to national welfare, so long as appropriate fiscal arrangements are put in place. A rent-based taxation system can be expected to maximise government revenue from foreign investment in mining projects. It therefore lends support to the objective of maximising the PSV of mineral development. Policy measures aimed at deriving nonfiscal benefits from mining projects, through integrating the projects more completely into the national economy, can affect adversely the amount of mining taxation revenue that can be collected. In evaluating the national welfare effects of these nonfiscal policies consideration needs to be given to the effect of the policies on mining taxation revenue.

Lease allocation

The method of allocating exploration and production rights to private interests can have profound effects on national welfare from minerals development. The relationship between methods of lease allocation and minerals taxation policy has been clearly identified only recently (Willett, 1983; Smith, 1984). This relationship has important policy implications: minerals taxation policy cannot be developed in isolation from policy on lease allocation if inadvertent erosion of the minerals taxation base by the system of lease allocation is to be avoided.

Methods of lease allocation used by governments can be classified as one of, or a combination of, the following: first-come-first-served; work-program bidding; and cash bidding. The first-come-first-served method allocates leases to the first applicant. Usually conditions are applied relating mainly to the amount, nature and timing of exploration work to be done by the applicant. Provision is made for withdrawal of all or part of the lease if these conditions are not satisfied within a specified time. The work-program bidding system involves competition between applicants on the amount, nature and timing of exploration, subject to minimum conditions being met on technical expertise and financial backing. Limits are imposed on the duration of leases. Both the first-come-first-served and the work-program bidding systems provide a large amount of discretionary power to the authorities in deciding what constitutes a desirable exploration program.

Cash bidding involves the auctioning of exploration rights. In its pure form, cash bidding is unconditional; the successful bidder, having made a lump-sum cash payment to the government, is left to choose his optimal exploration program, and has automatic development rights in the event of a commercial discovery. In practice, cash bidding is rarely unconditional. It usually takes on some of the restrictions of work-program bidding, particularly in relation to limits on the duration of leases and to associated requirements to relinquish some or all of the lease area after a specified time.

Owing to the insecurity of tenure given under the first-come-first-served and work-program bidding systems, there is a tendency for explorers to dissipate the expected rent of exploration acreage, thereby eroding the minerals taxation base. Explorers operating under these systems know that retention of the lease is conditional upon their fulfilling commitments entered into with the administering authorities on exploration programs. There can be no presumption that the conditions laid down by the authorities will maximise the expected rent value of the exploration acreage. Under the work-program bidding system, applicants compete with each other in terms of the exploration work they will offer to undertake and, under competitive pressure, will bid away all of the expected rent in the form of work-program commitments. These commitments will be greater than those that would maximise the expected rent value of the area; the expected rent will have been dissipated.

It might be thought that this problem of rent dissipation under the work-program bidding system could be ameliorated through the authorities relaxing their requirements for further work in the event of the first phase of exploration proving unsuccessful. But applicants, knowing that the authorities are sympathetic to their requests for a revision of commitments, submit unrealistically ambitious work programs in the first place. The authorities then have no meaningful way of selecting the best applicant.

The dissipation of expected rent through excessive and premature exploration under the first-come-first-served and work-program bidding systems means that the expected value of receipts from resource rent tax (under which exploration is a deduction) is reduced towards zero. Cash bidding does not have this effect, at least in its purer forms, because competing applicants bid away the expected rent in the form of offers of cash payments to the government, rather than dissipating it through additional work-program commitments.

As a charge against the expected rent value of leases, cash bids constitute a tax on anticipated resource rent. The Brown Tax or RRT are taxes on *realised* resource rent; they are conditional on the actual amount of rent realised. Cash bidding and conditional rent taxes therefore go hand in hand as the optimal package of charges for allocating community resource property. The question arises of the relative emphasis to be placed on the bid part and the conditional part of the rent tax regime. This is governed by the previously announced details of the conditional tax, especially the rate of tax. Participants in the auction for exploration rights will formulate their cash bids in the light of these details. The higher the rate of conditional tax, the lower will be the cash bid.

The socially optimal combination of bid and conditional tax is determined in part by the attitudes to risk of the government and the bidders. If total reliance is placed on the cash bid (such that there is no conditional tax), the private bidder takes all of the risk; if total reliance is placed on the conditional tax (through a Brown Tax at the rate of 100 per cent) the government takes all of the risk. Between these two extremes of zero and a 100 per cent rate of conditional tax, where some reliance is placed on the conditional tax and the remainder on the cash bid, risks are shared between the private bidder and the government. The optimal degree of risk sharing is influenced by the degree of risk aversion of the private bidder relative to the government. The greater the risk aversion of the private bidder relative to the government, the greater is the reliance on the conditional tax (that is, the higher is the rate of conditional tax) in the optimal arrangements (see Leland, 1978; Emerson and Lloyd, 1983; Emerson and Garnaut, 1984). It has been argued that on both theoretical and practical grounds heavy emphasis should be placed on the conditional rent tax, with cash bidding picking up any residual expected rent as well as serving as an efficient method of lease allocation (Emerson and Garnaut, 1984).

Public and local private investment
Governments often require some national public or private equity participation in mining projects. Often this is desired for symbolic rather than purely economic reasons. Local equity participation on less than fully contributing terms (free equity, or partially paid equity) is effectively another form of tax. Free equity, under which the government becomes a part-owner and is entitled to a share of the profits without contributing any equity capital, has an effect similar to the ordinary income tax.

An important consideration bearing upon the national welfare effects of policies requiring minimum levels of local equity is the direct cost of foreign equity relative to the opportunity cost of local equity. The direct cost of foreign equity is given by the IRR on the stream of financial flows associated with the foreign equity financing (the equity inflows minus the foreigners' share of the project's income in each period). If this IRR is greater than the opportunity cost of local equity, the socially optimal level of foreign equity is zero. If the IRR is smaller than the opportunity cost of local equity, the optimal level of foreign equity is 100 per cent. In practice, it is difficult to predetermine whether the IRR will be greater or smaller than the opportunity cost of local equity for a particular project, since the IRR will depend on the project's actual profitability. Nevertheless, it usually will be instructive to estimate the IRR on foreign equity capital flows for actual and planned projects, in relation to the opportunity cost of local equity, as a step in the analysis of the welfare effects of local equity requirements.

It should be remembered, too, that the revenue objectives of local equity participation on fully contributing terms can often be achieved through taxation. The direct economic justification for local equity participation is weakened when taxation could have collected a similar amount of revenue.

Domestic purchases of inputs

Governments often specify in mining contracts that preference be given by the mining company to domestic suppliers in purchasing inputs for the mine and the mine workforce. Usually there is some provision for recourse to foreign suppliers if local products are markedly inferior in quality, if their price is much greater or if their supply is highly irregular. Nevertheless, some element of protection of local products against imports is usually implicit in domestic purchase requirements.

Mining companies would purchase some products locally in the absence of these requirements. It is only the increase in purchases associated with domestic purchase requirements that is relevant to the analysis of the welfare effects of these policies. Estimating these welfare effects involves weighing the national benefit of the domestic purchase requirements against the national welfare losses associated with the purchase of products by mining companies, effectively at a higher cost.

The national benefit is the increase in the income of nationals resulting from the requirements. The national cost is the loss of mining taxation revenue and of the income of national shareholders in the mining company resulting from the increase in costs that arises from the requirements. At higher levels of cost some mine production may be deterred. The national public and private income forgone as a result of this reduction in mine output in that case is an additional source of national loss.

If income taxes or rent taxes are applied at high rates on mining projects, the host government bears a large proportion of any increase in mining costs resulting from domestic purchase requirements. For example, if the marginal rate of tax is 70 per cent, the government bears 70 per cent of any increase in costs (in the form of mining taxation revenue forgone).

If subsidisation of local suppliers is perceived as being desirable, a less distorting means of doing this is to use some of the taxation revenue to provide direct subsidies to the local suppliers, or to provide the subsidies directly from consolidated revenue. As this does not add to mining costs, mining decisions are unaffected by the policy.

There may be circumstances, however, in which domestic purchase requirements can be effective in increasing national welfare from mining. Very often the community in the vicinity of a mine site suffers social disruption as a result of the mine, and an anti-mine sentiment can emerge. This can be fuelled by perceptions that the main beneficiaries of the mine are outsiders, while the local community bears much of the cost. Local groups very often possess the means to disrupt mine construction and mining operations. These sorts of pressures can be reduced by linking the interests of the local community to the success of the mining project. This can be done through local purchase requirements involving the participation of the community in the immediate vicinity of the mine site. The mining company might be asked to formulate and implement a local business development program, designed to raise the capacity of the local community to supply goods and services to the mining project on reasonably competitive terms.

The national costs of government stipulations regarding domestic purchases and local business development programs are easier to estimate than the benefits. An estimate of the increase in mining costs associated with these requirements can often be obtained from mining companies. It is then relatively straightforward to examine the effect on the project's PSV of raising costs by the estimated amount.

As a general presumption, it seems likely that reasonable local business development programs will often generate net increases in national welfare, but that strict domestic purchase requirements, through their effects on mining taxation revenue and on the income of national shareholders in mining enterprises, will be costly to the nation.

Manpower training and localisation

Host governments, concerned about continued reliance on foreign companies to develop the country's mineral resources, often require that nationals be trained in mining operations and administration, and that the mine workforce be progressively localised.

Voluntary training of nationals by mining companies can represent a genuine positive externality. If the mining company bears the cost of generating human capital through training and this human capital is released at some stage for employment elsewhere in the economy, the PSV of the increased production resulting from this human capital should be credited to the mining project (see Lal, 1975: 60–5). However, when some of the training is not pursued voluntarily but is a result of government requirements, there is an additional national cost associated with producing the benefits of training. This cost is the national (public and private) share of mining income forgone as a result of increased training costs. The national benefit is the PSV associated with the externality of training, plus the increase in the income of the trained nationals during their employment in the mining project.

Requirements for localisation above the level and rate which would be pursued voluntarily by mining companies can produce complex welfare effects. First, these requirements produce a transfer of income to national workers. The share of this income that otherwise would have accrued to foreigners is a national benefit. On the cost side, there is the opportunity cost of the local labour employed in response to the localisation policy, and the PSV forgone as a result of any reductions in mining income.

In general, some training and localisation will be pursued voluntarily by mining companies, especially since the wages and salaries of local manpower are usually much lower than the cost of employing expatriates. Training and localisation also offers the advantage to foreign mining companies of improving their public image in host countries. Any host government requirements in excess of these levels can be evaluated by reference to the national benefits and costs of these policies.

Downstream activities

Policies aimed at encouraging the domestic processing of minerals may take several forms, including minimum domestic processing requirements, perhaps accompanied by a lower selling price for unprocessed mine output; taxation penalties and concessions; and subsidies on government-provided processing inputs.

If there is a genuine domestic divergence associated with the domestic processing of minerals such that the social cost of processing is lower than the private cost, the first-best policy to achieve the socially optimal level of processing is to subsidise the input for which the domestic divergence exists, while applying a rent tax on mining. Other policies, which can have the effect of lowering the production of raw minerals, involve greater cost to the nation in achieving a given level of processing input, and are therefore second-best.

To illustrate how the welfare effects of a second-best mineral processing policy might be taken into account, consider the case of minimum domestic processing requirements accompanied by a lower selling price for mine output.[11] The effect of a lower selling price on mine output is similar to that of a production-based tax. It will deter mining projects that are expected to be marginal at international mineral prices and it will prevent the full economic recovery of minerals from operating mines. Figure 3.1 represents the mining industry in the host country. The host country is unable to influence the world mineral price, P*. The long-run industry supply curve is LMC and optimal output is OQ*.

A lower price, P_1, reduces industry output to OQ_1. This results in a deadweight loss in producer surplus (resource rent) equal to the area of triangle B. Rectangle A is also lost. The national loss associated with the loss of rectangle A and triangle B is that part of the income represented by these areas which otherwise would have accrued to the host government as taxation and to host country nationals as dividends.

Once an estimate is made of the reduction in selling price resulting from selling mine output to the local processing plant, it is not difficult to estimate the effect on collections of the various taxes at the new lower output (OQ_1). To these estimates must be added estimates of the losses in mining taxation revenue associated with the contraction in mine output from OQ* to OQ_1. This is more difficult, since it involves estimating the elasticity of supply of mine output. A simpler but not highly reliable procedure would be to question mining company executives about the likely effect on mining decisions of a specified reduction in selling price. On the basis of their responses on output losses, a crude estimate of the associated revenue losses might be made.

A second example of second-best mineral processing policies is the application of an export tax on unprocessed minerals with no export tax on processed minerals. A mining company could avoid the export tax by establishing domestic processing facilities, the intention of the export tax policy. The same applies to the application of higher royalty rates on processed than on unprocessed minerals. The difference in royalty rates is effectively an export tax.

Any losses in export tax revenue associated with domestic processing should enter into the economic evalution of mineral processing projects.[12] These losses can be estimated by multiplying the export tax rate by the quantity or value of mine output which is processed domestically (depending on whether the export tax is quantity-based or value-based). On the other hand, mine output will tend to be larger when a company agrees to process its mine output domestically, since the company will not be subject to the distorting effects of the export tax. Then there will be some offset to the loss in export tax revenue. The offset will be equal to the national (private and government) share of any increase in mining income associated with the increase in mine output.

Summary and conclusions

The conceptually correct measure of national welfare from mining and mineral processing is present social value. However, governments often claim that their policies towards mining and processing have been formulated with a view to generating foreign exchange, value added, employment and domestic linkages. These purported benefits are partial, and are often misleading indicators of the national worth of projects. Policies aimed at obtaining them may therefore not be consistent with increasing national welfare from mineral development.

The methodology for the appraisal of mining and processing projects proposed in this chapter provides a framework for tracing more comprehensively the welfare effects of particular projects and policies relating to mining and minerals processing. These policies concern taxation, lease allocation, local equity participation, domestic purchases of intermediate inputs, manpower training, and downstream processing.

There is a close relationship between the method of mining taxation and the process of lease allocation. The combination that maximises national welfare is a tax based on realised resource rent combined with cash bidding for exploration leases. Local equity participation on concessionary terms is, in effect, just another form of tax. Local equity participation on fully contributing terms can offer advantages over taxation in some circumstances, but when it is required simply for the purpose of generating revenue, its objective can be served equally well through a rent tax. Requirements for the domestic procurement of intermediate inputs and for manpower training and localisation tend to raise the cost of minerals production. In evaluating the national welfare effects of these policies, consideration needs to be given to the effect that the policies have on mining taxation revenue. This also applies to policies aimed at promoting the domestic processing of locally extracted minerals, such as minimum processing requirements and differential royalty rates graduated according to whether or not mine output is processed domestically.

Much has been made in the past of the data limitations confronted in attempting to evaluate the welfare effects of minerals projects and policies. Often the problems have been overstated and recourse made to more convenient, short-cut methods of investigation. The short-cuts involved can be very costly when they lead to the adoption of policies which do not increase national welfare. At the very least, a nonempirical investigation utilising the concept of present social value as it relates to national benefit–cost analysis provides the policymaker with a broader view of the welfare effects of alternative minerals policies.

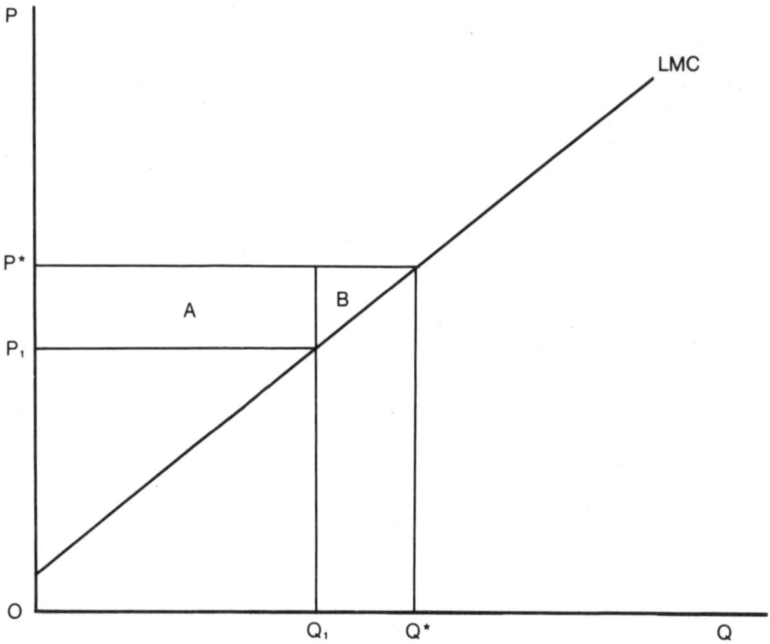

Figure 3.1

Notes

1 This can occur when the two projects involve different amounts of investment, and the larger project, although it offers a lower rate of return, offers a larger absolute discounted value of net benefits when evaluated at the appropriate discount rate.

2 For a discussion of retained value, see Mikesell (1971), and ESCAP/CTC (1981: ch. 1).

3 In the case of metallic minerals, the final product has usually been nominated as metal ingots, cathodes or wire bars, and not fabricated metal products.

4 This conclusion also applies to other short-cut methods, such as EPR and DRC. On the use of EPR, see Corden (1971); Balassa and Schydlowsky (1968; 1972); and Bhagwati and Srinivasan (1980). On DRC, see Lal (1974); Little and Mirrlees (1974); Pearson (1976); Srinivasan and Bhagwati (1978); and Warr (1983). Emerson (1983) provides a critical discussion of both methods.

5 Converting domestic prices into foreign exchange terms at the shadow exchange rate (rather than the official exchange rate) is equivalent to applying the SCF to the good; see Squire and van der Tak (1975: 33–6, 93–7, 128–30).

6 See the later discussion on the choice of a discount rate.

7 The notion of concessionary financing in this context has the meaning that foreign funds are being supplied at an effective rate that is below the host country's social discount rate.

8 This assumes that the distributional weights of all income groups are uniform; see the later discussion under 'Distributional issues'.

9 For a critical discussion of externalities in project evaluation, see Little and Mirrlees (1974: 335–49); Lal (1975: 74–6); and Squire and van der Tak (1975: 21–3).

10 Some benefit–cost theorists, including Harberger (1969) and Little and Mirrlees (1969), advocate attaching a premium to the discount rate to reflect the suboptimality of savings in the country. These ideas have been challenged by Warr and Wright (1981). The methodology suggested here does not attach any such premium.

11 Companies could expect a lower price on domestic sales compared with exports in any case, since they would usually face lower transportation costs. The situation described in the discussion, however, is one in which mining companies receive a price for domestic sales which is lower, even after adjusting for transport-cost differences.

12 There would be no export tax revenue losses if the mining company had processed its mine output domestically in the absence of government policies to promote processing. The following analysis deals with circumstances in which mine output otherwise would not have been processed domestically.

References

Anderson, J.R., J.L. Dillon and B. Hardaker (1977) *Agricultural Decision Analysis* Ames, Iowa: Iowa State University Press

Balassa, B. and D.M. Schydlowsky (1968) 'Effective Tariffs, Domestic Cost of Foreign Exchange, and the Equilibrium Exchange Rate' *Journal of Political Economy* 76, pp. 348–60

—— (1972) 'Domestic Resource Costs and Effective Protection Once Again' *Journal of Political Economy* 80, pp. 63–9

Bhagwati, J.N. and T.N. Srinivasan (1980) 'Domestic Resource Costs, Effective Rates of Protection and Project Analysis in Tariff-Distorted Economies' *Quarterly Journal of Economics* 94, pp. 205–9

Boadway, R. (1975) 'Benefit–Cost Shadow Pricing in Open Economies: An Alternative Approach' *Journal of Political Economy* 83, pp. 419–30

Bosson, R. and B. Varon (1977) *The Mining Industry and the Developing Countries* New York: Oxford University Press

Brown, E.C. (1948) 'Business-Income Taxation and Investment Incentives' in *Income, Employment and Public Policy. Essays in Honor of Alvin H. Hansen* New York: Norton; rptd in R.A. Musgrave and C.S. Sharp (1959) *Readings in the Economics of Taxation* Homewood, Illinois: Richard D. Irwin

Corden, W.M. (1971) *The Theory of Protection* Oxford: Clarendon Press

—— (1974) *Trade Policy and Economic Welfare* Oxford: Clarendon Press

Diaz Alejandro, C.F. (1976) 'International Markets for Exhaustible Resources, Less Developed Countries, and Multinational Corporations' Discussion Paper No. 256, Economic Growth Center, Yale University, New Haven, Connecticut

ESCAP/CTC Joint Unit on Transnational Corporations (1978) 'Transnational Corporations in Export-Oriented Primary Commodities: A General Conceptual Framework for Case Studies' Working Paper No. 1, United Nations, Bangkok

—— (1981) *Transnational Corporations and Primary Commodity Exports from Asia and the Pacific* Bangkok: United Nations, ESCAP/UNCTC Publication Series B, No. 1

Emerson, C. (1982) 'Mining Enclaves and Taxation' *World Development* 10, pp. 561–71

—— (1983) Minerals Policy: Taxation and Domestic Processing, Doctoral Thesis, Department of Economics, Research School of Pacific Studies, Australian National University, Canberra

—— (1984) *Mining Taxation in ASEAN, Australia and New Guinea* ASEAN–Australia Economic Papers No. 14, ASEAN–Australia Joint Research Project, Kuala Lumpur and Canberra

Emerson, C. and R. Garnaut (1984) *Mining Taxation in the ESCAP Region: Review and Proposals for Reform* Bangkok: United Nations, ESCAP/UNCTC Joint Unit on Transnational Corporations

Emerson, C. and P.J. Lloyd (1983) 'Improving Mineral Taxation Policy in Australia' *Economic Record* 59, 166, pp 232–44

Garnaut, R. and A. Clunies Ross (1975) 'Uncertainty, Risk Aversion and the Taxation of Natural Resource Projects' *Economic Journal* 85, pp. 272–87

—— (1983) *Taxation of Mineral Rents* Oxford: Clarendon Press

Gillis M. et. al. (1978) *Taxation and Mining* Cambridge, Mass.: Ballinger

Harberger, A.C. (1969) 'Professor Arrow on the Social Discount Rate' in G.G. Somers and W.D. Wood (eds) *Cost–Benefit Analysis of Manpower Policies* Kingston, Ontario: Industrial Relations Centre, Queens University; rptd in A.C. Harberger (1972) *Project Evaluation: Collected Papers* Chicago: Markham

Joshi, V. (1972) 'The Rationale and Relevance of the Little–Mirrlees Criterion' *Bulletin of the Oxford University Institute of Economics and Statistics* 34, pp. 3–31

Lal, D. (1974) 'Methods of Project Analysis: A Review' World Bank Staff Occasional Paper No. 16, Johns Hopkins University Press, Baltimore and London

—— (1975) *Appraising Foreign Investment in Developing Countries* London: Heinemann

Leland, H.E. (1978) 'Optimal Risk Sharing and the Leasing of Natural Resources, with Application to Oil and Gas Leasing on the OCS' *Quarterly Journal of Economics* 92, pp. 413–37

Little, I.M.D and J.A. Mirrlees (1969) *Manual of Industrial Project Analysis in Developing Countries, Volume II: Social Cost–Benefit Analysis* Paris: OECD

—— (1974) *Project Appraisal and Planning for Developing Countries* London: Heinemann

Mikesell, R.F., ed. (1971) *Foreign Investment in the Petroleum and Mineral Industries* Baltimore and London: Johns Hopkins University Press

Munasinghe, M. and J.J. Warford (1982) *Electricity Pricing: Theory and Case Studies* Baltimore and London: Johns Hopkins University Press, for the World Bank

Pearson, S.R. (1976) 'Net Social Profitability, Domestic Resource Costs, and Effective Rates of Protection' *Journal of Development Studies* 12, pp. 320–23

Pearson, S.R. and J. Cownie, eds (1974) *Commodity Exports and African Economic Development* Lexington, Mass.: D.C. Heath

Pouliquen, L.Y. (1970) 'Risk Analysis in Project Appraisal' World Bank Staff Occasional Paper No. 11, Johns Hopkins University Press, Baltimore and London

Reutlinger, S. (1970) 'Techniques for Project Appraisal Under Uncertainty' World Bank Staff Occasional Paper No. 10, Johns Hopkins University Press, Baltimore and London

Sen, A.K. (1972) 'Control Areas and Accounting Prices: An Approach to Economic Evaluation' *Economic Journal* 82, pp 486–501

Smith, B. (1984) Resource Rent Tax and the Allocation of Mineral Exploitation Rights (mimeo) Faculty of Economics, Australian National University, Canberra

Squire, L. and H.G. van der Tak (1975) *Economic Analysis of Projects* Baltimore and London: Johns Hopkins University Press

Srinivasan, T.N. and J.N. Bhagwati (1978) 'Shadow Prices for Project Selection in the Presence of Distortions: Effective Rates of Protection and Domestic Resource Costs' *Journal of Political Economy* 86, pp. 97–116

Thoburn, J.T. (1977) Primary Commodity Exports and Economic Development London: John Wiley

—— (1981) *Multinationals, Mining and Development: A Study of the Tin Industry* Westmead, UK: Gower

UNIDO (United Nations Industrial Development Organization) (1972) *Guidelines for Project Evaluation* New York and Vienna: UN, Sales No.E.72.II.B.11

—— (1980) *Mineral Processing in Developing Countries* New York and Vienna: UN, Sales No.E.80.II.B.5

Warr, P.G. (1977) 'On the Shadow Pricing of Traded Commodities' *Journal of Political Economy* 84, pp 865–72

—— (1982a) 'Shadow Pricing Rules for Non-Traded Commodities' *Oxford Economic Papers* 34, pp. 305–25

—— (1982b) 'Sub-Optimal Savings and the Shadow Pricing of Labour' Working Papers in Economics and Econometrics No. 77, Department of Economics, Research School of Social Sciences, Australian National University, Canberra

—— (1983) 'Domestic Resource Cost as an Investment Criterion' *Oxford Economic Papers* 35, pp. 302–6

Warr, P.G. and B.D. Wright (1981) 'The Isolation Paradox and the Discount Rate for Benefit–Cost Analysis' *Quarterly Journal of Economics* 96, pp. 129–45

Willett, K. (1985) 'Mining Taxation Issues in the Australian Federal System' in P. Drysdale and H. Shibata (eds) *Federalism and Resource Development: The Australian Case* Sydney: George Allen & Unwin

Zorn, S. (1978) Issues in Negotiation of Non-Renewable Resource Projects in Development Countries (mimeo) paper presented at the United Nations Center on Transnational Corporations Round Table on Negotiations with Transnational Corporations, Seven Springs Centre, New York (April)

4 ASEAN–Australia interests in mineral commodity associations and agreements

SAHATHAVAN MEYANATHAN

BEFORE dealing with the main topic of this chapter, introductory sections relating to mineral prices and intervention in commodity markets are provided.

Mineral prices

The concept of 'price' in the case of minerals has two allocations functions apart from the usual one of balancing supply and demand.[1] It also allocates consumption between present and future generations, and distributes the economic benefits of mining among producing countries, foreign investors and consumers. While the former function is concerned with conservation and income, the latter deals with the distribution of income between countries. Mineral prices may not fulfil such functions optimally, however, due to imperfections in market structures or knowledge. The distinguishing feature of a mineral resource is that it is exhaustible and the intertemporal sum of the services it provides is finite,[2] which raises the question of rationing consumption optimally among generations. The market mechanism may be viewed as a 'bargaining process' among a number of participants, each of whom has imperfect knowledge. As future generations cannot be present during the current bargaining process, there is no way for their preferences to be properly represented. Prices therefore fail to fulfil the function of rationing resource exploitation optimally among generations. It may be noted that mineral exploitation is itself founded on imperfect knowledge of the extent, availability and grade of deposit. The appeal to the theory of 'second-best' in such cases is of no help either, as there are no universally accepted criteria for choosing among suboptimal solutions. Therefore there is no generally accepted theory about the pricing of minerals (Bosson and Varon, 1977). Some, however, view resource owners as wealth maximisers who are attempting to produce the resource in a way that will maximise the present value of the asset.[3] This view requires assumptions about the physical attributes of the mine and the proper determination of the rate of discount. Given such assumptions, a price can be determined which satisfies the owner's risk-adjusted discount rate.

On the other hand, the role and determinants of mineral prices have been subjected to different interpretations. Under competitive conditions, prices have been viewed as a function of supply and demand in the short run, and as cost-determined in the long run. Another strand in the thinking precipitated by the price changes in the oil market has stressed the idea of the availability and cost of substitutes as the basis for pricing.[4] Yet another viewpoint has been that of the developing countries, whose concern with the inequality in the global distribution of income has induced them to re-examine the determination and function of prices, seeing prices essentially as means of resource transfer. What has been obvious over the past decade has been the unpredictable extent of mineral price changes,[5] undoubtedly reflecting bargaining by governments laying claim to minerals-related income or 'rent', as well as demand and supply shifts.

To a large extent, the 'market prices' of minerals are based on their market structures. These markets range from free commodity markets to those that are to a greater or lesser extent integrated vertically and horizontally where 'prices' are administered. Markets that are free, where prices are determined at arms' length, are the exception rather than the rule in minerals. There are some metals that approximate to such a structure and which are traded on the London Metal Exchange, notably lead and copper, and tin (except that there is a floor support price).

In the market for iron ore, a condition of bilateral oligopoly exists between the highly concentrated producers and the principal consumers. Some 40 per cent of trade is in the form of long-term contracts, where prices are negotiated at arms' length, some 20 per cent consists of intracompany transfers at administered prices, with only about 20 per cent of the trade occurring under 'free market' conditions—that is, short-term, arms-length contracts (Rodrick, 1982). The concentrated market structures of some minerals, such as molybdenum and nickel, have led to a very few producers dominating pricing. In bauxite, alumina and aluminium, where some 75 per cent of trade is between affiliated companies, the prices are notionally determined transfer prices. The crude oil market structure has been characterised as one of oligopoly with a competitive fringe, which has been able to control prices until recently, when the effect of a downward demand shift has begun to threaten its control.

To the producing countries, prices are the most important factor in determining the economic benefit to be derived from the exploitation of resources. Imperfect knowledge as well as imperfect access to knowledge on pricing are sources of potential conflict and obstacles to their planning. Countries heavily dependent on exports of a given mineral sometimes produce at high rates in order to earn foreign exchange to cover expanding imports and debt-servicing costs, even when prices may not cover full cash costs.

It is pertinent to note that mineral prices have historically been subject to control or influence. Major companies, producer nations and consumers have all had a role in this regard. While companies have unilaterally tried to influence prices, producing countries and consumers have more often than not tried to act through coordinated international action. whereas developed countries have sought security of supplies at reasonable, but low, prices, developing countries aspire to promote their mineral development and the extraction of mineral rents through higher prices. The remarkable success of OPEC generated interest among the developing countries in producer alliances for other commodities. OPEC became a model, radically strengthening belief in the effectiveness of producer–exporter cooperation as a means of promoting the interests of the Third World.

Intervention in commodity markets

The main aim of the producer associations has been to influence commodity markets and prices through collective action. Reasons for intervention in commodity markets, as articulated by the developing countries, have tended to cluster around the 'market failure' issue. Within the context of commodity markets, market failure is attributed to 'excessive' instability (Table 4.1).[6] This is seen as a negative externality of price fluctuations. The macroeconomic effects of price instability on the economies of developing countries have usually been cited as a justification for intervention at an international level.

The principal externality of commodity price instability is reflected in fluctuations in foreign exchange balances. Substantial adjustment costs may ensue, particularly if fluctuations in export earnings are exacerbated by a heavy concentration of export production on a limited range of commodities (Smith, 1978). These costs tend to be exhibited in the form of fluctuations in the national income, in the level of economic

Table 4.1 Mineral price variability, 1970–82

Copper	0.23
Bauxite	0.42
Iron ore	0.24
Lead	0.46
Zinc	0.37
Manganese	0.34
Petroleum	0.83
Tin	0.50
Tungsten	0.40

Note: Variability measured by coefficient of variation from data obtained from World Bank *Commodity Trade and Price Trends* (1983)

activity, and in exchange rates. To a certain extent, some of these costs could be reduced if a country had ample foreign exchange reserves or adequate borrowing facilities abroad. In the more likely situation of suboptimal reserves, the cost of instability could be greater (Brainard and Cooper, 1968; Massell, 1964; Smith, 1978).[7]

Related to the foreign exchange externality is the *prima facie* case that instability has detrimental effects on economic development and growth. Typically, a substantial fraction of government revenue is derived from export taxes. Instability in export income therefore creates problems for development planning, inducing problems for savings, investment, and economic growth (Tomek, 1969; Smith, 1978; Reynolds, 1963; Massell, 1964; Brainard and Cooper, 1968; Meier, 1976).[8]

One possible way of handling the foreign exchange earnings problem is to deal with it directly through the IMF's compensatory financing facility as the first-best policy rather than by stabilising prices. This involves the granting of loans to countries experiencing a shortfall in export proceeds (compared to some 'export norm'), the loan being repayable when export earnings improve. Despite the argument that direct transfers such as compensatory financing would be more efficient than price stabilisation, it is not an appealing device for developing country exporters on political grounds, since it involves a loan element (Behrman, 1979). Moreover, such an arrangement does not provide the developing countries with the capability of 'influencing' the trend of commodity prices. To the extent that such a capability exists, it has political as well as economic appeal.

The factors outlined here are only a few of the reasons why nations engage in formulating international agreements. As suggested above, it must be borne in mind that there can be equally important social and political elements that underlie such negotiations (Brown, 1980).

The organisation and development of associations of mineral commodity exporting countries has been a lengthy process, dating from about 1960. The first stage in the formation of these associations spanned 1960–72, while the second stage was from 1973 onwards.

The first association of exporter countries created was the Organization of Petroleum Exporting Countries (OPEC), established in 1960. Following this, copper exporting countries met at the Intergovernmental Copper Conference in Zambia in 1967 to establish the Intergovernmental Committee of Copper Exporting Countries (CIPEC).

In 1973, OPEC quadrupled the price of crude oil. This evoked the Third World's desire to control and exploit its natural resources in its own interests. The success of OPEC strengthened the ties between the developing countries and culminated in the adoption of the Declaration on the Establishment of a New International Economic Order at the United Nations in May 1974, and The Charter of Economic Rights and Duties of States in December 1974. In both documents, emphasis was given to the formation and functioning of producers' associations (Martner, 1979). Encouraged by such instruments, the International Bauxite Association was formed in 1974 and, in 1975, the Primary Tungsten Association was organised. After years of discussion, The Association of Iron Ore Exporting Countries was formed in 1977. Intergovernmental discussions also took place on lead, zinc and manganese around the same period, and in 1983 the Tin Producers' Association was formed.

The formation of producer–exporter associations has, however, not negated recent cooperation among producers and consumers under the aegis of the United Nations for the conclusion of commodity agreements. The various International Tin Agreements date from 1931. In fact, for tin, the producers' association was formed as a result of dissatisfaction in this direction.

This paper attempts to delineate ASEAN–Australian interests in the mineral commodity associations and agreements outlined above. In the next section, a brief review of the mineral industry in the region is undertaken. This is followed by an examination of the minerals in the region that are subjected to agreements. The agreements that are the focus of this study are those relating to intergovernmental producer arrangements and producer–consumer arrangements—that is, OPEC, IBA, CIPEC, AIEC and PTA. These are followed by a discussion of the special position of tin, which is affected by both the International Tin Agreement and the Tin Producers' Association. The final section then reviews the preceding sections, and Appendix 4.1 describes briefly the International Study Group on Lead and Zinc and the Intergovernmental Group of Experts on Manganese.

Minerals of the ASEAN–Australia region

The ASEAN–Australia region is richly endowed with an abundance of mineral deposits.[9] Together ASEAN and Australia accounted for 41 per cent of world tin reserves, 25 per cent of world nickel reserves and about 22.7 per cent of bauxite reserves in 1981. Energy resources in the area include coal, petroleum, natural gas and uranium. Tin, tantalum, antimony, nickel, cobalt, bauxite, copper and zinc are among the non-fuel minerals distributed in the ASEAN region. Australia, on the other hand, produces a diversified range of minerals, including bauxite, iron ore, copper, tin, zirconium, rutile, manganese, lead, zinc and silver.[10]

Malaysia is the world's largest exporter of tin, accounting for about one-third of global tin output. Malaysia is also a net energy exporter, of petroleum and natural gas,[11] and also produces moderate quantities of copper, bauxite, antimony and tantalum. Petroleum and natural gas are Indonesia's most significant mineral resources, with crude petroleum constituting over 50 per cent of its total exports. In addition, it is also a major producer of tin, an important source of copper and nickel and, to a lesser extent, of bauxite and manganese. In Thailand, apart from sizeable amounts of tin, tungsten, tantalum and iron ore, nonmetallic minerals such as fluorspar and potash are also found. The Philippines is one of the world's few producers of nickel. However, copper is its top mineral export, accounting for about 50 per cent of mineral export receipts. It is also an important producer of chromium, molybdenum, gold and cadmium, but its oil and gas reserves are comparatively small by ASEAN standards. Singapore lacks mineral resources and is totally dependent on imports of raw materials and energy. Australia, the sixth largest country in the world, produces a variety of minerals. It is the largest exporter of alumina, lead and mineral sands, and is an important supplier of coal,[12] iron ore, nickel and zinc. It also accounts for 30 per cent of global bauxite production. Australia's uranium resources account for about 20 per cent of the free world's proven economic reserves, and it also produces crude oil and natural gas.[13]

Interests and cooperation in associations and agreements

OPEC

In the late 1950s and early 1960s, the role of government policies in shaping the oil industry became more significant in both producing and consuming countries. In the developing countries in particular, governments became increasingly aware of the role the industry could play in achieving their economic development goals. They became more concerned with international prices, balance of payments, their petroleum reserves, and the location and ownership of refining facilities. The structure of the international oil industry at this time was characterised by the size of the seven major companies and their high degree of vertical integration. These characteristics affected the terms at which concessions were obtained in the producing countries and provided the oil companies with significant discretion in producing, pricing and marketing their crude oil (Siddayo, 1980).

The establishment of OPEC in 1960 was a turning point in the position of the major oil companies vis-à-vis the host countries. The formation of OPEC was largely a response to the reduction of posted prices and realised prices by the oil majors (see Siddayo, 1980; McKern, 1976). With the existence of OPEC, a drastic alteration ensued in the property rights arrangements between the oil companies and the host governments. The policy followed by OPEC has consisted of two phases. In the first phase, from 1960 to 1969, the organisation followed a cautious policy of information gathering,[14] observation and internal consolidation. As of 1970, it adopted a nationalist stance, together with a firm policy of market intervention. The five founding members of OPEC (Iraq, Iran, Kuwait, Saudi Arabia and Venezuela) were joined by Qatar in 1961, Indonesia and Libya in 1962, Abu Dhabi in 1967 (later replaced by United Arab Emirates) and Algeria in 1969 (Martner, 1979).

In 1964, OPEC managed to get the transnational companies to include the calculation of royalties in their production costs; and in 1968 the Declaration on Oil Policy in member countries defined the conditions governing exploitation, participation, prices and proposed direct state intervention, with the ultimate aim of placing every oil enterprise in each OPEC country under that country's control. Between 1970 and 1973, OPEC strengthened into a thirteen-member organisation with the inclusion of Nigeria in 1971 and Ecuador and Gabon in 1973. At about this time, some OPEC countries began to nationalise their oil resources—Algeria in 1971, Iraq and Libya in 1972.

As for prices, which had not changed for ten years, Algeria undertook measures in July 1970 to challenge the existing posted prices, by unilaterally raising the price of its oil. Other OPEC countries soon followed suit. The 1973 Arab–Israeli war and the embargo on the sale of oil led to a substantial price increase, and in October the OPEC countries fixed the price at $11.65 per barrel, and furthermore decided to adjust prices every quarter in line with inflation (Martner, 1979; *Petroleum Economist*, August 1983). OPEC thus succeeded in transferring powers of economic decisionmaking from companies to the governments of producing countries.[15] Between 1975 and 1976 there were moderate increases in prices, whereas the period 1977–78 was one of relative price stability. In 1979 and 1980, on the other hand, price increases ensued as a result of new OPEC pricing policies and shortfalls in Iranian output. At the Bali Meeting of December 1980, OPEC countries agreed that members could charge prices ranging from $32 to $41 a barrel for their oil. As a result, between 1977 and 1981, prices increased from $12.38 to $34.50 per barrel (Ooi, 1982).

The second 'oil shock' brought about the recession in the West. The increased use of alternative energy sources, coupled with conservation measures and increased production by non-OPEC countries, led to an oil glut. In 1981, Saudi Arabia sponsored an agreement under which a quota would be allotted to each OPEC member, to be reviewed regularly, and would, in theory, support the posted prices by restricting supply. For the year 1982, the quotas amounted to 17.5 million barrels per day. This system fell through in 1982 as Nigeria, Iran, Libya and Venezuela produced more oil than their shares, increasing OPEC production to 18 million barrels per day.

Prices thereafter weakened. Even Saudi Arabian light crude was being made available at $31 per barrel, compared to the official price of $34 per barrel. In 1982 OPEC also faced an internal crisis, with Iran challenging the Saudi Arabian leadership (Far Eastern Economic Review, *Asia Yearbook, 1983*). On 18 March 1983, for the first time since its formation in 1960, OPEC passed a resolution to lower the price of Saudi Arabian crude marker, which took the price from $34 to

$29 a barrel (see note 14). This reduction was a result of three main factors:

1 Non-OPEC suppliers (including those of the Eastern Bloc) surpassed the production of OPEC. Since 1973 the additional output of non-OPEC countries such as the United Kingdom, Mexico, Egypt, Norway and the USSR had reached about 5.5 million bpd. As a result, OPEC's share of world oil production was reduced by a considerable margin and OPEC had become a 'residual' supplier.
2 Increasing conservation and the use of alternative energy sources had decreased consumption.
3 Stockpiles of crude which were bought in the late 1970s were unloaded onto a stagnant market. These factors also changed the balance of power within OPEC. Saudi Arabia's power as a 'swing producer' was curtailed (*Petroleum Economist*, January 1981; April 1982).

Indonesia's position in OPEC

Indonesia is the only member of OPEC from the ASEAN–Australia region, although Malaysia and Brunei also are net petroleum exporters and base the price of their crude on OPEC prices. The other countries of the region are affected by OPEC, as importers of energy. Australia, while producing some 65 per cent of its own liquid fuel requirements, has adopted OPEC prices as the basis of domestic taxation and pricing policy.

The 1945 Constitution of Indonesia reaffirmed state control over its natural resources, and Law No. 44 of 1960 set the rules governing oil and gas extraction in Indonesia[16] (Siddayo, 1980). This legislation formed the basis for Indonesia's participation in OPEC in 1962, as the country shared the sentiments of the Middle Eastern producers. Although Indonesia had raised its oil prices several times in the 1960s after negotiations with oil companies, prices remained below the $2 per barrel mark until 1971, when Indonesian crude went up by $2.17 per barrel. When OPEC posted prices were raised to $11.60 a barrel in December 1973, Indonesia also increased its crude prices to $10.80 per barrel. During 1974, Indonesia raised its export prices twice: from $10.80 to $11.70 a barrel at the end of March 1974, and from $11.70 to $12.60 a barrel in July 1974. These increases were higher than those approved for other OPEC countries, as Indonesia had based its arguments on three factors: its crude was of better quality than Middle Eastern oil, its marketing and transportation costs were different from those of other OPEC producers; and its price was in the form of 'realised' prices rather than 'posted' prices. In 1975, Indonesia adopted a four-level pricing system reflecting the different qualities of its crude.[17] For the years 1976 and 1977, Indonesia increased its crude prices in tandem with OPEC increases. The second 'oil shock' of

1979–80 raised the average price of Indonesian crude by 139 per cent over the January 1978 average price (Ooi, 1982).

As an OPEC member Indonesia is unique, as it is both an importer and an exporter of oil. In 1982 it produced 2.4 per cent of total world output and ranked eleventh among the world's major producers and fourth among OPEC countries, accounting for slightly more than 7 per cent of total OPEC production (*Petroleum Economist*, November 1983). It benefited from OPEC pricing policies right up to 1981. Crude export earnings were $16.5 billion for the fiscal year 1981/82, and were an important source of finance for development. However, the situation changed in 1982 with declining world oil demand and quotas imposed on OPEC members. In 1982 Indonesia was significantly affected by the oil glut.[18]

Indonesia's OPEC quota was fixed at 1.3 million bpd, considerably lower than the 1.56–1.64 million bpd level the country had produced in the preceding years.[19] Due to a weak market and lower production, oil export earnings fell to $12.3 billion in 1982/83 and further to $11.9 billion in 1983/84. Indonesia, normally a staunch supporter of OPEC, was adversely affected by the quota system, and there were calls for a pull-out from OPEC. However, Indonesia has continued playing according to OPEC rules, even at the cost of a sales decline.[20]

CIPEC (le Conseil Intergouvernemental des Pays Exportateurs de Cuivre)

During the late 1970s, copper was third after steel and aluminium in terms of tonnage of metal consumed in the world. Copper has a wide range of industrial uses due to its resistance to corrosion and its high conductivity of electricity and heat. Canada and the United States are the major developed country producers, accounting for about 27 per cent of copper ore production and 24 per cent of refined copper production. Among the more important producers in the Third World are Chile, Zambia and Zaire. Indonesia, the Philippines and Australia are small producers by world standards. Copper is also recovered from scrap, both old and new, which makes up an important share (about 30 per cent) of overall world consumption.

Copper is usually traded in its refined state rather than as ore, although there is substantial arm's-length trade in concentrates. Prices of copper products tend to move together, and are quoted in terms of a differential from one of the copper products traded on the principal commodity exchanges (mainly electrolytic wirebar).

The market structure for copper has been characterised as 'a homogenous oligopoly with a large competitive fringe' (Labys, 1982: 81). Partly as a result of nationalisation in the developing countries, the copper producing industry is still highly integrated, but less concentrated than before (Rodrick, 1982; Nappi, 1980). State-owned firms own some 41 per cent of mine capacity and considerable smelting and refining capacity. When copper is traded in the unprocessed form as concentrates, smelting and refining charges are subtracted from the London Metal Exchange (LME) refined wirebar price to arrive at an appropriate price. Their role in determining these deductions gives buyers an advantage, it is argued, due to their superior technical knowledge and negotiating strength (Labys, 1982).

Since the early 1930s the international copper market has been segmented into two parts: the American market, where transactions are made at 'US producer prices' which are charged by copper refineries to their local fabricators; and the rest of the world, where transactions are mainly influenced by LME prices. As the bulk of the world's refined copper is sold directly by the refiners to the fabricators or handled through merchants, only small amounts are sold through the LME and the New York Commodities Exchange (COMEX).[21] Copper price quotations are based on refined metal, with transactions based on annual contracts.[22] Despite the small volume traded, the price of copper is largely determined by transactions on the London Metal Exchange (Labys, 1982: 84).

The international copper market has had a long history of controls aimed at influencing prices. Between 1886 and 1960 there were seven attempts at price fixing, which met with varying degrees of success (Zorn, 1978). The formation of CIPEC in 1967 was prompted by the sharp decline in copper prices and the desire of exporting countries to counter what they viewed as the excesses of the transnational corporations. The founder members were Chile, Peru, Zaire and Zambia, which in 1967 controlled 36 per cent of world exports (excluding Socialist countries). In 1975 Indonesia became a full member of CIPEC and Australia, Papua New Guinea, Mauritania and Yugoslavia became associate members, not bound by the decisions of the full members with respect to export and pricing policies (Mikesell, 1979). CIPEC has been concerned with questions of control and information as much as with price (Zorn, 1978). In the late 1970s, CIPEC countries (including associate members) accounted for over 45 per cent of the total exports of copper ores and concentrates.[23]

Since its formation, CIPEC's policy has changed several times. During the years 1967 to 1974, it was dominated by the governments of Chile, Peru and Zambia, which advocated government control of copper resources and managed to secure partial or complete nationalisations. The early period was one of administrative organisation and information gathering, and when prices declined in 1970 CIPEC members were still preoccupied with national control rather than decisive market action.

The formulation of a CIPEC producer price in 1968 and a resolution authorising joint action to halt the price slide in 1971 were never put into action, and instead the organisation was mainly concerned with evolving a means of supporting the Chilean government in its efforts to sell copper from the nationalised Kennecott and Anaconda mines.

In June 1974 (following the success of OPEC) a proposal for direct CIPEC intervention in the market was adopted, but this was not backed up with concrete plans of action. By late 1974, when prices had dropped by 60 per cent, compared with those current earlier that year, CIPEC took several steps to defend the market. These included an agreement with Japanese refiners to keep excess supplies of refined copper off the international markets. CIPEC member countries also resolved to cut exports by 10 per cent from the end of 1974. In April 1975, the Special Conference of Ministers held in Paris decided to reduce copper production by 15 per cent, but the actual production decline was only about 6 per cent below the 1974 level (Zorn, 1978). By November 1975 CIPEC had formally renounced unilateral market regulation, announced that it was willing to meet with consumers to discuss arrangements for stabilising prices, and later abandoned the policy of cutbacks.

The talks began in Paris in December 1975 as part of the Conferences on International Economic Cooperation (the North–South Dialogue), and also within the Intergovernmental Group of Experts convened by UNCTAD as part of the Integrated Programme for Commodities at the end of 1976. At the first meeting, in November of that year, the preparatory meeting was conducted in a politically charged atmosphere marked by prolonged debate along ideological lines. The Group of 77 and the UNCTAD Secretariat used these meetings to build a case for a copper buffer stock (Brown, 1980). Chile and Zaire were not interested in retaining copper under the Integrated Programme, and Chile withdrew from participation. Subsequent meetings within the Integrated Programme period (October, 1978) failed to achieve any substantial progress.

The failure of UNCTAD to achieve agreement on copper was due to the differences in the aims of the countries involved. The United States wanted market transparency (described below) and a buffer stock, at most, while Peru wanted a buffer stock and supply management. Canada, on the other hand, favoured market transparency and supply management but no export control. Australia wanted to start with market transparency only. Indonesia was of the view that the three aims should be incorporated into the agreement as follows:

1 market transparency (exchange of information on marketing practices, production, consumption, costs of production, costs of marketing, prices, use of copper, etc.) which could facilitate the working of other mechanisms;
2 buffer stock;
3 supply management (production controls, export controls, producer stocks, to stop overproduction when prices fall).

The agreement, in Indonesia's view, would be implemented in three stages, with two-year intervals approximately between the implementation of each aim (*South*, October 1980).

Market intervention measures were again discussed by CIPEC in 1982, and a study group meeting in Paris in August recommended coordinated intervention by the four founder members in the London and New York futures markets. Again, nothing concrete ensued.

The positions of ASEAN and Australia in CIPEC differ to some extent. While Australia is an associate member, Indonesia is a full member. The Philippines, on the other hand, is not a member of CIPEC. As Australia is a small producer of copper by world standards—producing about 3 per cent of the world's total—its interests in the organisation are not as strong as those of the major producers. As an industrial country, it is reluctant to support actions which its industrialised trading partners oppose. Thus, as an associate member it participates in the meetings of CIPEC but is not bound by the decisions taken. Although Indonesia's production levels are about the same as Australia's, it has taken a strong interest in the activities of CIPEC[24] and in the copper meetings of UNCTAD. This is mainly due to its overall policy on minerals, to gain the maximum financial contribution to its development efforts. Although the Philippines is a producer of copper, it is not a member of CIPEC, as it views the organisation as lacking cohesion and concerted policy actions. There are indications that it might join CIPEC as an associate member in the future.

The International Bauxite Association (IBA)

Bauxite is an ore containing a minimum of 40 per cent alumina and a maximum of 10 per cent silica, and is used to produce alumina and aluminium. There are several types of bauxite, which can be categorised into three main groups corresponding to the process employed for refining to alumina: monohydrated bauxite, found in France, Hungary, Italy, Yugoslavia, Greece and the Soviet Union (treated by the European Bayer process); trihydrated bauxite, found in Guyana, Surinam, Guinea, Ghana, Sierra Leone and Australia (treated with the American Bayer process); and a mixture of the two types, found in Jamaica, Haiti and the Dominican Republic (treated with a variant of the two processes) (Nappi, 1980a). Alumina, the product of the refining process, is the raw material for the smelting stage, which produces aluminium.

The bauxite–aluminium industry is characterised by classical vertical integration, with six multinational companies dominating the industry: Alcan, Alcoa, Alusuisse, Kaiser, Péchiney–Ugine–Kuhlmann, and Reynolds.[25] Even today, these firms control over 68 per cent of bauxite production capacity in the western world, 72 per cent of alumina capacity and 62 per cent of aluminium smelter capacity. The industry is almost completely dependent on the Bayer/Hall–Heroult process. Until 1978 there was no 'market price' for aluminium or fabricated products; there was only the catalogue price of the major producers, known as the 'producer price'. In 1978 the London Metal Exchange (LME) started trading in the spot (and then the forward) market for aluminium ingots and (later) bars. The free market trading volume is still small. There is a small spot market for bauxite, but virtually all alumina produced is transferred between affiliates of the major producers.

The vertical integration of the industry provided an ideal environment for 'transfer pricing', and the response to this by governments has been nationalisation of the relatively minor calcined bauxite industry in Guyana, acquisition of minority shares in Guinea, manipulation of export taxes, and partial nationalisation by Jamaica.

During the early 1970s talks were initiated by Jamaica on the formation of a bauxite producers' association, which was the recurrent theme of the Pan-West Indian/Pan-Caribbean movement of the 1960s. The Caribbean's five producers (three of them major ones) proposed a bauxite producers' association, with the aims of acquiring more control over foreign companies and increasing domestic processing of the ore. The first 'oil shock', which transformed their current accounts into deficits, spurred these countries into convening in March 1974 in Guinea to lay the foundation of an International Bauxite Association. Australia, Guinea, Guyana, Jamaica, Sierra Leone, Surinam and Yugoslavia signed, and a few months later Indonesia, Ghana, Haiti and the Dominican Republic also joined the Association (Phenix, 1983). With the entry of India in 1983, the IBA accounts for about 80 per cent of the world's bauxite output. Over 75 per cent of world bauxite production comes from the three major IBA producers—Australia, Jamaica and Ghana. Unlike OPEC, whose members are all developing countries, IBA has a mixed membership. IBA's objective is to promote the rational and orderly development of the bauxite industry, and to guarantee member countries that the exploitation, transformation and marketing of bauxite will assure them of a 'just and reasonable' real income. The objectives of the Association are general enough to let each country make its own interpretation.

Between the first two meetings of the IBA, a large number of member countries modified their mining policies and tax structures. Jamaica took the lead in imposing tax increases in May 1974. It levied a uniform royalty, which was almost double the previous levels, and a variable production levy based on the quality of bauxite extracted by aluminium companies and related to the ingot price. In November 1974 the government of Surinam increased its mining-related revenues by means of a retroactive tax on production. Following this, other countries—

Guyana, Haiti, the Dominican Republic, and to a lesser degree Australia—modified their tax structures.

The obligations in Article IV of the Association call for the exchange of information and research, the harmonisation of policies, the co-ordination of policies towards multinational enterprises, and—most importantly—steps to obtain national control of the industry. The recommendations of the Association require a two-thirds majority but are not binding on members. The IBA's first common action, in December 1977, was the agreement on a minimum price of US$24 per tonne of bauxite headed for the North American market. However, the action had little short-term effect, as the prices actually charged by all the suppliers of the United States market were higher than this base. When the price of aluminium increased by 120 per cent between 1973 and February 1978, the price of bauxite recorded a 128 per cent increase. This was partly due to the increase in the variable tax on West Indian bauxite production, which is based on the ingot price of aluminium.

The Association has had no standard pricing policy, and its activities concentrate on developing a data base and recruiting new members. The IBA continually stresses its desire to avoid confrontation and to engage in dialogue with its customers. The existence of identified bauxite reserves amounting to 28 per cent of the world total in non-IBA countries, the concentration of exploration in non-member countries (especially Brazil),[26] and the threat of possible recourse to non-bauxite sources of alumina (clays and shales) inhibit the actions of the Association (Nappi, 1980a).

At UNCTAD, disinterest in united action by the three major producers raised doubts about the gains to be made by producers from the inclusion of bauxite under the Integrated Programme. Countries such as Brazil and Australia were particularly concerned that any agreement designed to maintain prices might involve curtailment for their export expansion plans. Bauxite is considered a 'non-core' commodity in the Integrated Programme (Brown, 1980).

Australia, the leading exporter of alumina, the largest producer of bauxite and the third largest exporter of bauxite, is a power to be reckoned with. It holds a strategic position in the determination of all common policies of the IBA. Australia has not favoured price fixing but supports a range of prices for the commodity as guidelines for trade. Unlike other producers, Australia is not heavily dependent on bauxite export earnings, and therefore acts as a moderating influence to ensure that the Association does not adopt monopoly pricing practices (Barnett and Hemphill, 1983). In contrast to Australia, Indonesia is a small producer (although it has the potential to become an important producer in view of its reserves).[27] Accordingly, its role is not a dominant one in the Association. It would, however, prefer Australia to play a more assertive role on behalf of the producers in the organisation.

The Association of Iron-ore Exporting Countries (AIEC)

Iron ore is one of the more abundant minerals found in the earth's crust. The main producers are the USSR, Australia, Brazil, the United States, China, India and Canada. About 47 per cent of world production of iron ore is traded internationally. Brazil, whose market share was about 10 per cent in the early 1980s, is expected to increase its share to 30 per cent by the late 1980s and has emerged as the largest exporter after Australia (Labys, 1982). Much of the traded iron ore undergoes processing into one of several forms: calibrated lumps, concentrates, sinter, metallised agglomerate or pellets. The industry exhibits high levels of concentration on both the buyer and seller sides and is characterised as a weak bilateral oligopoly (Labys, 1982). Prices are determined through a wide variety of mechanisms, ranging from bilateral negotiations (long-term contracts) to inter-affiliate transfers (Rodrik, 1982: 543). Iron ore is not traded on any of the world's commodity exchanges, and over the long-term prices depend on long-term contracts.[28]

While the 1950s saw an unprecedented growth of iron ore demand in the industrialised countries as a result of postwar reconstruction, the 1960s, in contrast, were characterised by excess supply and weakening prices. It was at this juncture that some Third World countries tried to form an association. The first meeting, held in Caracas in 1968, paved the way for the establishment of the AIEC seven years later, by which time Australia and Sweden had also joined the initial group, which was composed of Algeria, Chile, India, Mauritania, Peru, Sierra Leone, Tunisia, and Venezuela. By 1977 Chile had also joined the Association (Nappi, 1980b; Martner, 1979). Although the objectives of the Association include assisting members to secure 'remunerative returns' from the iron ore industry, it essentially acts as a centre for the flow of economic and technical information. Under the draft charter, the AIEC will not take collective action or impose directives upon its members. The AIEC does not fix iron ore floor prices or export quotas, or maintain buffer stocks. The signatories account for only 50 per cent of world production (excluding the USSR) and about 35 per cent of world reserves. Any attempt to restrict production could initiate production increases in non-member countries, especially Brazil. As in the case of the IBA, the AIEC also consists of countries whose political ideologies differ too widely to support either sustained cooperation or concerted action (Nappi, 1980b).

As iron ore was included in the Integrated Programme as a 'non-core' commodity, the Association held various meetings to crystallise its position. At UNCTAD, iron ore was not considered suitable for buffer stocking because of the high cost of such a scheme, the absence of a normal market and the multiplicity and classifications of the ore. Differences surfaced between those interested in expanding their market shares and those intent on preserving what they had. India, falling under the latter category, envisaged a market-sharing arrangement which would ensure the country an outlet for its ore, traditionally threatened by high transportation costs. Brazil, on the other hand, would not discuss any arrangement that might hinder its export expansion plans.

The debate at UNCTAD, which began in the early 1970s, focused on two other issues. The first of these was the production of iron pellets, which some countries argued should be located in developing exporting countries. However, the use of sinter as an attractive blast-furnace feed has weakened the pellet market. The second issue centred on the pricing of ore exports on an fob rather than a cif basis.[29] The four iron ore conferences under the Integrated Programme, in comparison with these earlier ones, discussed no substantial issue (Brown, 1980).

Although the Philippines attended the meeting of Third World producers in 1968, it did not become a member of the AIEC, presumably because it is a minor producer of ore. Australia has opposed attempts to control the production and exportation of iron ore, and, by virtue of its export volume, has been a strong influence in the UNCTAD conferences.

The Tungsten Producers' Association (TPA)

Tungsten ores occur geologically as wolframite and sheelite, and are used to harden other metals and alloys. Nearly 40 per cent of tungsten is directed towards the special steels sector (high-speed steels, tool steels), while defence and military uses account for 5–10 per cent. As with other metals, its demand is correlated with industrial production in the developed countries.[30]

The main feature of the tungsten market is the concentration of supply in both the developed and developing worlds. The USSR, Canada and the United States together, produce more than 30 per cent of the world's tungsten. The remainder is produced by Australia, Burma, Bolivia, South Korea, Portugal, Thailand, China, Austria and Japan. World trade in tungsten is conducted in a market approximating a bilateral oligopoly between the industrial countries consuming tungsten and the producers, who are influenced by the Socialist producers. The world's largest tungsten producer, China, plays an unpredictable role in the world market.[31] It has the largest proven reserves of tungsten in the world and is a dominant supplier of high grade ores.

The weak market that persisted after 1981 resulted in production cutbacks in Thailand, Australia and other countries. Tungsten is Thailand's largest mineral export after tin, and in 1982 it accounted for US$26 million (*South*, June 1982).

During the years 1959 and 1964 Russia imported substantial quantities of tungsten ore from China. It retained Chinese ores of better grade for consumption and inferior Russian ores were exported at lower than the LME prices. This action, coupled with disposals from Czechoslovakia, Britain and China, depressed the London price, and the 'crisis' led to the closing down of many mines. The Bolivian miners' call for a remedy of the situation led to the establishment of the UNCTAD Committee on Tungsten (UCOT), which convened its first conference in January 1963 (Tan, 1979).

With little progress achieved at UNCTAD in the ensuing years, Australia, Bolivia, Peru, Portugal, Thailand, Zaire and Rwanda formed the Tungsten Producers' Association in 1975 (Martner, 1979). Since then, Spain and South Korea have also joined the Association. Representatives from China, Canada, Brazil, Mexico and France attend the Associations' meetings as observers. The main objective of the producers' association is to promote world consumption of tungsten and its components. It also focuses on research and surveys on tungsten mining, processing, smelting and refining, as well as the compilation and publication of relevant data.

It also attends meetings of the Tungsten Committee of UNCTAD.[32] At the Bangkok meeting of March 1982, producers, in the absence of an international commodity agreement for tungsten, pressed for 'rationalisation of production' and resisted selling at 'unrealistic' price levels. However, the meeting failed to formulate a united stand by producers. The Association is weakened by the absence of the USSR, Canada, the United States and, most importantly, China. Moreover, it is not in a position to counter stockpile releases by the United States, GSA, which between 1969 and 1970 released an amount equivalent to a year's production, with disruptive effects on prices. The United States stockpile in 1982 stood at a massive 40 000 tonnes, with 25 000 tonnes reserved for strategic purposes and 15 000 tonnes authorised for disposal. Tungsten pricing also faces recycling of scrap metal and competition from other metals in end uses.

Thailand, by virtue of the relative importance of the mineral, has strongly supported the initiatives of the Tungsten Producers' Association, both within the organisation and at UNCTAD. As tungsten constitutes a small sector of the Australian minerals industry, Australia has adopted a moderate stance in the Association. Both countries view the intransigence of the United States as a major reason for the lack of action at UNCTAD.

The International Tin Agreements (ITA) and the Association of Tin Producing Countries (ATPC)

Tin was one of the earliest metals known to man, but is relatively scarce compared with other elements such as lead and zinc. The principal deposits are scattered along a belt surrounding the Pacific Ocean, the deposits on the Asian portion being larger than those on the American side. Other deposits are found along the Eastern side of the Atlantic Ocean and in South and Central Africa (Wright, 1966; Harris, 1978).

The production of tin is concentrated in the developing countries, while its consumption is mainly in the developed world. The ASEAN region has the highest concentration of tin activity in the world, constituting about 63 per cent of the world's tin output (excluding centrally planned economies) (ITC, 1982). Other major producers include Bolivia, Nigeria, Australia, Zaire, Brazil, and South Africa. Malaysia is the largest producer, followed by Indonesia (which has replaced Bolivia as the second largest), and thirdly Thailand. Malaysia's share of total world production of tin in concentrates was about 30 per cent in 1981; Indonesia's was 17 per cent, and Thailand's 15 per cent.

Among the ASEAN countries, Indonesia's reserves are larger than those of Malaysia and Thailand (which have about the same level of reserves). Australia has less than 5 per cent of world reserves.[33] Bolivia has some 10 per cent of global reserves, and the People's Republic of China and the USSR have important reserves. For the period 1978–80, the average Malaysian production was about 5 per cent per annum of its reserves, higher than that of the other major producing countries (1.9 per cent for Indonesia and 2.7 per cent for Thailand).

The world tin market involves trade in concentrates as well as refined metal (ingots, slabs), but most of the tin produced is traded as metal. In 1980 Malaysia exported 36.8 per cent of total world exports of tin metal, followed by Thailand (18 per cent), Indonesia (13.8 per cent) and Singapore (7.3 per cent). Thus ASEAN as a whole accounts for 75.9 per cent of the world's tin exports. Australia is the fifth major exporter, mainly of concentrate. Although the United Kingdom is not a producer, its ore imports are sufficient to maintain its position as a major smelter, accounting for about 11 per cent of world production. The major importers of refined tin include the United States (28.3 per cent in 1980), Japan (19.1 per cent), the EEC (29.0 per cent) (the Federal German Republic accounting for the largest share in that grouping), and the USSR (8.7 per cent). A small proportion of tin is supplied by secondary sources (about 5 per cent).

Tin has no single marketing structure. For example, some tin is marketed through supply contracts between governments (or state corporations) and consumers. This is the case with Indonesia's state-produced tin, which is sold by the State Corporation through its offices in London and Antwerp. Thailand's output is marketed through a single commercial outlet, and the Malaysia Mining Corporation sells to consumers directly through a London agent. A substantial amount of tin is marketed through tin merchants (brokerage firms which deal directly with the LME) or through brokers who represent consumers. Despite the existence of different consumer–producer arrangements, most consumers deal directly with a smelter or have contracts with merchants. The merchant usually buys the tin ore or metal, ships it and finances the entire transaction until it reaches the consumer (Labys, 1982). The smelting companies in Penang, the foundation of the Penang Market, act as intermediaries for the miners and the buyers of metal on the market. The price is quoted ex-smelter, with shipment within 60 days at the smelter's option. The two other markets where tin is quoted are the London Metal Exchange and the New York market. Prices in the other two markets are closely related to the Penang market. In addition to the physical (cash) market, the LME operates a forward market where prices are quoted for three months ahead.

The tin market structure is generally considered a bilateral oligopoly (Labys, 1982). Multinationals and governments dominate the supply side. The concentration of smelting capacity is high, with the four largest firms (two in Malaysia, one in Indonesia, and one in Thailand) accounting for 52 per cent. On the demand side, the high concentration among tinplate manufacturers gives rise to a potentially strong bargaining position. Tin-can manufacture is even more concentrated, with large firms dominating the market (Thoburn, 1982b).

While four of the five ASEAN countries have interests in tin, three of them (Malaysia, Indonesia and Thailand) are members of the International Tin Agreements (ITA); Australia is also a member. The formation of the Association of Tin Producing Countries is partly related to the problems associated with the functioning of the Tin Agreement and with negotiations for the Sixth Agreement.

The depression of 1930 triggered the quest for control of the tin market. In March 1931 the first International Tin Agreement laid the foundation for this control over the next ten years. Its general principle of regulation of production by a system of quotas was subsequently maintained by the Second Agreement of 1934–36, the Third Agreement of 1937–41 and the nominal Fourth Agreement of 1942–46. From the Second Agreement, the absorption of stocks also became an important objective. The signatories of the First Agreement, in 1931, were Malaya, Nigeria, Bolivia and the Netherlands East Indies, which together produced no less than 85 per cent of the world's tin. To safeguard their share of production, other producers were invited to join the agreements. Thailand was a party from 1931 until 1941, French Indochina, Cornish tin companies and Portugal from 1934 to 1936, and the Belgian Congo from 1934. These countries were subjected to lenient terms regarding export restrictions. However, other producing countries such as China, Burma, Japan and Australia did not join the agreement (Fox, 1974).

The Korean War in 1950 postponed the urgency for the negotiation of a postwar agreement. However, the first agreement under the guidelines of the United Nations was ratified in June 1954 by the more important producing and consuming countries, except for the United States and Indonesia. Indonesia joined in July 1954, making it possible for the Tin Council to start operations. The producing and consuming countries held the same number of votes in the Council, though each member's vote varied according to its percentage share in the total exports and imports. A buffer stock of 25 000 tonnes was contributed by producing countries, and was controlled by a director whose market intervention was aimed at maintaining the price between a ceiling and a floor level. The Council could impose export restraints only so long as the buffer stock amounted to at least 10 000 tonnes.

The Second Agreement was ratified in mid-1961 and had the same objectives; mainly to avoid price fluctuations and to ensure 'adequate' supply at 'reasonable' prices. The liquidation of the American strategic stockpile by the GSA, coupled with increased supply by new producers, led to price declines in 1966, and against this background producers signed the Third International Tin Agreement on 1 July 1966. Buffer stock purchases helped to keep prices above the floor, and between 1969 and 1972 prices were relatively stable. The discontinuation of GSA commercial sales further strengthened market stability from mid-1968 to mid-1973.

The Fourth International Tin Agreement (July 1971) retained the principles of earlier agreements but attempted to increase flexibility. The agreement was to be administered by the International Tin Council (ITC), where both producers and consumers had equal weight. Among the seven producing countries, Malaysia had the largest voting share with 47 per cent, while Japan had 24 per cent on the consuming side. The elected executive president could suspend intervention if prices reached established limits or if the move was justified by exchange-rate variations. The Council could, by a two-thirds majority, institute export controls regardless of the quantity of tin in the buffer stock. Third, in the case of scarcity the Council could pressure producers to distribute available supplies to member consumers. The International Monetary Fund agreed during this period to help in financing buffer stocks. This Agreement (like the earlier ones) saw the absence of the United States as a member (Fox, 1974), but in 1975 the United States decided to join the Fifth Agreement, which took effect in 1976.

Apart from maintaining the buffer stock and export controls, the Agreement introduced some modifications. Producing countries were required to contribute the equivalent of 20 000 tonnes to the buffer stock (in tin, currency, or both, depending on the Council's decision). The Agreement for the first time imposed on the consuming countries an additional contribution amounting to the equivalent of 20 000 tonnes of tin metal; this was, however, on a voluntary basis. With the United States (accounting for 25 per cent of world consumption) joining the Agreement, and the doubling of the buffer stock size, the Fifth Agreement was expected to be effective in curbing price fluctuations.

The early period of the Fifth Agreement marked a dramatic

improvement in the tin market. The higher demand was due to the improvement in the industrial climate—leading to stock replenishment and increased consumption. The ITA export controls were relaxed, and were finally lifted on 30 June 1976. Sales of tin from the buffer stock reached a high level, and when news of the almost exhausted buffer stock holdings was released, market demand strengthened so sharply that the price rose above the ITA ceiling. The second oil shock, in 1979/80, caused a fall in demand, and in April 1982 export control was announced by the ITC in an attempt to prevent a further price decline. However, producers were doubtful of its effectiveness, as the Agreement was due to expire on 30 June 1982 (Bank Negara, Annual Report). The Fifth ITA enjoyed the widest participation since the Tin Agreements came into being after the second world war. As mentioned above, the United States (under the Carter Administration) had joined the Agreement for the first time. The United States and West Germany held enough votes jointly to veto decisions requiring a two-thirds majority of both consumer and producer votes on key issues, especially when determining the buffer stock price range. These two countries also resisted proposals for increases in the buffer stock price range, and were critical of policies in the producing countries.

Dissatisfaction on the producers' side centred on the difficulty and delay in the revision of the price range. For the period 1977–79, the market price was around the ceiling price and the buffer stock was depleted, rendering the Agreement ineffective. The Fifth Agreement also saw the formation of the Economic and Price Review Panel. This body was important in persuading Bolivia, the highest cost producer, to join the Agreement. With the revision in October 1981 of the price range, Bolivia's cost of production was above the ceiling price, while that of the ASEAN countries was not far above the floor price (Thoburn, 1982b).

The Fifth Agreement, which was to expire in 1981, was extended to 1982. Four negotiating sessions were held at the United Nations Tin Conference in Geneva between April 1980 and June 1981 to negotiate the Sixth Agreement, which came into effect on 15 July 1982. The protracted series of negotiations was due in part to the consuming countries' positions. The United States maintained that the export control measure should be scrapped in favour of a much larger buffer stock (70 000 tonnes). The producers conceded to an extent and a buffer stock size of 30 000 tonnes was agreed to, with another 20 000 tonnes to be financed through borrowings. It was also agreed that export controls would be activated only after 35 000 tonnes of tin had been accumulated in the buffer stock. Producers became disenchanted and bitter with the United States, however, when it refused to join the Agreement after concessions to its views had been made.[34]

Membership under the Sixth ITA comprised six producing countries, which included the three ASEAN producers and Australia. Producers accounted for 82 per cent of global tin production. There were seventeen consuming countries, representing 51.6 per cent of world consumption, as members. This was in contrast with the seven producing countries and 22 consuming countries under the Fifth Agreement. Notable absentees were Bolivia and the United States. As a consequence of the reduced membership, the size of the normal buffer stock was substantially reduced from the originally agreed quantum of 30 000 tonnes to only 19 666 tonnes. In addition to the normal buffer stock, which was to be financed by member governments, the Agreement provided for a supplementary buffer stock of 20 000 tonnes to be financed by borrowings, bringing the total buffer stock size to 39 666 tonnes, close to the buffer stock size of 40 000 tonnes under the previous Agreement. In contrast to the Fifth Agreement, where contributions were compulsory only for producers while consuming members contributed on a voluntary basis, contributions to the Sixth Agreement were shared equally among producing and consuming members. Another significant change concerned the trigger point for the imposition of export controls. While the previous Agreement had stipulated that export controls should be imposed when buffer stock holdings amounted to 10 000 tonnes, the Sixth Agreement provided that such controls be enforced only when the buffer stock holdings reached 70 per cent of the maximum allowable volume—that is, 20 766 tonnes. This amount of tin was transferred from the Fifth ITA to the new Agreement to allow export controls imposed under the Fifth ITA to continue under the Sixth ITA (Bank Negara, Annual Reports).

Interest in the Tin Agreement among ASEAN producing countries stems from the belief that the Agreement would support a price adequate to achieve satisfactory profits for mines and taxes for governments. Australia has participated in all six International Tin Agreements—in the first three as a consumer, and in the second three as a producer—and its position in the Agreements has been the same as that of the ASEAN producers. From 1957 to 1977 the buffer stock was self-financing, and made a profit for its participants partly from investment income and partly from tin transactions over a period of rising prices. The price has fallen below the floor price only once (in 1958), while the ceiling has been exceeded for months and even years at a time, so that the Agreement has worked asymmetrically. The Tin Council has lacked the resources for a larger buffer stock, and consumers have resisted a revision of the price range in such situations, so that the Council has always lagged behind the market.

Smith and Schink (1976) found in their study that, to have kept price fluctuations over the period from 1956 to 1974 within plus or minus 15 per cent of the trend, a buffer stock of 100 000 tonnes would have been required, a figure confirmed by the World Bank (1978). On the effectiveness of the Agreements in stabilising prices, Smith and Schink claim that the dominant influence in the market was the GSA and not the Council. For the period 1970–82, the variability in the price of tin was the highest of the non-fuel minerals reviewed in this paper (see Table 4.1). Producers regard export control as a more effective instrument in the working of the Agreement than the existence of large stocks overhanging the market, despite the costs of mine closures which export controls necessitate.

In late 1985 the future of the Tin Agreement was in great doubt, as a result of the failure of the Council to honour its commitments to pay for some 68 000 tonnes of tin, purchased at a cost of around US$600 million in order to support the price. The stockpile had reached 130 000 tonnes, half of which had been purchased with borrowed funds. Trading in tin was suspended when the price fell to £stg 8140 per tonne, and there were doubts about whether member governments would provide the funds needed to pay for these purchases. A failure of the tin market threatened the survival of even the London Metal Exchange (*The Economist*, 2–8 November 1985).

The agreement establishing the Association of Tin Producing Countries (ATPC) came into force on 16 August 1983. The four founding members—Malaysia, Indonesia, Thailand and Bolivia—account for about 90 per cent of world tin production, and the ASEAN producers together account for 74 per cent. The move to form the producer association gained momentum in 1982, following producer countries' disenchantment with the troubled Fifth ITA.[35]

The ATPC was conceived at a time when the tin market was near collapse; prices had declined sharply, huge surpluses were overhanging the market, and the GSA was stepping up its sales. The newly formed ATPC shares a common feature with the prewar producer pacts, in that both were conceived at a time when market conditions were unfavourable.

Malaysia was the first proponent of a tin producers' pact. As the world's largest tin producer, Malaysia has always played an important role in producer–consumer negotiations. However, in view of developments in the world tin market since late 1981, Malaysia has explored new approaches, including the need to form a tin producers' association. The ATPC proposed by Malaysia gained momentum with time and the support of the other producing countries, especially Thailand and Indonesia. Nevertheless, the Agreement took time to take shape, as there were disagreements over voting rights, functions and control measures.

The objectives of the ATPC are to ensure remunerative and equitable returns to tin producers and adequate supplies to consumers, to cooperate in tin marketing, and to encourage research and development. Among its four main objectives, the ATPC focused attention on research and development, and marketing. The pricing objective has for the time being been held in abeyance due to the existence of the Sixth ITA.

Since its formation, the Association of Tin Producing Countries has emphasised research and development to increase the use of tin and to reduce production costs at the mines. The South East Asian Tin Research and Development Centre in Malaysia concentrates on production research, while the producer-funded International Tin Research Institute in the United Kingdom focuses on consumption research. The Institute is financed by the five major tin producers, Indonesia, Malaysia, Nigeria, Thailand and Zaire, and a voluntary contribution is made by Australia. Another area to which the ATPC has paid attention is marketing. The problems that surfaced during 1981 and 1982 at the London Metal Exchange led to the formulation of an alternative marketing system through the reform of the Penang Physical Tin Market and the establishment of tin futures trading on the Kuala Lumpur Commodity Exchange. The Association has also been concerned with tin ore smuggling in the ASEAN region, which, given the high price of tin, has meant a considerable loss of revenue.

Concluding remarks

This paper has sought to trace ASEAN–Australian interests in mineral producer and commodity agreements. The producer associations reviewed include those relating to petroleum, copper, bauxite, iron ore, tungsten, and tin. The International Tin Agreement is the only international producer–consumer agreement on minerals, and is of special significance to the ASEAN–Australia region, which accounts for a substantial part of world production. The major aim of the producer associations cited has been to influence commodity markets and prices in their favour through collective action. To the minerals exporting countries, prices represent the most important factor in determining the economic benefit from the exploitation of mineral resources. Interest in producer alliances was accelerated by the success of OPEC in appropriating mineral rents. However, outside OPEC, the structural conditions facing the other minerals associations have negated successful collective intervention. Additionally, some associations (such as the IBA and the AIEC) seem to lack political cohesiveness because of widely differing political ideologies. Despite this, they have focused attention on other aspects of the commodity, such as information exchange, the promotion of processing, marketing and transportation, and the modification of fiscal regimes. They have also attempted to crystallise the producers' positions in the UNCTAD negotiations, but with little success, except in the case of tin.

The analysis earlier in this chapter indicates that the interests of ASEAN and Australia in the minerals associations have been conditioned by the following factors: the relative importance of the particular mineral in the country's economy; the importance of the individual country to world trade in its minerals; its policies towards import and export trade; and the perceived 'capability' of the respective associations. In Australia, tin and tungsten are relatively small sectors, so that the state has supported producer actions in the ITA and has taken a 'reasonable' stance in the TPA. In the case of the ATPC, Australia participated in all the preliminary meetings but took time to sign the accord, presumably waiting to see what action followed as the agreement came into force. In the case of the other minerals—copper, bauxite and iron ore—these are large, competitively strong sectors for Australia, and producers have generally viewed intervention with disfavour. Australia's position in these associations has thus been one of passivity, a position it feels confirmed by the perceived ineffectiveness of the organisations at market intervention. It has no doubt been in Australia's interests to adopt such a position in order to protect its market expansion plans for these commodities. On the ASEAN side, Thailand, Indonesia and Malaysia have generally been strong proponents of collective action, in tin and tungsten for Thailand, in petroleum and tin for Indonesia, and in tin for Malaysia, as these are highly important sectors in their economies. In respect of bauxite and copper, Indonesia has taken a similar stand, in keeping with its overall policy stance, although these are comparatively smaller sectors. Little concern has been generated in the case of the Philippines in the iron ore and copper associations, as it perceives the benefits to be derived from these as minimal.

Thus the positions adopted by the four ASEAN minerals producers are often at variance with the position of Australia, reflecting differences of opinion about the optimal mechanism for maximising their returns from minerals exploitation. Australia has seen fit to participate in those agreements where its individual position is perceived to be weak, while abstaining from active participation where its bargaining position is strong. Its pre-eminent position in bauxite and iron ore has enabled it to be less active in those sectors, whereas its interests are more closely allied to the collective interests of Malaysia, Thailand and Indonesia in the case of tin. Fortunately, the distribution of strengths in the minerals industries between the two regions is such that there are few areas where the interests of Australia and ASEAN in commodity agreements are in direct opposition.

Appendix 4.1

International associations on lead and zinc, and on manganese

The International Lead and Zinc Study Group
Australia is a founder member of this body, which was formed in 1959. The group meets once a year and comprises all the major producers and consumers of the mineral. It conducts intergovernmental consultations on lead and zinc production without resorting to economic measures. The Secretariat in London collects and publishes material on the nature and structure of the industry, including smelters and secondary recovery.

Australia is the most important producer of zinc after Canada, and is next in importance in the production of lead worldwide after the United States (*Asian Research Bulletin*, January 1984; Robinson, 1975). The Australian industry and government view the work of this group in a favourable light.

The intergovernmental group of experts on manganese
Manganese is part of the Integrated Programme of UNCTAD and is exported by several developing countries, including Brazil, Gabon, India, Mexico and Ghana. IN 1979 Australia's production of the mineral was about 8 per cent of the world's total. The Philippines, Malaysia and Thailand also export small quantities of manganese. There is no satisfactory substitute for the metal, and it is essential in high-grade steel manufacture. There is an absence of production in the major consuming countries, such as the United States. The diversity of economic and political attitudes makes it difficult for countries such as the USSR, South Africa and China to participate effectively in the group. The UNCTAD meetings have reviewed issues such as pricing policy, processing of manganese ore into ferromanganese, problems of decline in price and consumption, reliability of ore supplies, access to markets, research and development, transparency of markets and transportation (Brown, 1980; UNCTAD *Monthly Bulletin*, various issues).

Notes

This chapter owes much to discussions with government officials from the ASEAN countries and Australia.

1 Much of this section is based on Bosson and Varon (1977).
2 This abstracts from the question of recycling and uncertainty; see Dasgupta and Heal (1979) for details.
3 See Hotelling (1931); Solow (1974); Dasgupta and Heal (1979: ch.11); Griffin and Teece (1982).
4 The technology for producing a substitute product is referred to as 'backstop technology'. The price of the substitute places a ceiling on future price increases of the mineral (Dasgupta and Heal, 1979). The difference between this price and that based on costs of production can be taken to represent the intrinsic value of the mineral in the ground, which should ideally accrue to the producing country (Bosson and Varon, 1977).

5 In the short run mineral prices are influenced by a multitude of factors such as supply and demand, inflation, production costs, degree of competition among producers, protection, and exchange rates. These are compounded by manipulations and political considerations. Over the longer term, structural developments affect prices. The long-run price for minerals generally approximates to the long-run cost trend (Bosson and Varon, 1977).

6 Instability is demonstrated by the amplitude of changes and the frequency of such changes in the variables of interest (Tomek, 1969). It may be considered excessive when fluctuations in the market price are out of phase with more constant factors such as domestic retail prices, productivity, or population growth (Brown, 1975) and when prices exceed the limitations of a price range believed to be normal.

7 Similar arguments could be made for the developed consuming countries (Tomek, 1969).

8 Informational externality is yet another issue commonly associated with commodity instability. Excessive price instability may misguide producers into reallocating resources (in the form of overinvestment or underinvestment in primary commodities) when longer run equilibrium considerations argue otherwise. Spot prices are therefore not regarded as useful guides to future production and investment, creating uncertainty about long-run profitability. Certainly, investors associate price instability with risk, and may consequently underinvest. They may also seek portfolios of investments to reduce uncertainty.

9 Much of this section is drawn from BMR (1983).

10 Although not covered in this chapter, Brunei (which joined ASEAN in 1984) is dependent on oil and gas exports. It is not, however, a member of OPEC.

11 Petroleum accounted for about 27 per cent of the total exports of Malaysia in 1982 (Malaysia, *Economic Report, 1984/85*).

12 Reserves of coal in ASEAN are not large, and are generally of poorer quality than those in Australia.

13 Exports of minerals and fuels accounted for 40 per cent of Australia's total exports in 1981/82 (BMR, 1983).

14 Information gathering was the first major project undertaken by OPEC with the aim of preventing further cuts in the posted prices of oil (Zorn, 1978). The main function of posted prices is to determine the tax and royalty payments from oil companies to the exporting countries (Robinson, C., 1975).

15 Western acquiescence and the ability of multinational companies to pass the price increase on to customers were major factors in the success of OPEC's efforts to raise oil prices (Zorn, 1978: 223).

16 Indonesia created two national oil companies in 1957, and in 1960 nationalised its oil industry. In 1968 the two companies were merged into Pertamina (Noreng, 1982). In 1982 Pertamina was responsible for about 5 per cent of crude output, while the rest was produced by the 37 or so foreign contractors (Ooi, 1982).

17 The four grades were made up of: Rantau (1st grade); Minas, Attaka, Sepinggan, Duri, Bekapai, Badak, Bunyu, Lirik (2nd grade); Cinta, Jitabarang, Handil (3rd grade); and Walio, Pamugian, Juanta/Tarakan (4th grade). Minas crude is the representative crude, or the price-reference

('marker') crude, from Indonesia, with an API gravity of 35 (Ooi, 1982).

18 Between 1978 and 1981 world oil demand declined by about 8.5 per cent, while demand for alternative fuels increased by 11 per cent (Far Eastern Economic Review, *Asia Yearbook, 1983*).

19 Like most OPEC members, Indonesia produced more oil than its quota; in 1983/84 it produced 1.33 million bpd, and in 1984 1.48 million bpd. The excess over the OPEC quota helped the country over a threatening debt problem in 1983, which arose as a direct result of the world oil price drop (*Far Eastern Economic Review*, August 1984).

20 Indonesia is part of OPEC's Monitoring Committee, set up to supervise the operations of the London production-sharing agreement. The Committee is made up of the oil ministers of the United Arab Emirates, Algeria, Venezuela and Indonesia. The Indonesian Minister of Mines was Chairman of OPEC in 1981.

21 COMEX is the New York Commodity Exchange. Although less copper is traded on this market than in London, oriented mainly towards the international market, COMEX offers commercial agents a wider variety of the metal. The 'US producer price' has tended to be close to the COMEX price in the recent past (*South*, October 1980).

22 The risk of supply cuts (or ore shortages), and of losing markets in times of gluts, have resulted in long-term contracts which employ a variety of formulas for pricing copper, based on the LME prices (Rodrick, 1982; Mikesell, 1979).

23 CIPEC members have a large role in world trade of blister copper (75 per cent of the total) than in world production of copper ore (30 per cent).

24 An Indonesian held the position of Secretary-General of CIPEC in 1982.

25 It is a most concentrated and vertically integrated industry, with only a minute share of trade taking place at 'arms length'.

26 The threat posed by Brazil is important: the Trombetas project alone has a planned capacity of 3.5 million tonnes per year.

27 Its reserves stood at 700 million tonnes in 1979.

28 According to Labys (1982), about 20 per cent of international iron ore trade is captive; of the remainder, 40 per cent or more is transacted under contractual arrangements of ten to fifteen years' duration, and 40 per cent as short-term contracts with both prices and quantities renegotiated annually.

29 An issue not openly discussed in the meetings was the increasing frequency of breaches of the terms of long-term contracts by importers. While exporters continued to observe the terms to the letter, buyers had increasingly come to regard the contract as a general framework (Brown, 1980). This issue is considered an important one for Australia, as the country's iron ore mines rely heavily on Japanese steel mills' imports, which are tied to long-term contracts. Between 1975 and 1980, 'Japanese imports of Australian iron-ore fell short of base contract levels by roughly 70 million tons. Valued at current prices, this represents a loss in revenue to Australian mines of something like US$1.2 billion' (*Far Eastern Economic Review*, 8 January 1982, p. 41).

30 With the onset of the recession in 1981, the automotive, mining and oil drilling equipment industries were adversely affected and the demand for tungsten plummeted. Industrial recycling and increased use of alternative metals (uranium, carbide, molybdenum and coated ceramics) also lowered

demand. The price in 1977 was US$170 per tonne, but in June 1982 it was US$107 per tonne (Tan, 1979; *South*, June 1982).

31 An inverse relationship between Chinese supply and tungsten price was observed between 1962 and 1975 (Tan, 1979).

32 At its fifteenth session, held in December 1983, the Committee discussed proposals on the question of stabilising the tungsten market. However, it has yet to convene a negotiating conference. The Committee also reviews the tungsten market situation and discusses statistical issues (UNCTAD *Monthly Bulletin*, various issues).

33 See Department of Trade and Resources (Australia), 1983.

34 Producing countries have found it hard to gauge the refusal of the United States to join the Agreement. Presumably it felt that the concessions made were not attractive enough for it to participate. It may be noted that, in the case of the International Natural Rubber Agreement, producers had to do away with export controls so that the United States would participate.

35 For details, see the *Asia Research Bulletin*, May 1983.

References

Al-Chalabi, F.J. (1983–84) 'In Favour of a New Sharing of the Oil Market' *OPEC Bulletin* December–January, pp. 3–6

Department of Trade and Resources (Australia) (1983) *Survey of Major Western Pacific Economies* Canberra: AGPS

Bank Negara (Malaysia) (various years) *Annual Reports*

BMR (Bureau of Mineral Resources) (1983) *Australian Mineral Industry Annual Review* Canberra: AGPS

Barnett, D.W. and J.P. Hemphill (1983) 'The Minerals Industry in Australia 1970–1980: A pattern for Growth' Paper prepared for the ASEAN–Australia Minerals Project (February)

Behrman, J.R. (1979) 'International Commodity Agreements: An Evaluation of the UNCTAD Integrated Programme' in W.R. Cline (ed.) *Policy Alternatives for a New International Economic Order* New York: Praeger

Bosson, R. and B. Varon (1977) *The Mining Industry and the Developing Countries* New York: Oxford University Press

Brainard, W.C. and R.N. Cooper (1968) 'Uncertainty and Diversification in International Trade' *Food Research Institute Studies* 8, 3, pp. 257–85

Brown, C.P. (1975) *Primary Commodity Control* Kuala Lumpur: Oxford University Press

—— (1980) *The Political and Social Economy of Commodity Control* London: Macmillan

Dasgupta, P.S. and G.M. Heal (1979) *Economic Theory and Exhaustible Resources* Cambridge University Press

ESCAP (1982) *Transnational Corporations and the Distribution of Gains in the Tin Industry of South East Asia: An Updated Study* Bangkok: ESCAP (September)

Fox, W. (1974) *Tin, the Working of a Commodity Agreement* London: Mining Journal Books

Griffin, J.M. and D.J. Teece (1982) 'Introduction' in J.M. Griffin and J.D. Teece (eds) *OPEC Behaviour and World Oil Prices* London: George Allen & Unwin

Harris, K.L. (1978) 'Tin' *Mineral Commodity Profiles* Washington, DC: US Bureau of Mines

Hotelling, H. (1931) 'The Economics of Exhaustible Resources' *Journal of Political Economy* 39, 2 (April)

ITC (International Tin Council) (1982) *Tin Statistics 1971–1981* London: ITC

Labys, W.C. (1982) 'The Role of State Trading in Mineral Commodity Markets' in M.M. Kostecki (ed.) *State Trading in International Markets* London: Macmillan

McKern, R.B. (1976) *Multinational Enterprise and Natural Resources* Sydney: McGraw–Hill

Martner, G. (1979) Producers–Exporters Associations of Developing Countries Geneva: UNAPEC

Massel, B.F. (1964) 'Export Concentration and Fluctuations in Export Earnings: A Cross Section Analysis' *American Economic Review* 54, 2 (March)

Meier, G.M. (1976) *Leading Issues in Development Economics* Oxford University Press, 1964.

Mikesell, R.F. (1979) *The World Copper Industry: Structure and Economic Analysis* Baltimore: Johns Hopkins University Press

Ministry of Finance (Malaysia) (1984) *Economic Report, 1984/85* Kuala Lumpur

Nappi, C. (1980a) 'IBA and the International Bauxite Market' *Commodity Market Controls* Lexington Books

—— (1980b) 'AIEC and the International Iron-Ore Market' *Commodity Market Controls* Lexington Books

Noreng, O. (1982) 'State Trading and the Politics of Oil' in M.M. Kostecki (ed.) *State Trading in International Markets* London: Macmillan

Ooi, J.B. (1982) *The Petroleum Resources of Indonesia* Kuala Lumpur: Oxford University Press

Pearce, D.W. and J. Rose (1975) 'Editorial Introduction' in D.W. Pearce and J. Rose (eds) *The Economics of Natural Resource Depletion* London: Macmillan

Petroleum Economist (January 1981, April 1982, July, August and November 1983)

Phenix, P. (1983) The Political Economy of the Bauxite and Aluminium Industry with Special Reference to Jamaica, mimeo

Reynolds, C.W. (1963) 'Domestic Consequences of Export Instability' *American Economic Review* (Papers and Proceedings) 53, 2

Robertson, W. (1982) *Tin: Its Production and Marketing* London: Croom Helm

Robinson, C. (1975) 'The Depletion of Energy Resources' in D.W. Pearce and J. Rose (eds) *The Economics of Natural Resource Depletion* London: Macmillan

Robinson, I.M. (1975) 'Zinc' in C. Payer (ed.) *Commodity Trade of the Third World* London: Macmillan

Rodrick, D. (1982) 'Managing Resources Dependency: The United States and Japan in the Markets for Copper, Iron Ore and Bauxite' *World Development* 10, 7

Siddayo, M. (1980) *The Supply of Petroleum Reserves in South-East Asia* Kuala Lumpur: Oxford University Press

Smith, G.W. (1978) 'Commodity Instability and Market Failure: A Survey of

Issues' in F.G. Adams and S.A. Klein (eds) *Stabilizing World Commodity Markets* Lexington Books
Smith, G.W. and G.R. Schink (1976) 'The International Tin Agreement: An Assessment' *Economic Journal* 86 (December)
Solow, R.M. (1974) 'The Economics of Resources or the Resources of Economics' *American Economic Review* (Papers and Proceedings, May)
Tan, C.S. (1979) 'Trade Strategy under Bilateral Oligopoly: A Case Study of China's Tungsten Trade' *Philippine Review of Business and Economics* XVI, 1 (March)
Thoburn, J. (1982a) *Multinationals, Mining and Development* London: Gower
—— (1982b) 'Tin: Agreement or Cartel?' *Resources Policy* (September)
Tomek, W.G. (1969) 'Stability for Primary Products: Means to What Ends?' Occasional Paper No. 28, USAID Prices Research Project, Department of Agricultural Economics, Cornell University (September)
World Bank (1978) 'The World Tin Economy: An Econometric Analysis' Staff Commodity Paper No. 1 (June)
—— (1983) *Commodity Trade and Price Trends, 1983* Baltimore: Johns Hopkins University Press
Wright, P.A. (1966) *Extractive Metallurgy of Tin* Amsterdam: Elsevier
Zorn, S.A. (1978) 'Producers Associations and Commodity Markets: the Case of CIPEC' in F.G. Adams and S.A. Klein (eds) *Stabilizing World Commodity Markets* Lexington Books

5 The minerals industry in Australia

DONALD W. BARNETT

AUSTRALIA is the sixth largest country in the world. It is five times the area of the European Economic Community and almost as large as the United States of America. Australia is well endowed with natural resources and is a technically advanced country, yet it has a population of barely fifteen million.

These endowments have resulted in Australia's being self-sufficient in most raw materials and a leading exporter of many. In agriculture Australia is the world's largest exporter of wool and meat, its second largest exporter of sugar, and a leading exporter of grain, fruit and dairy products. In minerals Australia is an equally important supplier. It is the world's largest producer and exporter of bauxite, alumina, rutile and zircon; it vies with Brazil as the largest exporter of iron ore; it is second only to the United States as an exporter of coal; and it is overtaking Canada as the leading exporter of nonferrous minerals. The locations of its principal mining centres are given in Figure 5.1.

Australia became a significant minerals producer as soon as gold was discovered in 1850 in New South Wales and Victoria. Major discoveries of silver, lead, zinc, copper, tin and more gold were made in the succeeding decades of the nineteenth century, and the great mining towns of Broken Hill and Kalgoorlie were established. But in the first half of the twentieth century Australia gradually lost its position of prominence. The combination of vast distances, small domestic demand, small-scale production technology and, at times, poor government policy gradually choked the initial development boom. Australia had to wait until the early 1960s for a resurgence in minerals production. By then, the 'tyranny of distance' could be offset by the economies of scale produced by employing large production units and developing huge mines. The rapidly developing Japanese economy provided a 'natural' market for the Australian output. Also, Australian government policy had changed to a pro-mining, free-enterprise stance that encouraged local and foreign firms to explore and develop the country's mineral wealth.

In the 1950s the production of heavy mineral sands expanded rapidly. In the 1960s, largely to meet demand from the blossoming Japanese economy, large-scale iron ore, bauxite, coal and manganese mines were developed. In the ten-year period to 1970, Australia rose from insignificance as an exporter of these commodities to world prominence, and by the start of the 1970s Australia had also developed an integrated nickel industry.

Discoveries of oil and natural gas in the 1960s and early 1970s substantially reduced Australia's oil imports from 100 per cent of requirements to about 30 per cent. Australia is now set to become an exporter of liquefied natural gas (LNG). Large uranium deposits were discovered in the five years after 1968 in the Northern Territory, South Australia and Western Australia. By 1982 two of these deposits were being mined, and more are to be developed later in the 1980s, thereby ensuring Australia's position of influence in the uranium fuel-supply cycle.

The growth in Australia's minerals production throughout the 1970s is shown in Figure 5.2. Production increased from about $A1500 million (US$1665 million) in 1970 to $A7436 million (US$8480 million) in 1980. In real terms the value of mine production increased over this period by 72.1 per cent, or 5.6 per cent a year. With the discovery of oil and gas in Victoria, production became more evenly distributed within Australia, about 90 per cent being about equally divided between New South Wales, Queensland, Victoria and Western Australia. The Northern Territory, because of new uranium production, should increase its share of total mineral output in the 1980s. Western Australia, with its new LNG exports, and Queensland and New South Wales, with the expansion of their coal production, should maintain their relative shares. Victoria's position is wedded to the price of petroleum products, and in the longer term to success in new oil exploration. Towards the end of this century South Australia should become an important producer of copper, uranium, and possibly oil, although its total production in the 1980s is likely to be limited. Tasmania's prospects continue to wane.

The growth in mineral export earnings has been even more rapid than the growth in production. Export revenue increased from $A1250 million (US$1390 million) in 1970 to $A6843 million (US$7800 million) in 1980. In real terms exports increased by 83.5 per cent over the decade, an average annual increase of 6.3 per cent. In 1980 Australia's mineral and mineral commodity exports equalled 92 per cent of the value of mine output.

The percentage of production exported for each mineral and the principal destinations are shown in Figure 5.3. Japan remains the dominant market for Australia's mineral exports. However, Japan's share has fallen from 54 per cent of exports in 1970 to 42 per cent in 1980, while exports to other Asian countries have doubled over the decade, from 7 per cent to 15 per cent (Table 5.1). Over the same period black coal, iron ore and aluminium ores and metal have consistently been the three largest earners of mineral export revenue (Table 5.2).

The value of mineral imports increased from about $A260 million (US$287 million) in 1970 to about $A2131 million (US$2340 million) in 1980 (Table 5.3). The major mineral import is crude oil, and the increase in the value of imports since 1970 is largely attributable to the escalating price of oil, rather than to an increasing reliance on foreign mineral supplies. Certainly in 1970 the Middle East supplied 48 per cent of Australia's mineral imports, and by 1980 this had increased to 64 per cent. However, in the late 1960s Indonesian oil constituted nearly 20 per cent of Australia's mineral imports, but with increased production of light Victorian crude the need for Indonesian imports quickly waned; throughout most of the 1970s Indonesia supplied about 11–12 per cent of Australia's mineral imports. Oceania is Australia's third largest supplier, but the value has fallen from 10 per cent of Australia's imports in 1970 to 6 per cent in 1980. Imports from Oceania consist mainly of rock phosphate for fertiliser production.

The minerals industry now accounts for 5.7 per cent of Australia's gross domestic product and about 36 per cent of its export income. In 1970 mineral exports represented about 25 per cent of Australia's export income.

Production and trade in the major minerals

Coal

Coal is Australia's most important mineral. In 1980, 76.3 million tonnes of saleable black coal were produced with a value of US$2165

Table 5.1 Destination of Australia's mineral exports, 1970 and 1980 (as percentage of total)

Country	1970	1980
Japan	54	42
Other Asian and Pacific countries	7	15
Total Asian and Pacific countries	61	57
EEC excluding United Kingdom	10	11
United Kingdom	12	10
Total EEC	22	21
United States	12	13
Other	5	9
	100	100

Source: Adapted from BMR

Table 5.2 Australia's mineral exports, 1970 and 1980 (US$ million)

Mineral product	1970	Per cent	1980	Per cent
Black coal	218	15.6	1925	24.7
Iron ore and pellets	361	25.9	1327	17.0
Alumina and aluminium	167	12.0	1236	15.8
Lead	112	8.1	644	8.3
Nickel	23	1.7	489	6.3
LPG	47	3.4	381	4.9
Zinc	68	4.9	288	3.7
Copper	88	6.3	286	3.7
Pig iron and steel	79	5.7	149	1.9
Tin	13	1.0	134	1.7
Uranium	0	0	112	1.4
Gold	20	1.4	103	1.3
Rutile	39	2.8	101	1.3
Tungsten	9	0.6	60	0.8
Salt	11	0.8	50	0.6
Zircon	16	1.1	40	0.5
Other	121	8.7	174	6.1
Total	1392	100.0	7800	100.0

Source: Adapted from BMR

Table 5.3 Australia's mineral imports, 1970 and 1980 (US$ million)

Mineral product	1970	Per cent	1980	Per cent
Crude oil and other refinery feedstock	180	62.7	1950	80.3
Phosphate rock	28	9.8	127	5.2
Gem diamonds	7	2.4	57	2.4
Sulphur	7	2.3	52	2.1
Others	65	22.7	244	10.0
Total	287	100.0	2430	100.0

Source: Adapted from BMR

million, and 32.9 million tonnes of saleable brown coal were produced with a value of US$122 million.

Brown coal is produced and consumed entirely in Victoria, mainly to produce electricity. A small amount of brown coal is briquetted for home heating. Production is expected to more than double by 1990, but reserves are adequate for several centuries.

Australia's black coal output increased from 48.4 million tonnes in 1970 to 76.3 million tonnes in 1980, with a record 95.4 million tonnes in 1981. Nearly 95 per cent of Australia's black coal comes from New South Wales and Queensland. New South Wales coal production increased from 35.5 million tonnes in 1970 to 42.7 million tonnes in 1980 and 56.2 million tonnes in 1982. Queensland coal production increased from 10.3 million tonnes in 1970 to 28.5 million tonnes in 1980 and 36.8 million tonnes in 1982. Queensland produces largely good-quality hard coking coal, with steam coal making up about one quarter of the output. Production is generally from large strip mines. New South Wales produces, in approximately equal quantities, hard coking, soft coking and steam coal. However, production of hard coking coal will lag behind that of the other two types throughout the 1980s. The proportion of production from opencut mines in New South Wales is steadily increasing, although some 70 per cent of production was still from underground mines.

About half of the black coal produced is consumed in Australia, electricity production being the major use. In 1980, 36.4 million tonnes of coal were consumed in Australia, with electricity production accounting for 67 per cent of this consumption, the iron and steel industry 23 per cent and alumina production 3 per cent. About 65 per cent of all black coal used domestically is produced and consumed in New South Wales.

Exports of coal have been fairly evenly divided between New South Wales and Queensland. In 1982 New South Wales exported 25.0 million tonnes, of which 11.4 million tonnes were steam coal, while Queensland exported 24.8 million tonnes, of which 23.5 million tonnes were coking coal. Japan remains the dominant export market for Australian coal (Figure 5.4). In 1982 Japan took 65 per cent of Australia's exports, consisting of 25.7 million tonnes of coking coal and

Table 5.4 Prices and Federal excise levies for Australian crude, first half of 1981 ($A per barrel)

Field size	Location	Producer return	Levy	World parity equivalent
Parity crudes, new oil				
	Bass Strait	30.23	0	30.23
	Barrow Island	30.61	0	30.61
Parity crudes, old oil				
Small fields	Bass Strait	27.23	3.00	30.23
(less than 2 million b/yr)	Barrow Island	27.61	3.00	30.61
	Moonie	28.81	3.00	31.81
Medium fields	Bass Strait	14.80	15.43	30.23
(2–15 million b/yr)	Barrow Island	14.97	15.64	30.61
Large fields	Bass Strait	11.54	18.69	30.23
Non-parity crudes				
	Bass Strait	2.57	27.66	30.23
	Barrow Island	3.12	27.49	30.61

Source: Australian Burean of Mineral Resources, 1982

6.6 million tonnes of steam coal. Korea is the second largest importer, taking 4.3 million tonnes in 1982. Taiwan is the third largest importer, taking 2.2 million tonnes in 1982.

It is very likely that exports will double to over 100 million tonnes by 1990, with steam coal representing nearly half these exports (Barnett, 1983). Most of the new mines will be opencut mines utilising draglines, or draglines and shovel-truck combinations for removing overburden. However, longwall mining equipment will play an increasing role in underground production. These high-productivity technologies will ameliorate the problem of obtaining sufficient labour for the mines.

Australia has adequate coal reserves to cope with the anticipated expansion in production. Black coal reserves are probably over 1 teratonne (1 teratonne equals 10^{12} tonnes). Probably more than 90 per cent of these reserves are in New South Wales and Queensland. New South Wales alone has 490 gigatonnes of *in situ* reserves (1 gigatonne equals 10^9 tonnes). Australia's measured and indicated reserves of black coal total 50.4 gigatonnes, 22.7 gigatonnes in New South Wales and 26.9 gigatonnes in Queensland. The New South Wales reserves winnable by opencut mining total 3.5 gigatonnes, while 'underground' reserves are 19.2 gigatonnes. Coking coal reserves total 4.8 million tonnes, and reserves of steam coal 17.9 million tonnes. Queensland's reserves consist of 6.5 gigatonnes of opencut coal and 20.4 gigatonnes of 'underground' coal. More than half (14.2 gigatonnes) of Queensland's reserves are coking coal.

The reserves currently regarded as marketable are much smaller than those cited above. New South Wales contains 8.6 gigatonnes of marketable reserves, 5.4 gigatonnes of coking coal and 3.2 gigatonnes of steam coal. Queensland's marketable reserves total 12.5 gigatonnes: 5.5 gigatonnes of coking coal and 7.0 gigatonnes of steam coal.

Petroleum and natural gas

In terms of dollar value, petroleum and natural gas are about equal with iron ore as the second most important Australian mineral. In 1980 Australia produced 22.2 million cubic metres of crude oil valued at US$860 million, and 9.4 giga cubic metres of natural gas valued at US$200 million.

Although crude oil had been produced in Australia from the early 1960s it was not until 1970, when the Victorian field in Bass Strait offshore from Gippsland went into full production, that it was produced in large quantities. Production in 1970 was 10.3 million cubic metres, of which the Gippsland field produced 7.5 million cubic metres. Barrow Island in Western Australia produced 2.7 million cubic metres, and the Queensland onshore field at Moonie produced 0.3 million cubic metres. Australian production peaked at 25.5 million cubic metres (158.5 million barrels) in 1978; by 1980 it had fallen to 22.2 million cubic metres, or 140 million barrels (Figure 5.5).

The higher world price for oil and an assurance by the government that Australian producers would be able to sell newly discovered oil at world parity prices resulted in greatly increased exploration. Since the late 1970s a number of small fields offshore from Gippsland and onshore in northern South Australia have been discovered, and Australia should be able to maintain production at the 1980 level until at least the end of the 1980s.

The recoverable reserves of crude oil at the end of 1980 were estimated at 315 million cubic metres. The Gippsland fields contained 88 per cent of these reserves, all of which are low-sulphur, high-API crude.

The reserves are adequate only for 65 per cent self-sufficiency in petroleum, necessitating imports. In 1980 Australia imported 15.5 million cubic metres of petroleum products, including 8.1 million cubic metres of crude oil, 2.9 million cubic metres of enriched crude oil and other refinery feedstock, and 2.7 million cubic metres of residual fuel oil. Australia exported 2.0 million cubic metres of petroleum products in 1980 of which almost half was liquefied petroleum gas (LPG).

Australia has large reserves of oil shale, especially in Queensland. In the late 1970s it was thought by some that these reserves could be quickly developed to assure Australia of petroleum self-sufficiency. However, the cost of production, especially the capital component, proved too high, and these projects have been shelved. Similarly, the prospects for producing oil or synthetic natural gas from Australia's large coal reserves have dimmed.

The Gippsland field in Victoria is also Australia's largest supplier of natural gas. In 1970 Australia consumed 1.5 giga cubic metres of natural gas. The Victorian Gippsland field supplied 41.6 per cent of this gas, the South Australian Gidgealpa–Moomba field in the north of the state supplied 43.4 per cent and the Roma field in southern Queensland 14.3 per cent. In 1980 Australia consumed 9.6 giga cubic metres of natural gas. The Gippsland field supplied 47.5 per cent of this gas, the Gidgealpa–Moomba field 39.9 per cent, and the Dongara field in onshore Western Australia 9.2 per cent (Figure 5.6). In 1980 production declined for the first time.

There is no national interconnecting pipeline grid. The Gippsland fields supply Melbourne, while the Gidgealpa–Moomba fields supply Adelaide and Sydney. The Roma field supplies Brisbane, while Perth is presently supplied by the Dongara field. In the near future most of Perth's gas will come from the offshore fields on the North West Shelf, about 1400 km north of Perth. In Melbourne and Sydney most of the gas is used for domestic and industrial heating with only minor amounts used to generate electricity, whereas in South Australia about half the gas goes to produce electricity. In Perth the largest single use for the gas is in the production of alumina, other industrial uses and domestic heating taking the remainder. In Queensland the largest single use for natural gas is the production of fertiliser.

The remaining recoverable reserves of natural gas, unlike those of petroleum, more than doubled over the decade. By 1980 the recoverable reserves of natural gas were 844.5 giga cubic metres. However, about 50 per cent of the reserves occur in the North West Shelf, offshore from Western Australia. A pipeline is under construction to bring a limited amount of the gas to Perth, but most of the gas is dedicated for export as LNG. Japan is to be the market for the LNG, but contract negotiations dragged on for nearly ten years and were concluded only in late 1985.

Exploration for commercial reserves of methane, which might be extracted from the coal fields in the Sydney Basin, has continued for several years. However, production is not likely to begin from this source before the end of the 1980s.

Uranium

In 1980 Australia exported about 1130 tonnes of uranium oxide (U_3O_8) valued at US$112 million, a mere 1.4 per cent of Australia's mineral export income. However, Australia has 19 per cent of the world's low-cost uranium reserves and no domestic usage, and thus has the potential to be a very important supplier of uranium. In the late 1970s it seemed possible that uranium sales could rival those of iron ore. Even now it is possible that uranium exports by the early 1990s could represent 10 per cent of Australia's mineral export income.

The richest and largest reserves are located 200 km east of Darwin in the Northern Territory. Two mines are presently producing yellowcake, and a third company has obtained approval to develop a mine.

Australia's uranium resources consist of 294 000 tonnes of 'reasonably assured reserves' recoverable at below US$80 per kg, and 264 000 tonnes of 'estimated additional reserves' recoverable at below US$80 per kg, plus small amounts of higher cost reserves (BMR, 1982). The grade of the deposits in the Northern Territory averages about 0.35 per cent U_3O_8 (Barnett, 1979).

Iron and steel

Australia has a raw steel capacity of about 10 million tonnes per year and an iron ore capacity of about 120 million tonnes per year. Iron ore production in 1980 was 95.5 million tonnes, a remarkable figure considering that production in 1965 was 6.5 million tonnes (Figure 5.7). Raw steel production in 1980 was 7.6 million tonnes (Figure 5.8). Iron ore exports earned 17 per cent of Australia's mineral export income in 1980 and exports of pig iron and steel earned a further 1.9 per cent.

About 95 per cent of Australia's iron ore is produced in Western Australia, and the great bulk of this output comes from the Pilbara region, 1200 km north of Perth. Hamersley Iron is the largest producer with a mine capacity of 46 million tonnes. In 1980 production was 38.9 million tonnes, averaging 63.5 per cent iron. Mount Newman is the second largest producer with a mine capacity of 40 million tonnes. Production in 1980 was 27.6 million tonnes of 63.2 per cent iron. Cliffs Robe River is the third largest producer and in 1980 produced 15.1 million tonnes of coarse ore and 13.9 million tonnes of fines. However, unlike the other iron ore producers, who produce a haematite ore body, the Robe River ore body is limonite and averages 57.5 per cent iron. Robe River had the largest pellet plant in Australia, but it was closed in April 1980. The pellet grade was 63.5 per cent iron. The fourth largest producer, Goldsworthy Mining, produced 5.4 million tonnes of ore averaging 62.2 per cent iron ore in 1980.

Dampier Mining produced 2.8 million tonnes of iron ore from the Yampi Sound area 2000 km north of Perth. The ore averaged 68.5 per cent iron. Dampier produced a further 1.8 million tonnes of 63 per cent iron ore at Koolyanobbing, about 500 km east of Perth.

Iron ore was also mined from the Middleback Ranges in South Australia, with production of 2.5 million tonnes in 1980, and from Savage River in Tasmania, with 2.1 million tonnes produced in 1980 (BMR, 1982). The Savage River ore is magnetite, and all ore is exported as pellets.

Exports of iron ore grew from 0.1 million tonnes in 1965 to 79.8 million tonnes in 1980 (see Figure 5.7). Japan is Australia's largest customer, taking 72 per cent of 1980 iron ore exports, a reduction from the late 1960s when Japan was taking 90 per cent of Australia's iron ore exports. Korea bought about 5 per cent of Australia's iron ore exports in 1980 and Mainland China 6.6 per cent.

Production and exports of iron ore declined as the world steel recession deepened. It is unlikely that new mines will be developed, or that further expansion of existing mines will be undertaken, until there is clear evidence that the world economy is once again growing.

Australia's iron ore reserves are estimated to be 39 gigatonnes at an average grade of 54 per cent. Western Australia's reserves are placed at 33 gigatonnes at 55 per cent average grade (BMR, 1982). If all these reserves were economically recoverable, mine life would be about 400 years at the present production level.

Iron and steel is produced by the Broken Hill Proprietary Co. Ltd. Pig iron facilities are located in New South Wales at Port Kembla (3.7 million tonnes output in 1980), and Newcastle (2.0 million tonnes), at Whyalla in South Australia (0.9 million tonnes) and at Kwinana, south of Perth in Western Australia (0.5 million tonnes in 1980). Basic oxygen steel plants are located at Port Kembla (2.6 million tonnes production in 1980), Newcastle (2.2 million tonnes) and Whyalla (1.0 million tonnes). Port Kembla also has open-hearth steel furnaces, and in 1980 open-hearth steel output was 1.6 million tonnes.

Most steel production is consumed in Australia (see Figure 5.8). The world steel recession is forcing the Australian steel industry to become more efficient. The small plants outside New South Wales are being phased out, as are the older plants in New South Wales. No expansion in capacity is foreseen, and costs have been sharply pruned.

Lead and Zinc

Throughout the 1970s Australia was the world's third largest producer of lead (providing 11 per cent of world mine output), the largest exporter of lead ores, concentrates, bullion and refined metal, the fourth largest mine producer of zinc (supplying 7.5 per cent of world output) and, after Canada, the second largest exporter of zinc. In 1980 Australian lead and zinc exports totalled US$932 million, equal to 12 per cent of mineral export income.

Production of lead and zinc was relatively stable throughout the decade (Figure 5.9). In 1980, mine production of lead was nearly 400 000 tonnes and primary refined lead production was 200 000 tonnes. Mine production of zinc in 1980 was nearly 500 000 tonnes, and primary refined metal production was 300 000 tonnes (BMR, 1982).

The Mount Isa mine in northern Queensland is the largest single producer of lead and zinc, producing approximately 40 per cent of Australia's lead and 25 per cent of its zinc. The Broken Hill deposit in western New South Wales has a greater total production than does Mount Isa, but the production comes from four mines and in 1980 supplied 53 per cent of Australia's zinc. The remainder of the output comes from small mines in New South Wales, Tasmania and Western Australia.

As can be seen from Figure 5.9, about 50 per cent of Australia's lead and 60 per cent of its zinc are refined domestically. Broken Hill Associated Smelters operates the world's largest lead smelter with a capacity of 230 000 tonnes per year. It is located in Port Pirie and produces refined lead, electrolytic zinc, antimonial lead alloys, gold, silver, sulphuric acid, cadmium and copper matte. It is the only producer of refined lead in Australia. Lead concentrates from Broken Hill make up 85 per cent of its feed stock.

Most of the Mount Isa lead concentrate is smelted into bullion at Mount Isa and then shipped to England for refining. Australian Mining and Smelting owns the Imperial Smelter near Newcastle with a rated capacity of 30 000 tonnes of lead–silver bullion and 75 000 tonnes of refined zinc.

The Electrolytic Zinc Company operates a zinc refinery at Risdon in Tasmania. The refinery has a capacity of 210 000 tonnes per year, and is the third largest in the world; however, it is now very old. The Broken Hill Associated smelter at Port Pirie has a zinc capacity of 45 000 tonnes of electrolytically refined zinc.

Only 25–30 per cent of the refined metal is consumed domestically. In 1981 the amount of lead consumed in Australia totalled 45 984 tonnes. The production of lead-acid batteries consumed 50 per cent of the metal, 20 per cent was used in alloying and related uses, 12 per cent in the production of pigments and chemicals, 11 per cent for the production of pipe and sheet, and 7 per cent for cable sheathing. In 1981 Australia consumed 95 018 tonnes of zinc. Galvanising used 71 per cent, diecasting 9 per cent, and brass and bronze manufacture 8 per cent; 4 per cent was used in the production of oxides, 3 per cent to produce rolled zinc, and 5 per cent for other purposes.

On a per capita basis, Australians in 1980 consumed 5.3 kg of lead and 7.2 kg of zinc. World demand for lead and zinc is dominated by the industrial nations of Europe and by North America and Japan. Consumption in Asia and the developing nations grew rapidly through the 1970s at an average annual rate of 5 per cent, double the world's growth rate. However, by 1980 the developing nations' consumption of lead and zinc was still only 15 per cent of the world total.

The proportion of sales to the developing nations, and to Asia in particular, increased rapidly in the 1970s. Australia's exports of refined lead to Britain fell from 58 000 tonnes in 1970 to 16 000 tonnes in 1980, and those to the United States dropped from 59 000 tonnes to 7000 tonnes during the same period. However, exports to India and Asia (excluding Japan) rose from 24 000 tonnes to 88 000 tonnes. Most exports of concentrates and bullion went to industrialised countries.

Sales of refined zinc showed a similar trend. Sales to Europe and North America fell during the 1970s, whereas sales within the Asian and Pacific area rose.

Australian production should grow modestly through the 1980s. The Elura mine in New South Wales is scheduled to start production in 1985 with a production of 70 000 tonnes of contained zinc. Mount Isa Mines plan to produce an additional 60 000 tonnes of lead concentrate and 90 000 tonnes of zinc concentrate during the decade.

Australian reserves of lead at the end of 1980 were assessed at 19.15 million tonnes of contained lead, of which 15.83 million tonnes were classified as demonstrated economic resources. The economic reserves give a 40-year mine life at present production levels. The zinc reserves were assessed at 49.24 million tonnes of contained zinc, of which 22.84 million tonnes were classified as demonstrated economic resources (BMR 1982). The economic reserves of zinc give a mine life of 45 years.

Nickel

The first commercial discovery of nickel ore in Australia was made in 1966 at Kambalda, Western Australia. Exports of concentrates commenced in 1967 and production of refined nickel began in 1970. Production of ore and concentrates peaked at 85 900 tonnes of contained metal in 1977, but by 1979 had fallen to 79 700 tonnes. Production in 1980 recovered slightly, to 74 300 tonnes. The amount of processed material has increased steadily over the decade and in 1979 stockpiles of concentrates were also used, so that the amount of nickel processed exceeded mine production by 11 800 tonnes (Figure 5.10).

Nickel is produced from nickel sulphide ores scattered around Kalgoorlie, Western Australia, and from the oxide mine at Greenvale, Queensland. In 1980 the Queensland operation produced concentrates containing 28 314 tonnes of nickel. The concentrate is railed to Townsville for processing at the Yabulu refinery. The main product is sintered rondelles of nickel oxide containing about 90 per cent nickel. The product is Class II and is used primarily in steel making.

The Kambalda concentrator, owned and operated by the Western Mining Corporation (WMC), produced concentrates containing 37 823 tonnes of nickel metal in 1980. Most of the ore came from WMC's own mines. The concentrate is either flash-smelted in WMC's smelter in Kalgoorlie to produce a nickel matte, or is railed to Kwinana to be refined in WMC's refinery. The Kwinana refinery also refines some of the matte. The refinery produces Class I products suitable for plating, alloying and producing stainless steel, and for use in the chemicals industry.

Australia's domestic consumption of nickel is approximately 4000–5000 tonnes per year, so exports were as high as 82 000 tonnes in 1977, and in 1980 were 72 100 tonnes of contained nickel metal. Most exports go to Japan, North America and Europe.

Nickel producers worldwide have suffered for many years from excess capacity, so no significant increase in Australian capacity is likely for the foreseeable future.

The BMR estimates that at the end of 1980 Australia had reserves equal to 8.47 million tonnes of contained nickel, of which 2.15 million tonnes were classified as demonstrated economic reserves. The economic reserves are sufficient for about 25 years of mine life at the present rate of production.

Copper

Australia has a mine capacity of 250 000 tonnes of contained copper per annum, and produces about 4 per cent of world output. Australian production of concentrates in 1980 contained 244 000 tonnes of copper metal, three-quarters of which was smelted domestically; most of the remainder was exported to Japan. Approximately 20 per cent of the blister copper produced in Australia is exported to Japan for refining, while the remainder is refined in Townsville or Port Kembla. In 1980 Australia exported 121 300 tonnes of copper, of which 50 800 tonnes were refinery shapes, 47 400 tonnes were contained in ores and concentrates, and 18 100 tonnes were blister copper.

Since 1973 Australian production has been very stable. The copper contained in concentrates has varied between 210 000 tonnes and 250 000 tonnes, blister copper production has varied between 163 000 tonnes and 196 000 tonnes, and refined copper production between 138 000 tonnes and 166 000 tonnes (Figure 5.11). The Mount Isa Mine in northern Queensland yields more than 70 per cent of mine production and 85 per cent of Australia's primary refined copper. Output at Mount Isa increased from 110 000 tonnes in 1970 to 155 000 tonnes in 1985. Mount Isa Mines is mining 4.6 million tonnes of copper ore per year, averaging 3.4 per cent copper. The ore body is contiguous with, but discrete from, the lead–zinc–silver ore body, and is mined as a totally separate operation. The concentrate is converted to blister copper at Mount Isa and refined in Townsville into wirebar. Since July 1979 anodes for the company's copper refinery at Townsville have also been produced at Mount Isa. The refinery capacity is 155 000 tonnes per year.

Output from other producers in New South Wales, the Northern Territory, Tasmania and Western Australia fluctuated with world economic conditions. A number of new mines which had been opened in the late 1960s or early 1970s were closed pending a rise in the price of copper.

Two mines, Woodlawn in New South Wales and Teutonic Bore in Western Australia, both producing significant quantities of copper and other base metals, have been opened recently. Exploration during the 1970s located an enormous copper–uranium deposit at Olympic Dam in South Australia and a high grade copper–zinc deposit at Benambra in northeastern Victoria. Exploitation awaits an improvement in world markets.

Australian Mining and Smelting, which owns part of the Woodlawn mine and all of the Cobar copper–lead mine in New South Wales, also owns a copper refinery at Port Pirie. The refinery is rated at 35 000 tonnes of blister copper and 60 000 tonnes of refined copper per year.

Consumption of primary copper in Australia in 1980 was estimated at 91 400 tonnes. It is estimated that a further 20 000 tonnes of secondary copper were also consumed.

Six export markets accounted for 87 per cent of the total export revenues from copper in 1980. Japan accounted for 29 per cent of this income, the United States for 15 per cent, Germany for 14 per cent, Belgium–Luxembourg for 13 per cent, the United Kingdom for 8 per cent, and France for 8 per cent. Japan takes most of Australia's copper concentrate exports, while in 1980 Belgium–Luxembourg took most of its blister copper exports. Over recent years Europe has taken about 80 per cent of Australia's refined metal exports.

Australia could substantially increase its copper output. The Olympic Dam prospect in South Australia has the potential to rival Mount Isa as Australia's leading copper producer. But the project needs a very large production to be viable, and it is doubtful that world markets will be able to absorb the additional output for many years.

At the end of 1980 Australia had 15.6 million tonnes of copper reserves, of which 5.5 million tonnes were classified as 'economic demonstrated' (BMR 1982). The economic demonstrated reserves give a mine life of 22 years.

Tin

Australia's tin production increased from 3 849 tonnes or 2.5 per cent of western world output in 1965, to nearly 12 000 tonnes or 6 per cent of western world output in 1980. This increase was due to expansion by two groups: Renison Goldfields increased capacity by 4 900 tonnes per year and Aberfoyle increased capacity by 1 400 tonnes per year.

During the same period the proportion of tin concentrates smelted in Australia fell from 85 per cent to 50 per cent, even though the rated refinery capacity increased from 3 500 tonnes to 7 500 tonnes per year. Concentrate production fluctuated, partly in response to cost pressures and exchange rate movements and partly due to production quotas imposed by the International Tin Council. In both cases, as profitability fell high-cost producers, mostly small operations, suspended mining.

Most of Australia's output of tin comes from hard-rock mines, which, by world standards, are large, efficient and highly mechanised. The Renison Bell mine in Tasmania, for example, is the world's largest single tin mine and has a tin recovery rate of 71 per cent. Cleveland, another Tasmanian hard-rock mine, owned by Aberfoyle, has a recovery rate of only 50 per cent. An Aberfoyle hard-rock mine in New South Wales had a capacity of about 2 000 tonnes of tin metal, but its recovery factor fell from 74 per cent to 61 per cent, reducing output in 1980 to 1700 tonnes of tin. Queensland is also an important tin producer, with production in 1980 of 2 800 tonnes of tin, about two-thirds of which was alluvial.

Australia has two tin smelters. That in Sydney is operated by Associated Tin Smelters. It has a rated capacity of 15 000 tonnes of high-grade concentrates a year, equivalent to about 6 000 tonnes of tin per year from the concentrates available in Australia today. The other smelter is operated by Greenbushes Tin, primarily to allow that company to remove the maximum amount of tantalum from its tin concentrates. The smelter capacity is rated at 1 000 tonnes of tin per annum. The increased receipts for tantalum production more than compensate Greenbushes for its capital investment and the high unit cost of operating such a small smelter.

The production of tin concentrates and refined tin during the period 1970–80 is summarised in Figure 5.12.

In 1980 Australia consumed about 3 000 tonnes of tin. The manufacture of tinplate took 50 per cent, the production of solders 30 per cent, and tinning, bronze and brass, bearing and type metals, and plastic bottle production consumed most of the remainder. Australia's tin consumption fell steadily from about 4 000 tonnes per year in the early 1970s. This was partly due to the world recession, partly to direct substitution, and partly to a decline in the Australian canning industry. The high price of tin was an important factor in the reduced consumption.

Through the 1970s the pattern of Australia's tin trade altered significantly. Concentrates are now sold almost exclusively to Malaysia, whereas in the early 1970s substantial tonnages were sold to European smelters. Milling practice improvements eliminated the production of low-grade flotation concentrates, which at one time could be smelted only in Europe. Exports of refined tin have varied between 1 100 tonnes and 3 300 tonnes per year since 1970, the principal customers being the United Kingdom (531 tonnes in 1980), the Netherlands (242 tonnes), New Zealand (225 tonnes) and the United States (135 tonnes). Tinplate is exported to many countries in Asia, including Indonesia, Malaysia, the Philippines and Singapore. The total quantity rose from 44 000 tonnes in 1970 to almost 100 000 tonnes in 1980. Australia has generally imported sizeable quantities of tinplate, especially in 1974, when 75 000 tonnes were imported. However, in 1975 imports fell to 6 700 tonnes and in 1980 to a mere 6 tonnes. Australia also imports small quantities of tin metal, peaking for the decade at 222 tonnes in 1977, with 1980 imports at 44 tonnes.

Australian producers have been reluctant to build a new tin smelter and thereby to increase Australia's efficiency and capacity as a tin refiner. However, additional mine capacity in the 1980s will increase the pressure to add to refining capacity. Several prospects are being evaluated for large-scale openpit and underground mining. One of these, at Emmaville in New South Wales, could supplant the Renison mine as the world's largest producer, and would provide the incentive to raise Australia's smelting and refining capacity.

At the end of 1980 Australia's tin resources were estimated at 764 000 tonnes of contained metal, of which 209 000 tonnes were classified as demonstrated economic and a further 127 000 tonnes as inferred economic (BMR, 1982). The demonstrated economic reserves give a mine life of eighteen years, and the total 'economic' reserve a mine life of 30 years, assuming the 1980 rate of production.

Government and the management of the Australian minerals industry

Although the minerals industry vies with agriculture for the position of Australia's largest export earner, governments have focused relatively little political energy upon it. By contrast, the agricultural industry is a very strong political force. It is represented by its own political party, originally called the Country Party but now known as the National Party. This party is in coalition with the Liberal Party, and the coalition has formed the government of Australia for about 30 of the last 35 years. Not surprisingly, a considerable amount of Federal government policy formulation has been aimed at assisting the farming sector.

Federal agencies act as monopoly sellers of Australia's wheat and wool output, and the Australian Wool Corporation actually sets the floor price for wool on a daily basis. The government sends trade delegations overseas, often at the ministerial level, to sell agricultural output. The Federal government has given rural producers favourable tax treatment, and regulates the maximum price of various non-labour factors of production that are important to rural communities.

The place of minerals in government priorities

The minerals industry, by contrast, has been taken largely for granted. The Federal government has, by and large, operated as if the industry will function adequately in a virtual policy vacuum. Initially this 'no policy' policy was very successful at encouraging the development of an export-oriented raw materials sector. However, it gave no encouragement to enhanced domestic processing capabilities, and, by supplying a growing foreign processing industry, made it even more difficult to establish additional domestic processing. Additionally, in times of less vibrant growth in major markets such as Japan, the 'no policy' policy left the industry vulnerable to the pricing policies of alleged monopsonistic buyers. The lack of government support made it easier for 'no growth' advocates such as environmental groups and land-rights activists to erode the minerals industry's power base by withdrawing exploration or development rights.

However, it must be said that when the Whitlam Labor Government tried to take a more active role in minerals between December 1972 and 1975, industry leaders strongly opposed government actions. When the Liberal–National Party Deputy Prime Minister tried overtly to enforce price controls in key mineral exports in the late 1970s he was thwarted by the Premiers of Queensland and Western Australia rather than by industry leaders. He also apparently lacked the support of the Prime Minister (Bambrick, 1980).

In the 1960s and early 1970s, the State governments also adopted a basically neutral policy towards the minerals industry. However, by the late 1970s most States had become much more concerned about increasing their share of the economic rent earned by the minerals industry. The governments of New South Wales and Queensland have recently increased royalties and the charges for government services to an extent which implies that they must believe the minerals industry is insensitive to the level of government imposts.

The Australian Constitution has encouraged the Federal government to pay less attention to the minerals industry than perhaps it should have done. The Constitution lists the specific areas of Federal authority, which are largely related to foreign or interstate commerce, defence and immigration. The founding fathers intended a wide area of reserve authority to be left to the States so that Federal interference should be minimised. Mineral resources are not specifically mentioned in the Constitution. Thus they remain under State sovereignty except for offshore or Federal territorial areas. However, the Federal government's fiscal superiority, given to it 'almost accidentally',

Australia's limited domestic consumption of minerals, and the Federal government's control over 'trade and commerce with other countries, and among the states', provided it with powerful 'back door' controls (Barnett, 1979).

The Federal government's most powerful minerals policy tool is given through Customs (Prohibited Exports) Regulations. The exportation of certain raw or semiprocessed minerals, metals, fuels and petroleum products is prohibited unless permission is granted by the Minister for Trade and Resources or an authorised person. The government has only infrequently tried to impose export prohibitions, and in most cases it has not been able to make the actual or threatened ban achieve the policy objective, such as maintaining price levels. On the few occasions that export bans have been fully implemented the results have been negative. The most infamous example is the 33-year ban on the export of iron ore implemented in 1938. The rationale for the ban was that the Commonwealth Geological Adviser believed that Australia possessed sufficient iron ore reserves to supply only its own industry (Raggatt, 1968). The effect of the ban was to delay full-scale exploration for iron ore by about 30 years, to slow down the development of Western Australia, to entrench the monopoly of the single local steel producer, and to remove any possibility of Australia's establishing an export-oriented steel industry. Shorter bans on the export of LNG and uranium also appear to have decreased rather than increased the net benefits to Australia.

The most important power held by the States over minerals is ownership. They have generally retained title to minerals even when the land is privately owned. Historically, exploration leases were granted on a 'first come' basis, and mining leases were awarded only to the holder of the exploration lease. Now, some States require competitive bidding for the exploration lease, and the government reserves the right to award the mining lease to a company other than the one holding the exploration lease.

The States also use their minerals leasing power to obtain extra minerals processing facilities in their territories. Diversification in the iron and steel industry away from New South Wales was aided by iron ore lease conditions imposed by South Australia and Western Australia (Barnett, 1979).

The States have been development-oriented and favour private ownership. Except in the electricity sector, the States have not taken any active role in the marketing of minerals or in the planned development of mineral resources. In energy production, however, the States display a very active direct interest. Virtually all electricity in Australia is generated by State-regulated utilities. In Victoria the electricity authority mines over 30 million tonnes of brown coal a year. South Australian sub-bituminous coal is mined by the State utility, and the New South Wales State utility operates a number of black coal mines.

Until the mid-1970s royalties were low and the State governments were satisfied with the politically valuable benefits such as increased employment flowing from mineral development. Development was favoured for its own sake. The increased constraints imposed on State borrowing by the Federal government in the late 1970s, and their very limited taxing powers, resulted in the States significantly increasing the level of royalties and other charges to the minerals industry. These higher charges have often been applied without a full appreciation of their impact upon the short-run profitability and longer term growth of the industry. At the same time, the State governments imposed much tougher environmental laws and required the industry to pay for many items of social infrastructure previously provided by the government. The increased royalties and charges will not produce the revenues the States had counted on, and indeed could well produce less than would come from a better planned set of policies.

The problem is that the States have traditionally ignored the economics of the minerals industry. They do not possess an adequate data base or framework within which to analyse the impact of their policy decisions. These factors, together with the recession in the world economy, led to a sharp reduction in the growth of the minerals industry.

Taxation, royalties and other contributions

The corporate income tax rate paid by mining companies is the same as that paid by non-mining companies. At the start of the 1970s the rate was 47.5 per cent of taxable income. The Whitlam Labor government lowered the rate to 42.5 per cent for all companies, but eliminated the deduction from income of a number of capital expenditure categories and decreased the annual depreciation allowance on most other mining capital expenditure categories. When the Liberal-National Party regained office the corporate income tax rate was raised to 46 per cent and the depreciation rates were liberalised, but not to the levels that existed before the Whitlam government gained office. The mining companies' corporate income tax increased from US$87 million in 1970 to US$776 million in 1979.

Since all minerals except uranium are owned by the States, only the States can impose royalties on the production of these minerals. However, the Federal government can levy export-based and certain production-based duties.

Each State is free to set its own royalty on each mineral product produced within its borders. As one could expect, the actual royalties differ considerably between minerals and between States. Unit royalties are probably the most common, but ad valorem royalties also exist.

For instance, petroleum products often incur a royalty of about 10 per cent of the wellhead value. Profit royalties are rare, but New South Wales imposes a severe profit-type royalty on the lead-zinc-silver producers at Broken Hill and a lesser one on the Cobar copper-zinc mine. The maximum royalty rate at Broken Hill is 50 per cent of 'net profits' once net profits equal or exceed US$9.6 million. The Cobar mine faces a maximum royalty of 8 per cent of net profits. Tasmania also has provision to charge a profit-type royalty with a maximum rate of 7 per cent.

Royalties collected by the Australian States increased from $A38 million in 1969–70 to $A88.8 million in 1973–74, jumped to $A153.9 million in 1974–75, and then steadily to $A74.6 million in 1979–80[1]. The jump in the 1974–75 take occurred because Queensland and New South Wales increased their royalty rates. When the price of crude oil rose sharply in 1979, the Victorian royalty moved in harmony. Western Australia, South Australia and Tasmania maintained a constant royalty take in real terms.

Contrary to the predictions of some analysts in the early 1970s, the net taxes and royalties paid by the Australian mining industry steadily increased in real terms throughout the decade. In fact, between 1969–70 and 1978–79 the net taxes and royalties, in 1980 US dollars, increased from US$337 million to US$1150 million, a rise of 340 per cent. Over the same period the net taxes paid by Australian industry as a whole showed no substantial growth (Figure 5.13). It can also be seen from Figure 5.13 that, while the all-industry dividend payout virtually equalled the all-industry net tax collections throughout the decade, the mining industry dividend payout fell to less than half the mining industry tax and royalty collections. If the mining industry figures are compared with the all-industry figures, it is clear that the relative net tax burden faced by the mining industry increased substantially over the decade.

In addition to normal income tax and royalties, the mining industry is subject to special levies imposed by the Federal government. The two most notable of these levies are the coal export levy and the crude oil production duty. The coal export levy was introduced by the Whitlam government in an attempt to capture part of the economic rent generated by the Utah Development Corporation in Queensland. The Australian Constitution prohibits the Federal government from discriminating against a particular State, so the levy legislation had to be framed in a 'national' context. The government stipulated that the maximum levy of $A6 per tonne would be payable by producers of specific high-quality coking coal. All of Utah Development's coal fell into this category, and a few other Queensland and New South Wales producers were also ensnared. The unfairness of the levy (for example, the fact that the New South Wales mines eligible for the levy were all relatively high-cost underground producers) has resulted in the levy being gradually reduced and constrained in its impact. By 1982 the levy was payable only on high-grade coking coal produced from a depth of less than 60 metres in open-cut mines established before July 1980. The levy was reduced to $A3.5 per liable tonne.

The crude oil production levy is even more complicated. Again it is directed towards capturing part of the economic rent generated by crude oil production, and is especially aimed at the established low-cost producers. To circumvent the nondiscriminatory requirements of the Constitution, a uniform levy is imposed on all producers at the maximum rate, coupled with a scale of rebates to less profitable producers. The Constitution does not prohibit discriminatory 'hand-outs' to the States, or operations within a State.

The amount of the levy is revised half-yearly and is pegged to the import parity price of Saudi Arabian light crude. Recent discoveries (that is, fields discovered after August 1976) receive the full parity price but by year-end 1982 only the Fortesque field and a small field on Barrow Island qualified as recent discoveries. Part of the production from the Bass Strait and Barrow Island fields is classed as 'non-parity'. For this oil the producer receives the 1975 price. In the first half of 1981 the government levy amounted to $A27.50 per barrel. The unit returns to the producer and the government for this 'class' of oil and for 'parity' crudes in early 1981 are shown in Table 5.4. Variables are the date of discovery and the production rate. The introduction of a Resource Rent Tax in 1984 complicated the picture even further.

The coal levy grosses a maximum of about $A100 million per year for the Federal government, whereas the oil levy grosses about $A3000 million per year. Since the levies are tax deductible, the net benefit to the Treasury from the oil levy is about $A1600 million per year (US$1850 million at the 1980 exchange rate). By way of comparison, company tax receipts from all sources in 1979–80 totalled $A3200 million (Harris, 1980).

Since the 1960s mining companies have paid for most of the social and transport infrastructure required for their operations. The States have adopted different approaches to the provision of infrastructure. In the Pilbara region of Western Australia the iron ore companies operate their own railways and ports. In Central Queensland the railways are operated by the State government, although the coal companies must pay for their construction. Even in New South Wales, where for coal at least the rail and town network already existed, companies are being required to fund an increasing proportion of the social and transport infrastructure. Virtually all railways in New South Wales continue to be State-operated.

Queensland, and to a lesser extent New South Wales, use the railways to collect a further portion of the economic rent from coal and copper production. In Queensland the average charge for railing coal is two to three times the full cost to the Queensland government of operating the service. In the first half of 1982 the Queensland Treasury gained by way of 'excess rail freight' more than 2.5 times as much from the coal export industry as it collected from coal royalties. The New South Wales railway is not as profitable as the Queensland system. It is older, the terrain is more difficult, and it is much less efficient. Nonetheless, the New South Wales government still earns a sizeable profit from railing coal (Barnett, 1983).

When account is taken of all the taxes, near-taxes and social infrastructure generated or paid for by the Australian minerals industry, one must disagree with the statement in the Fitzgerald Report, sponsored by the Whitlam Labor government, that 'mining is a net burden on the economy' (Fitzgerald, 1974).

Pricing policies

Export Prices

Both Labor and Liberal–National Party governments have been concerned about the apparent cartel-like behaviour of certain buyers of Australia's minerals exports. The Labor Minister for Minerals and Energy introduced export controls on a wide range of mineral commodities in an attempt to place a floor under prices. These controls were removed when the Fraser government took office. Subsequently, however, Mr Anthony, the then Deputy Prime Minister and Minister for Trade and Resources, told the House of Representatives on 24 October 1978 that:

> The Australian Government's concern is that individual Australian companies face buyers who are co-ordinated or who have a high degree of consultation and who, as a result, can and do successfully play one seller off against another. The result is prices, terms and conditions which do not reflect a fair and reasonable return ... We are concerned about buying techniques which ... are designed to depress the prices received for these basic raw materials.

To counter this, Mr Anthony reinstated export controls on iron ore, coal, bauxite and alumina. The aim was to set minimum price levels for all grades of product. Mr Anthony told the House that:

> Exporters who wish to enter into negotiations under new or existing contracts will be required to obtain specific approval before making any offers or responding to any offers or entering into any commitments ... I will determine the parameters within which a company or companies will be authorized to negotiate.

Although New South Wales coal producers generally supported Mr Anthony's action, Queensland coal producers did not. Certainly the Premiers of Queensland and Western Australia were strongly opposed to the guidelines, as were a number of Federal Members of Parliament. Apparently, the Prime Minister also opposed Mr Anthony's actions.

On 12 June 1979 Mr Anthony announced that the guidelines would still apply to coal, bauxite and alumina, but not to iron ore (Barnett, 1981b).

In October 1979, Mr J. Scully, Secretary to the Department of Trade and Resources, also alluded to the price-depressing effects of certain buyer groups. Mr Scully stated that 'Australia would develop new projects, financed by the guarantee of "bankable" 20 year contracts at prices higher than today's ruling prices ... Australia sought world parity pricing, underlining a strong but not exclusive reliance upon market forces. It would not accept a cost-plus formula'.

By 1983 the government was obliged to accept one-year contracts at fob prices for coal and iron ore barely above the actual cost of production. Both the coal producers and buyers claimed that one important reason that the ruling prices did not provide for adequate profits was that the State governments were charging unreasonably high rates for rail and port services. However, in Western Australia, where railways and ports are privately operated and perhaps the lowest cost in the world, the ruling prices still did not provide an adequate return to the producers.

The Australian government set a floor price for uranium exports of about US\$30 per pound U_3O_8 plus various escalation provisions. It will not authorise exports unless the contract price at least equals the current floor price.

Domestic prices

State and Federal governments set or influence the prices of most domestic energy raw materials.

As we have already seen, the Federal government sets the price for crude oil on the basis of production rate and date of discovery. In 1965 the Federal government set the price of domestic crude oil at \$A3.14 per barrel, 67 cents above the landed cost of equivalent foreign oil, to give an exploration incentive. But when the Bass Strait fields commenced production in 1969 (these fields still provide two-thirds of Australia's oil requirements), the Gorton Liberal government lowered the price for Bass Strait oil to \$A2.06 per barrel. In September 1975, Labor increased it to \$A2.33—despite world oil having been in excess of \$A10 for nearly two years. In August 1977, the Fraser Liberal government introduced world parity prices for oil discovered after 14 September 1975, and a year later set the price for the consumer at world parity. The producer price was raised in increments, so that by 1981 half of production was at world parity (Barnett, 1979; Folie, 1980).

The Federal government also set the price for LPG. A primary concern of government is to minimise the impact of any price increases on rural consumers. Consequently the domestic price of LPG is often below the export parity price. The maximum price for retail petrol is also set by Federal and State legal agencies.

State government agencies set the wellhead price for natural gas. Again the price is generally set below the parity equivalent, although because of the transport component in natural gas it is not easy to establish what is the crude oil parity price for gas. However, it is clear that the Gas and Fuel Corporation of Victoria, the monopoly buyer of gas from the Gippsland offshore field, has held the price the producer receives below crude oil parity.

In 1969 the Gippsland producer signed a 20-year contract to supply gas to the Victorian market. The maximum price for the producer at the processing plant was 28.5 cents per gigajoule for the first three years and 39.5 cents per gigajoule thereafter. The minimum price for the entire twenty years was set at 24.5 cents per gigajoule. The actual price is dependent on volume, but it must lie between the two limits. Later the Gas and Fuel Corporation negotiated additional purchase contracts at prices a little above the initial price, and with provision for prices to rise at half the rate of costs in the community as measured by the consumer price index. This rate of increase is many times below the rate of increase of oil price rises since 1974, and has led to a rapid growth of gas consumption by industry (Barnett, 1979).

Predictably, the producer has concentrated upon searching for crude oil rather than natural gas, and the Gas and Fuel Corporation is now concerned about the security of supply beyond 1990. However, it has refused to increase the producer price.

South Australian natural gas was also underpriced, but recent price settlements have been considerably higher.

Various State agencies charged with the task of supplying electricity also set prices. They control the price of most of the steam coal sold on the domestic market in that they are virtually monopoly buyers. The electricity authorities also generally operate captive coal mines. Since the issuing of mining leases is controlled by the States, there is always the possibility that State electricity authorities will be given preferential treatment in the allocation of reserves.

Processing

A feature of the Australian minerals industry is the low percentage of production that is processed. Only 1 per cent of Australia's 1980 production of rutile and zircon, 2 per cent of tungsten, 8 per cent of ilmenite and alumina, 12 per cent of iron ore, 25 per cent of manganese and 42 per cent of tin were processed locally. With the exception of nickel and bauxite, it is only the traditional metals that are highly processed. In 1980, 68 per cent of zinc, 77 per cent of copper and 95 per cent of lead were processed to a metallic stage. About 95 per cent of the nickel output was also processed to a metallic form, and 77 per cent of the bauxite production in 1980 was converted to alumina. When domestic consumption is removed from the statistics, virtually all of Australia's exports of coal, iron ore, alumina, manganese, rutile, zircon, tungsten and ilmenite are exported unprocessed, while only about 25 per cent of tin and 50 per cent of zinc are processed before exportation (Figure 5.14). One can expect that the level of processing of uranium yellowcake will also be extremely low. Before the 1950s a much higher proportion of mineral output was processed in Australia. Recent mineral developments, with the exception of nickel and bauxite, have involved the supply of unprocessed raw materials to the established industrial centres of Europe and the growing centres of Asia.

Over the period 1974–80, investment in new fixed capital expenditure in all Australian industry varied between US$9000 million and US$12 000 million per annum. Investment in mining averaged 12.3 per cent of the total, while investment in metal extraction and production facilities averaged only 5.5 per cent of the total new capital expenditure.

State government policies have favoured the establishment of further processing of mineral raw materials. But each State government is interested only in developments within its own boundaries. In the case of iron and steel, the threat of withholding lease extensions from BHP, the sole producer, led to the establishment of a blast furnace in Western Australia and an integrated steel mill in South Australia. However, this diversification ultimately inhibited BHP from matching the world trend towards larger and more concentrated production facilities, and it is an important reason why BHP steel became uncompetitive (IAC, 1979). More recently, these far-flung plants have been closed or severely scaled back.

In the 1960s and 1970s State governments were generally not prepared to withhold the granting of a development lease to the discoverer, often a large international concern, in case it discouraged other firms from exploring in their State. Royalties were generally low, so that a royalty holiday had little attraction. In short, while the States desired further processing they had little bargaining power. They have thus been ineffective in maximising the amount of processing of Australia's minerals production.

The Federal government has been, and probably still is, preoccupied with its attempt to gain access for its agricultural products into the European Economic Community (EEC) to compensate for the markets Australia lost when the United Kingdom joined the EEC. It has not paid much attention to the further processing of its minerals output. For example, when Mr John Howard, then the Minister for Special Trade Negotiations, addressed the EEC in October 1977, fifteen of the thirty paragraphs of his submission mentioned agriculture and mineral raw materials but the exportation of processed minerals was mentioned only once. The statements on minerals emphasised the high grade and abundance of the reserves and the sizeable past investment in mining. Virtually no attention was paid to the tariff barriers imposed on processed minerals imported into the EEC, Japan and the United States.

To some extent the Federal government's lack of interest in further processing reflected that of the minerals industry. In the 1960s and 1970s the industry was preoccupied with developing mines and finding markets for minerals output. The Australian economy was growing rapidly and unemployment was virtually nonexistent. Mining was believed to produce greater profits than the processing of the raw materials. So long as new mines could be opened, this was probably true.

However, by the 1980s the opportunity to develop new mines had decreased and the mining industry has also learned that selling basic raw materials can be a high-risk, low-profit venture. The industry is now more receptive to the idea of increased local processing. Mr John Macleod, Chief Economist with Conzinc Riotinto Australia, argues that the scope for processing in Australia is 'very, very large' (1983: 116).

Foreign ownership

In the 1960s there were virtually no constraints upon foreign companies investing in Australia. With the minerals boom of the late 1960s and early 1970s the degree of foreign ownership in the Australian minerals industry rose rapidly. Between 1963 and 1975, foreign ownership of the minerals industry increased from 27 per cent to 52 per cent. Foreign ownership was highest in black coal, crude oil and iron ore (McKern, 1976).

By the 1970s the 'foreign takeover' of Australian resources was seen as a political issue. The Whitlam government introduced ownership guidelines whereby ownership had to be 100 per cent Australian for uranium mines and at least 50 per cent Australian for all other minerals. The Fraser Liberal–National Party government reduced the minimum Australian ownership of uranium developments to 75 per cent but the requirement for other minerals remained at 50 per cent Australian ownership. The Fraser government further liberalised foreign ownership rules by stating that if adequate Australian finance could not be obtained, a project might be permitted to proceed with reduced Australian ownership. That government also introduced a policy whereby, although a company may be predominantly foreign-owned, if it announced intentions of becoming at least 51 per cent Australian-owned, was already at least 25 per cent Australian-owned and had a board of whom a majority of the members were Australian, it would be termed a naturalising company and given the status of a naturalised company.

A naturalised company was able to proceed with new projects in its own right, in partnership with an Australian company, a naturalised company or a naturalising company. However, in the absence of special circumstances it was precluded from undertaking a project as a joint venture with a wholly overseas-owned company, as this would involve a departure from the 50 per cent guidelines.

The full extent of the changes by the present Labor government to these regulations remains to be seen. In keeping with its pragmatic approach to economic policy, it has further relaxed ownership requirements.

The environment

The Federal government's authority over exports allows it to insist upon the completion of environmental impact statements for new mines or mineral-related projects, even though this is usually the jurisdiction of the States. The Federal government can also institute public enquiries. The Federal Environmental Protection (Impact of Proposals) Act became law in 1974, and has been used to hold lengthy enquiries into uranium mining in the Northern Territory and the mining of mineral sands on Fraser Island, Queensland. The result of the former enquiry was a guarded go-ahead for uranium, whereas the latter resulted in the cessation of mining (McColl 1980).

By the late 1970s the Federal government had largely withdrawn from this area, and left the responsibility of requiring and evaluating mining environmental impact statements to individual State governments.

Aboriginal land rights

In the 1970s the Federal government paid increasing attention to the issue of Aboriginal land rights. On Australia Day 1977, the Aboriginal Land Rights (Northern Territory) Act was proclaimed. The Minister for Aboriginal Affairs stated that the Land Rights Act confers upon traditional Aborigines 'inalienable freehold title to land outside reserves'. By April 1978, 18 per cent of the Northern Territory was classified as Aboriginal reserves and 23 per cent as inalienable Crown Land. Land claims by traditional Aborigines could in time involve more than half the area of the Northern Territory (Bambrick, 1979).

The Land Rights Act stipulates that the consent of the traditional owners must be obtained before exploration or development can occur on their lands. There is provision for the decision of the traditional landowners to be overruled 'in the national interest', but 'national interest' is not defined. In the case of non-agreement over financial matters the compensation will be determined by an arbitrator.

The Aboriginals have consented to uranium mining at Ranger, Nabarlek and Jabiluka. The Ranger partners agreed to pay Aboriginal groups a total of 4.25 per cent of gross revenues. No provision was made to pay royalties to the Northern Territory, which greatly upset that fledgling State's government. In Western Australia, a Labor government has moved to restrain the impact of Land Rights claims.

Development and exploration

There is a widely held belief in the minerals industry that government policy, especially taxation and foreign ownership laws, has a direct effect upon the level of industry activity. The industry is therefore quick to respond to changes in government policy. A corollary of this view is that socialist governments discourage private exploration and development.

The Whitlam Labor government, which held office from 2 December 1972 to 13 December 1975, introduced rather radical taxation, ownership and investment changes, and has been accused of causing the downturn in the Australian mining and petroleum exploration and development industries. For example, Derry, Michener and Booth stated 'that the degree of uncertainty which now hangs over every aspect of mining and exploration in Australia has virtually halted significant exploration and mining development' (1974: 48). Three years later Mullins, Lawrence and Deschamps wrote that 'Australia is rapidly gaining in stature following the return to power of the Liberal–National Party coalition' (1977: 104). However, the statistics do not unequivocally support the above statements; in fact, the evidence casts doubt on the veracity of these claims.

Expenditure on new fixed private non-agricultural capital investment in Australia over the period from 1974 to 1980, measured in 1980 Australian dollars, climbed steadily and continuously from $A7785 million to $A10 969 million. The proportions of this investment for mining, metals extraction and production were 21.4 per cent in 1974, 22.0 per cent in 1975, 14.7 per cent in 1976, 13.9 per cent in 1977, 15.9 per cent in 1978, 16.3 per cent in 1979, and 20.1 per cent in 1980 (BMR, 1982). That is, after five years of a non-Labor government, the proportion of new fixed capital investment in the mining industry had not reached the level of the last year of the Labor government. It is likely that much of the development during Labor's term of government stemmed from previous discoveries, but clearly development did not 'virtually halt', although exploration remained low throughout the mid-1970s.

However, one cannot lay the blame for the downturn in exploration solely at the Labor Party's door. Minerals exploration expenditure, measured in 1980 Australian dollars, fell from $A422 million in 1970–71 to $A161 million in 1975–76 (Figure 5.15). But 70 per cent of this decline occurred before the Labor Party gained office. Metals exploration expenditure would probably have dropped in the early 1970s regardless of the party in office. By 1971 the stock market mining boom was ending and much of the money had been wasted on dubious exploration. Also, so much iron ore, bauxite and nickel had been found that there was little point in funding further exploration, particularly as the world was entering a period of low metal prices and general recession. But Labor did drastically curtail exploration for uranium and petroleum (Barnett, 1981b).

By December 1972, all the currently known uranium deposits in the Northern Territory had been found. Labor's ban on the development of these deposits froze exploration. Labor initially banned development because the Minister for Minerals and Energy believed that uranium prices would rise rapidly, and that Australia would benefit from the delay. However, when uranium prices did rise Australia was enmeshed in the antinuclear debate. The Fraser Liberal–National Party government also vacillated on the uranium issue, and the prohibition on new development remained in force until the late 1970s. By this time the uranium market was softening and Australia's export window had all but closed.

During Labor's term in office real petroleum exploration expenditure, measured in terms of 1980 Australian dollars, more than halved from $A235.8 million in 1972–73 to $A88 million in 1975–76. This was probably because Labor refused to grant developers world parity prices for either new or old crude oil, and banned the exportation of LNG from the North West Shelf. When the Fraser Government permitted a modified world parity pricing system for oil and lifted the ban on LNG exports, petroleum exploration started to increase (see Figure 5.15). Labor banned LNG exports from the North West Shelf because it wanted to build a trans-Australia pipeline to bring the gas to the populous eastern states. The oil companies wanted to convert the gas to LNG and export it because they felt this would give them a quicker return (Barnett, 1976). Although the Fraser government quickly gave permission for the export of 53 per cent of the reserves, it was not until 1985 that the North West Shelf consortium was able to finalise export contracts.

The Australian experience of the 1970s indicates that mineral companies, provided they are permitted to earn profits that are internationally competitive, may not be as sensitive to socialist or nationalistic governments as is often believed. The Saskatchewan uranium experience of the 1970s also lends support to this conclusion (Anderson and Barnett, 1983).

Summary

The Australian minerals industry experienced sustained rapid growth over the 1970s. The value of mineral production increased from US$1665 million in 1970 to US$8480 million in 1980, an increase in real terms of 72 per cent. More than 90 per cent of the value of production is exported, and over the period the value of exports increased in real terms by 83 per cent. Minerals represent about 36 per cent of the total value of Australia's exports. The three most important exports continue to be black coal, iron ore and alumina bauxite. Together they constitute nearly 60 per cent of the value of mineral exports. Crude oil and natural gas are the fourth most valuable mineral commodity group, but virtually all production is consumed domestically. However, LNG exports should become an important export earner by the end of the 1980s. The main future growth areas should be coal, aluminium, LNG, and uranium.

Australia is still very reliant upon the Asian market, and about 60 per cent of the value of mineral exports goes to Asia (mainly Japan). The dominance of Japan as a buyer declined from 54 per cent of the total value of Australian mineral exports in 1970 to 42 per cent in 1980.

The ten years from the late 1960s was possibly the period of greatest political change in Australia since Federation. During this period Australia had five Prime Ministers, each of whom had a distinct political style and philosophy. Prime Minister Holt was the last of the 'traditional' Liberal Prime Ministers. Prime Minister Gorton, who took over leadership of the Liberal–National Party government, was a nationalist and introduced a number of practical and social changes. In December 1972 Australia elected its first Labor government in two decades. This government's term was ended in 1975 and the policies affecting the minerals industry also underwent significant changes. Thus, Australia's mineral policy went from basically laissez-faire to nationalistic with, at times, severe restrictions on foreign ownership and investment. With the election of the Fraser Liberal–National Party government in December 1975, the policy ostensibly returned to a more free-market approach. But even then many of the nationalistic policies remained in force. Yet throughout this period, mineral production and exports steadily climbed. Exploration expenditure certainly declined in the early to mid-1970s but the cyclical nature of exploration is well known. Moreover, the downturn was due also to the sudden end of the 'nickel' mineral boom, a world recession and the continuance of a policy of low prices for crude oil. It is likely that had the Labor government allowed the price of newly discovered crude oil to rise with international prices the exploration downturn would have been much more moderate. Similarly, had it not been for the ban on exporting LNG and uranium, investment in mineral developments would probably have continued to increase throughout the period.

It seems that the government policies which exerted a pronounced effect upon exploration and development in the 1970s were not those affecting taxation or foreign ownership and investment, but those affecting price and market regulations. The implication is that, despite their rhetoric, industry leaders will tolerate a wide range of political persuasions and policies provided they have a reasonably free hand in the marketing of their output. In the Ps and Qs of international mining, profits outweigh policies.

Notes

This chapter is based in part on Barnett and Hemphill (1983).

1 The jump in royalties is somewhat masked if US dollars are used because of the widely fluctuating exchange rate: the Australian dollar went from US$1.19 in 1972 to US$1.50 in 1973, and then gradually slid back to US$1.10 in 1977, and to US$0.70 in 1985.

References

Anderson, D.L. and D.W. Barnett (1983) 'Taxation of Uranium Mining Ventures in Saskatchewan: A Policy Assessment' *Resources Policy* (September)

Bambrick, S. (1979) *Australian Minerals and Energy Policy* Canberra: Australian National University Press

—— (1980) 'Australian Mineral Commodity Marketing' *Resources Policy* 6, 2, pp. 166–78

Barnett, D.W. (1976) 'North West Shelf Oil and Gas in the Australian Energy Scene' *Bulletin of the Australian Society for Exploratory Geophysics* 7, 2, pp. 75–9

—— (1979) *Minerals ad Energy in Australia* Sydney: Cassell

—— (1981a) 'Oil and Gas in Australia' Earth Resources Foundation, University of Sydney (rptd in *Thompson's Oil and Gas Year Book* Melbourne: Thompson)

—— (1981b) 'Australia's Minerals Policy' in *National and International Management of Mineral Resources* London: Institute of Minerals and Metallurgy

—— (1983) 'Coal's Place in the Australian Economy' in *Proceedings, Coal and Society* Sydney: Earth Resources Foundation, University of Sydney, and Australian Coal Association

Barnett, D.W. and J.P. Hemphill (1983) 'The Minerals Industry in Australia, 1970–80: A Pattern for Growth' Paper presented to the Conference on International Trade in Minerals, Energy and Processed Products in ASEAN and Australia, Bangkok (February) (pub. Canberra: AGPS)

BMR (Bureau of Mineral Resources) (various years) 'Geology and Geophysics' *Australian Mineral Industry Annual Review* Canberra: AGPS

Derry, D.R., C.E. Michener and J.K.B. Booth (1974) 'World Mining Exploration Trends—Politics of Prospecting' *World Mining* (March) pp. 46–50

Fitzgerald, T.M. (1974) *The Contribution of the Mineral Industry to Australian Welfare* Canberra: AGPS

Folie, M. (1980) 'An Economic Appraisal of the Australian Oil and Gas Industry' *Resources Policy* 6, 2, pp. 125–42

Harris, S. (1980) 'Resources Policies in Australia' *Resources Policy* 6, 2, pp. 179–91

IAC (Industries Assistance Commission) (1979) *The Iron and Steel Industry, Part A: Basic Iron and Steel Products* Canberra: AGPS (draft report)

Joint Coal Board (1982) *Black Coal in Australia, 1981–82: A Statistical Year Book* Sydney: Joint Coal Board

Macleod, J. (1983) quoted in 'Australia: An Economic Overview' *Forbes* 132, 2, p. 116

McColl, G.D. (1980) 'The Mining Industry and the Natural Environment' *Resources Policy* 6, 2, pp. 153–65

McKern, R.B. (1976) *Multinational Enterprise and Natural Resources* Sydney: McGraw–Hill

Mullins, W.J., R.D. Lawrence and D.W. Deschamps (1977) 'Explorationists Seek U_3O_8, Porphyry, Silver, Sulphide Deposits' *World Mining* (Catalog, Survey and Directory) pp. 104–111, 235

Raggatt, H.G. (1968) *Mountains of Ore* Melbourne: Lansdowne Press

Figure 5.1

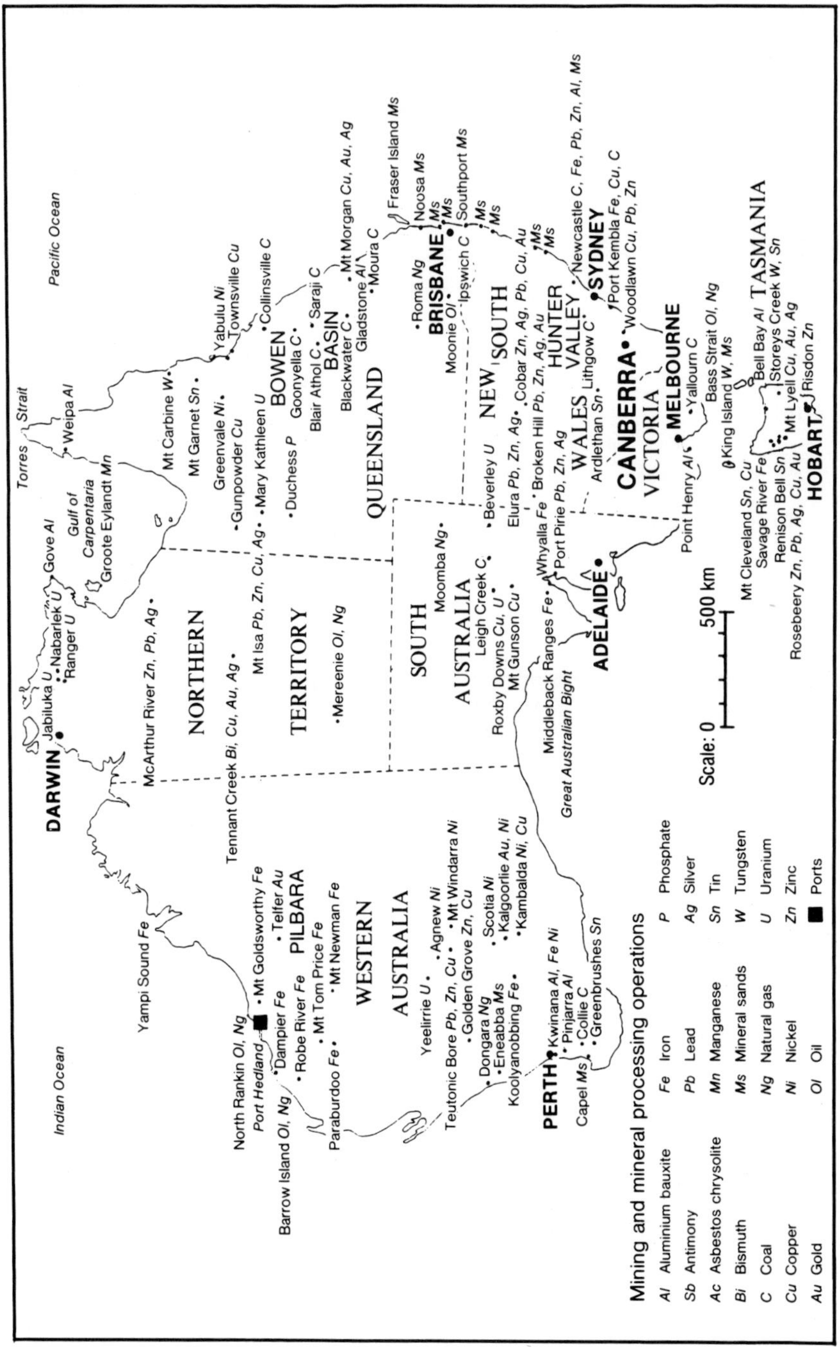
Mining and mineral processing operations
Al Aluminium bauxite
Sb Antimony
Ac Asbestos chrysolite
Bi Bismuth
C Coal
Cu Copper
Au Gold
Fe Iron
Pb Lead
Mn Manganese
Ms Mineral sands
Ng Natural gas
Ni Nickel
Ol Oil
P Phosphate
Ag Silver
Sn Tin
W Tungsten
U Uranium
Zn Zinc
Ports
Scale: 0 500 km
Indian Ocean
Pacific Ocean
Torres Strait
Gulf of Carpentaria
Great Australian Bight
DARWIN
PERTH
ADELAIDE
MELBOURNE
CANBERRA
SYDNEY
BRISBANE
HOBART
NORTHERN TERRITORY
WESTERN AUSTRALIA
SOUTH AUSTRALIA
QUEENSLAND
NEW SOUTH WALES
VICTORIA
TASMANIA
PILBARA
BOWEN BASIN
HUNTER VALLEY
Jabiluka U
Nabarlek U
Ranger U
Gove Al
Groote Eylandt Mn
Weipa Al
McArthur River Zn, Pb, Ag
Tennant Creek Bi, Cu, Au, Ag
Mt Isa Pb, Zn, Cu, Ag
Mereenie Ol, Ng
Mary Kathleen U
Gunpowder Cu
Greenvale Ni
Mt Garnet Sn
Mt Carbine W
Duchess P
Yabulu Ni
Townsville Cu
Collinsville C
Goonyella C
Saraji C
Blair Athol C
Blackwater C
Gladstone Al
Mt Morgan Cu, Au, Ag
Moura C
Fraser Island Ms
Noosa Ms
Roma Ng
Moonie Ol
Ipswich C
Southport Ms
Moomba Ng
Leigh Creek C
Roxby Downs Cu, U
Mt Gunson Cu
Beverley U
Elura Pb, Zn, Ag
Cobar Zn, Ag, Pb, Cu, Au
Broken Hill Pb, Zn, Ag, Au
Whyalla Fe
Port Pirie Pb, Zn, Ag
Middleback Ranges Fe
Lithgow C
Ardlethan Sn
Newcastle C, Fe, Pb, Zn, Al, Ms
Port Kembla Fe, Cu, C
Woodlawn Cu, Pb, Zn
Point Henry Al
Yallourn C
Bass Strait Ol, Ng
King Island W, Ms
Bell Bay Al
Storeys Creek W, Sn
Mt Lyell Cu, Au, Ag
Risdon Zn
Mt Cleveland Sn, Cu
Savage River Fe
Renison Bell Sn
Rosebeery Zn, Pb, Ag, Cu, Au
Yampi Sound Fe
North Rankin Ol, Ng
Port Hedland
Barrow Island Ol, Ng
Dampier Fe
Robe River Fe
Paraburdoo Fe
Mt Tom Price Fe
Mt Newman Fe
Mt Goldsworthy Fe
Telfer Au
Yeelirrie U
Agnew Ni
Mt Windarra Ni
Teutonic Bore Pb, Zn, Cu
Golden Grove Zn, Cu
Dongara Ng
Eneabba Ms
Koolyanobbing Fe
Scotia Ni
Kalgoorlie Au, Ni
Kambalda Ni, Cu
Kwinana Al, Fe Ni
Pinjarra Al
Collie C
Greenbrushes Sn
Capel Ms

Figure 5.2

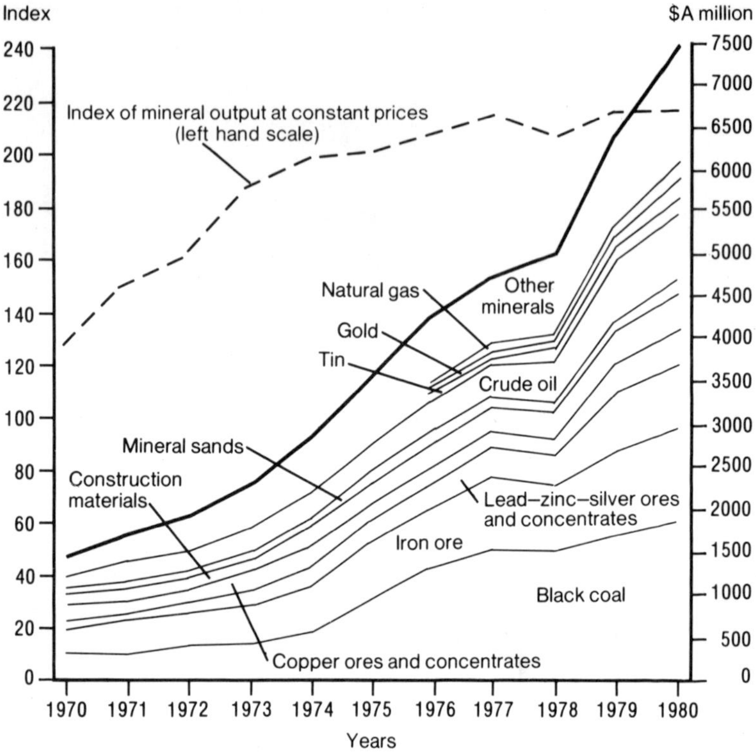

Source: BMR (1982)

Figure 5.3

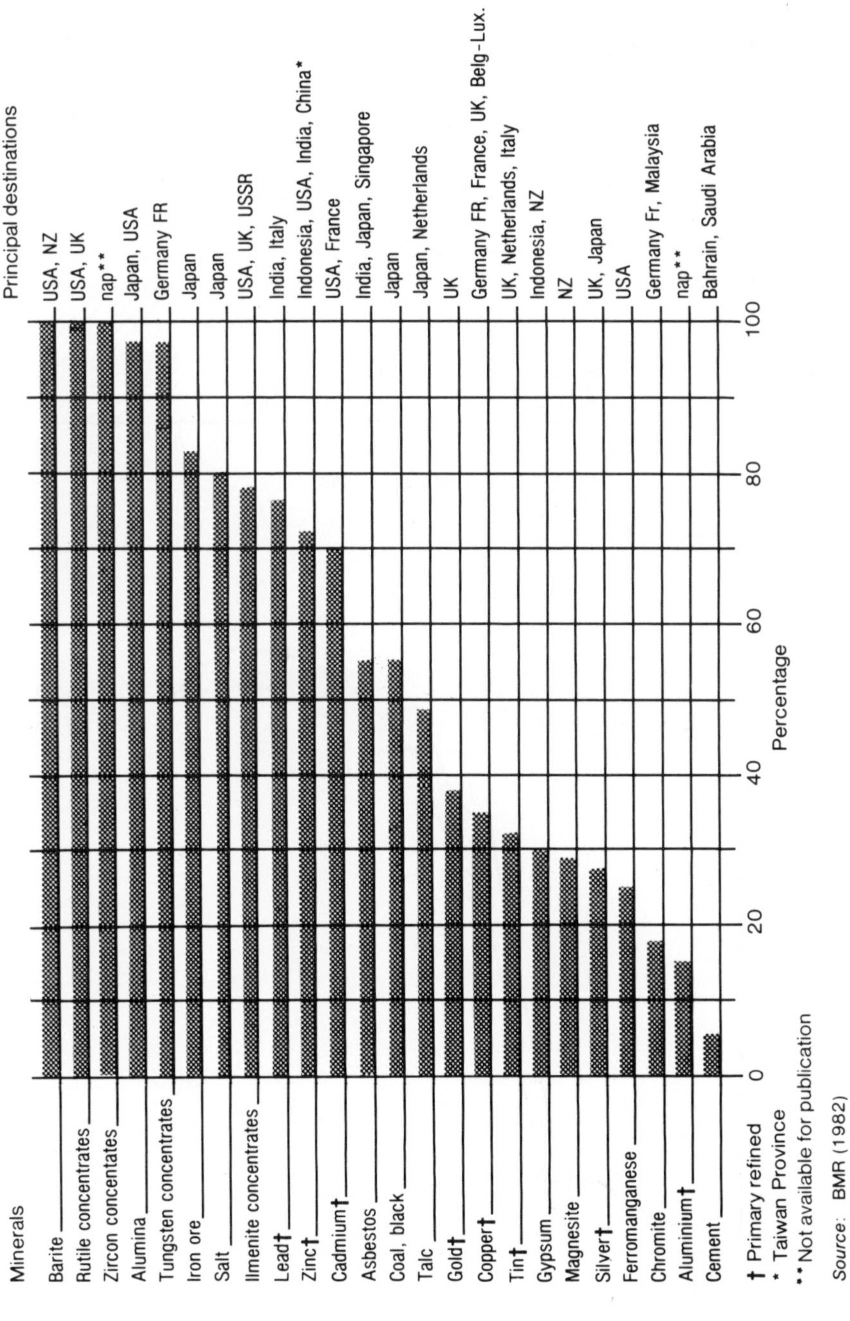
Minerals
Principal destinations
Barite
USA, NZ
Rutile concentrates
USA, UK
Zircon concentates
nap**
Alumina
Japan, USA
Tungsten concentrates
Germany FR
Iron ore
Japan
Salt
Japan
Ilmenite concentrates
USA, UK, USSR
Lead†
India, Italy
Zinc†
Indonesia, USA, India, China*
Cadmium†
USA, France
Asbestos
India, Japan, Singapore
Coal, black
Japan
Talc
Japan, Netherlands
Gold†
UK
Copper†
Germany FR, France, UK, Belg-Lux.
Tin†
UK, Netherlands, Italy
Gypsum
Indonesia, NZ
Magnesite
NZ
Silver†
UK, Japan
Ferromanganese
USA
Chromite
Germany Fr, Malaysia
Aluminium†
nap**
Cement
Bahrain, Saudi Arabia
0
20
40
60
80
100
Percentage
† Primary refined
* Taiwan Province
** Not available for publication
Source: BMR (1982)

Figure 5.4

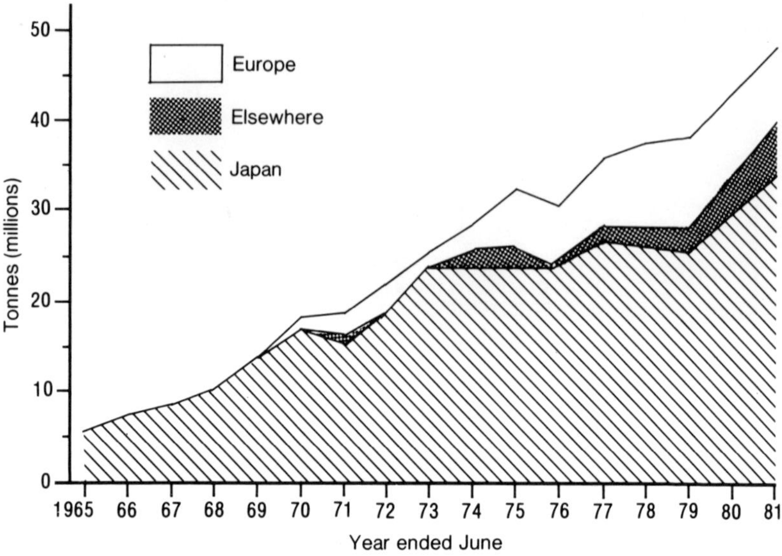

Source: Joint Coal Board (1982)

Figure 5.5

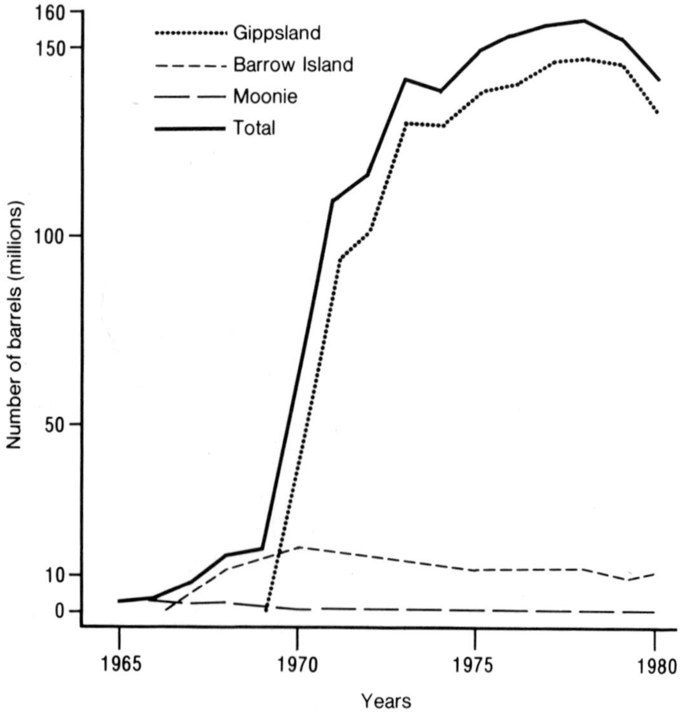

Source: Barnett (1981a)

Figure 5.6

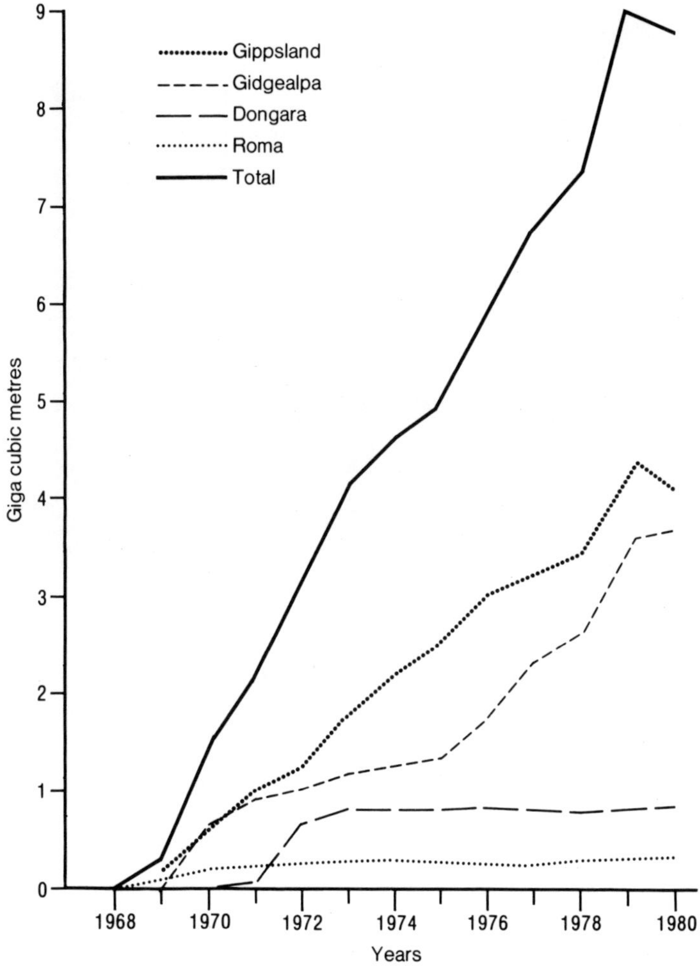

Source: Barnett (1981a)

Figure 5.7

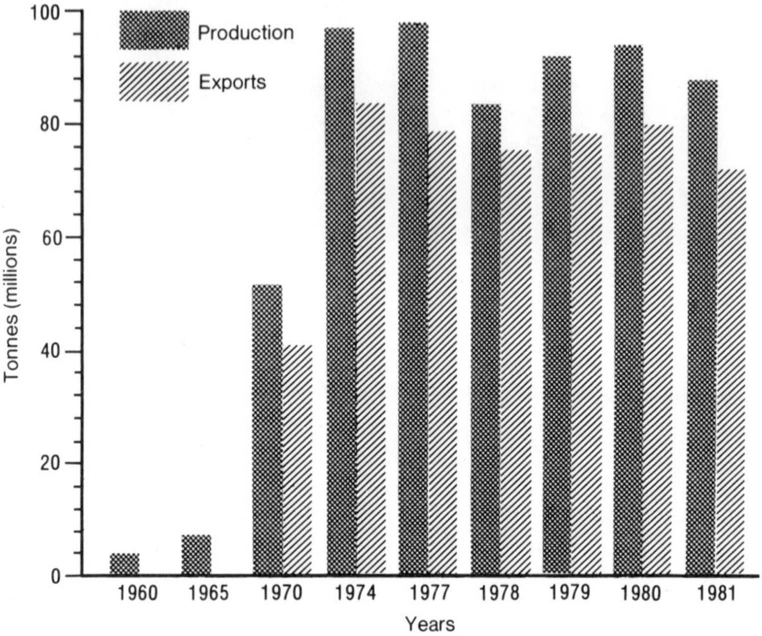

Source: Author's compilation.

Figure 5.8

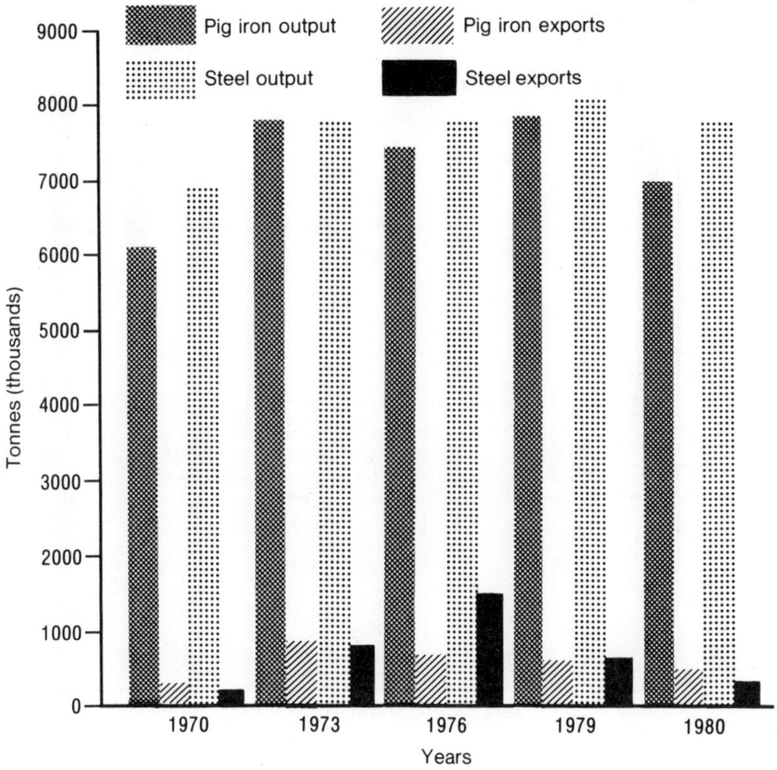

Source: . Author's compilation.

Figure 5.9

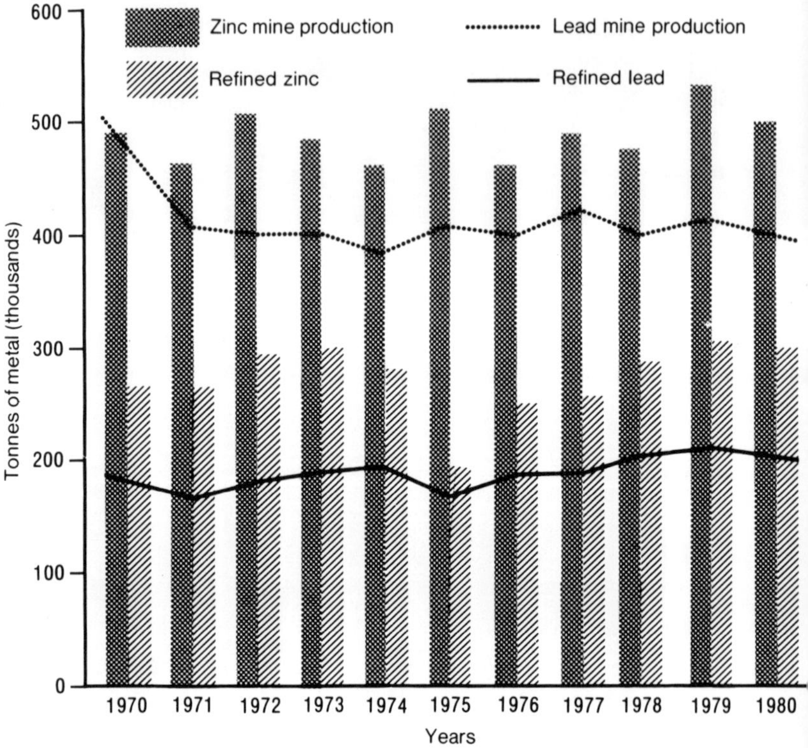

Source: Author's compilation.

Figure 5.10

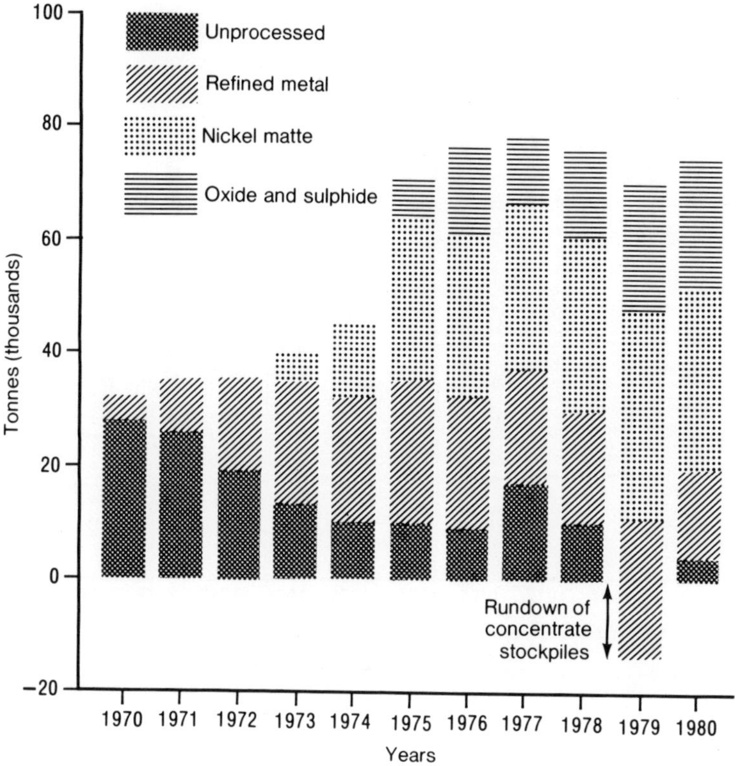

Source: Author's compilation.

Figure 5.11

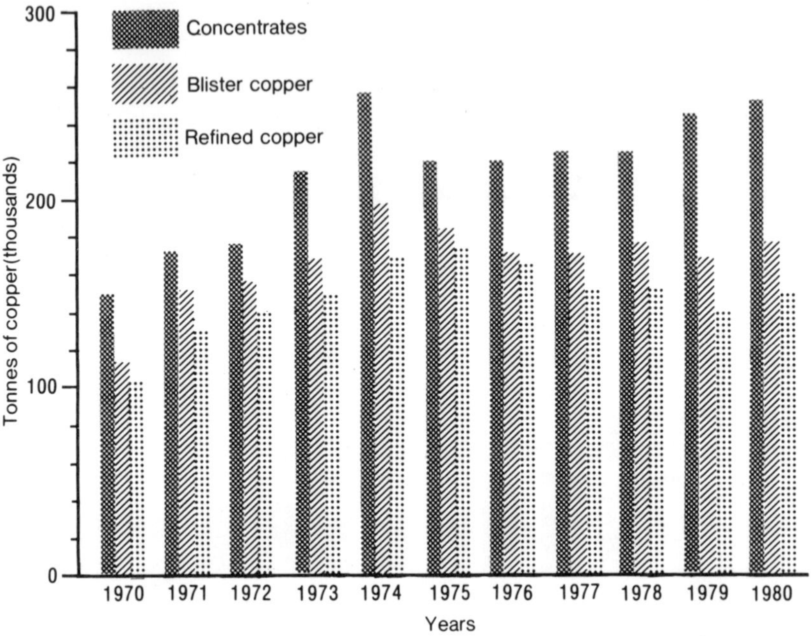

Source: Author's compilation.

Figure 5.12

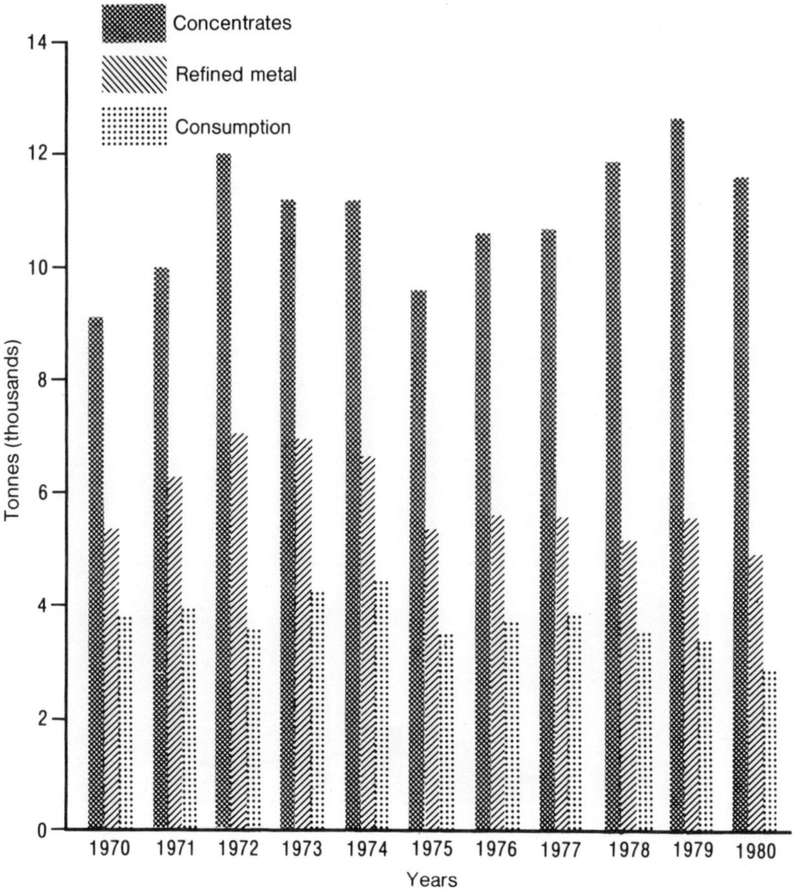

Source: Author's compilation.

Figure 5.13

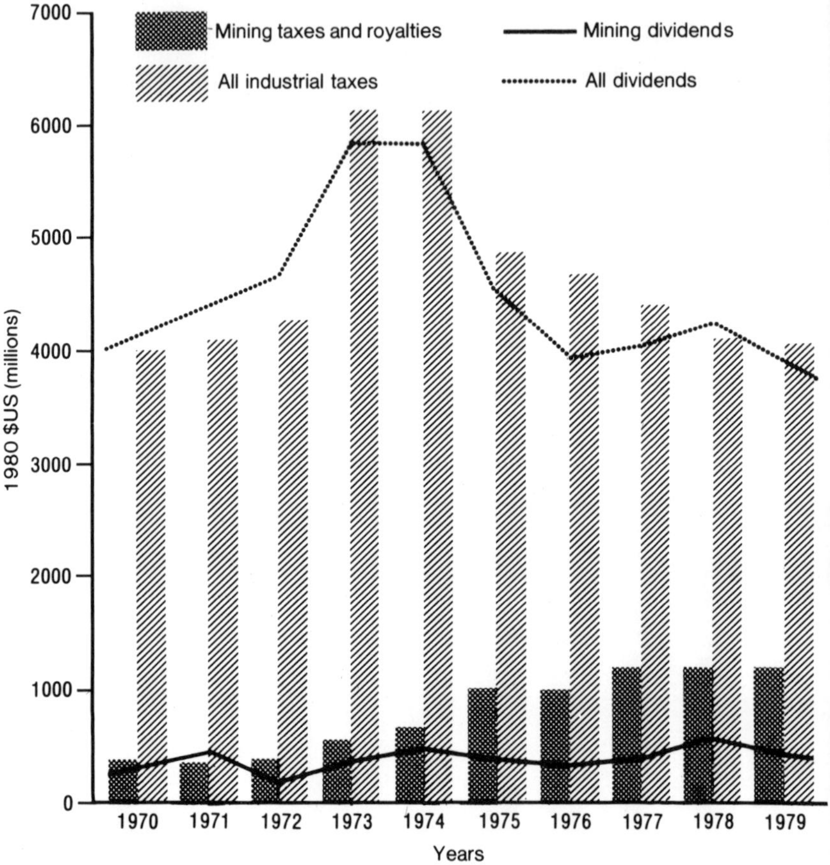

Source: Author's compilation.

Figure 5.14

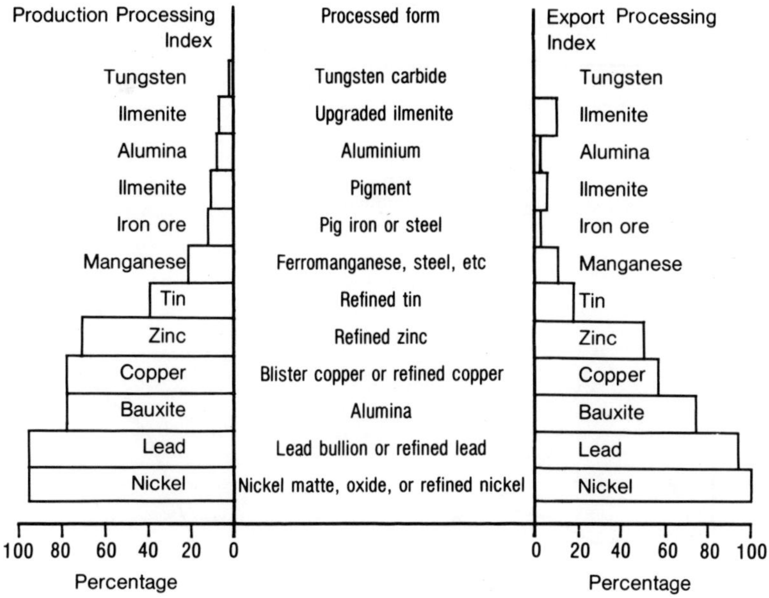

Notes: The production processing index is the percentage of production processed in Australia. The export processing index is the percentage of exports processed in Australia.

Source: BMR, (1982)

Figure 5.15

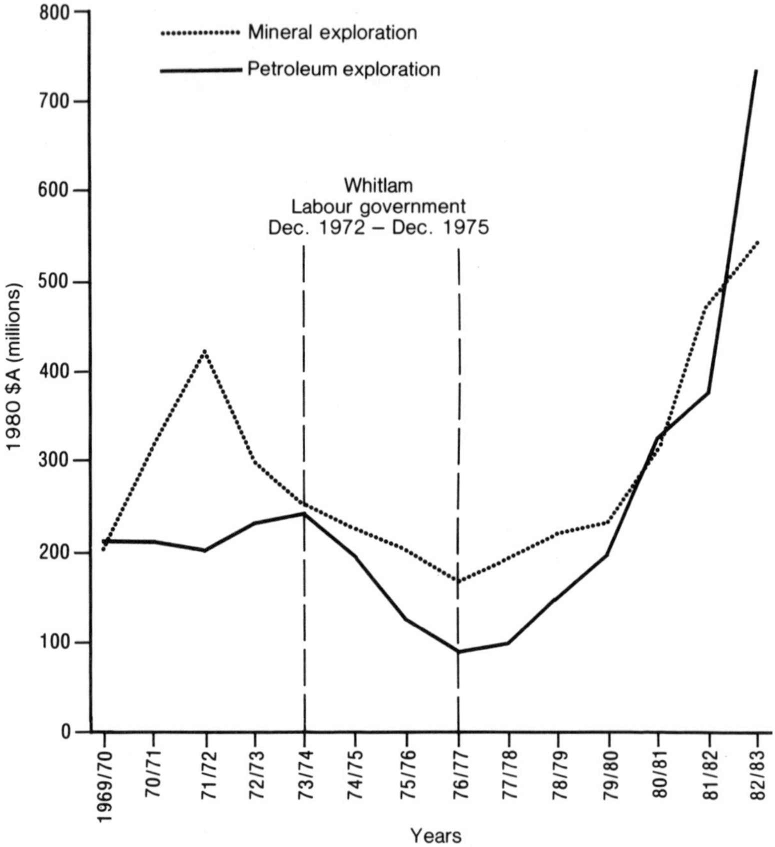

Source: Author's compilation.

6 Mineral and energy development in Indonesia

HADI SOESASTRO AND BUDI SUDARSONO

THE minerals sector plays an important role in Indonesia's economy, especially as a source of export earnings. Most of its minerals have been produced since just before the second world war. Production of tin started in the eighteenth century, that of coal in 1849, manganese ore and petroleum at the end of the nineteenth century, and finally nickel and bauxite in the 1930s. Production of other minerals, particularly iron and copper, has developed only in the 1970s.

Although the period between 1940 and 1966 saw a decline in the production of coal, tin and manganese ore, since 1970 there has been a general upsurge in minerals production, especially of nickel, gas, tin, coal and petroleum. As a consequence Indonesia has re-emerged as the second largest tin producer in the world. Furthermore, since 1980 it has replaced Algeria as the world's leading liquefied natural gas exporter, supplying at present around one-third of the market.

The contribution made by the mining sector of the economy to Indonesia's GDP increased dramatically in recent years from 1.6 per cent in 1966 to 25.7 per cent in 1980 (Table 6.1). At the same time, the contribution of minerals to Indonesia's total exports has increased from 39.3 per cent in 1966 to 84 per cent in 1982 and 78 per cent in 1983 (Table 6.4). This has been due largely to the increased contribution of petroleum (both crude oil and petroleum products), which rose from 33.4 per cent in 1966 to 74.7 per cent in 1983. Similarly, exports of LNG, which began only in 1977, were worth over US$2.5 billion annually from 1981 to 1983, or 11–13 per cent of total exports. In 1980, the mineral sector of the economy employed some 370 000 people, around 8 per cent of the total workforce.

The development of new mineral commodities and new mines has been encouraged since 1966 under the so-called 'new order' government. This government restored macroeconomic stability, liberalised the economy and provided incentives for both foreign and domestic

Table 6.1 Indonesia: sectoral contributions to GDP, 1981–83 (per cent, based on current market prices)

	1981	1982	1983
Agriculture	25.3	26.3	26.4
Mining	24.0	19.6	19.4
Industry	10.8	12.9	12.5
Utilities	0.5	0.6	0.4
Construction	5.8	5.9	6.2
Transportation and communication	4.4	4.7	4.7
Trade, banking and other services	29.3	30.0	30.1
Total	100.0	100.0	100.0
(billion rupiahs)	54 027.0	59 632.6	71 214.7

Note: Percentage share of mining (constant 1973 prices): 1977, 12.0; 1978, 11.0; 1979, 10.3; 1980, 9.3; 1981, 8.9; 1982, 7.6; 1983, 7.4.
Source: Central Bureau of Statistics *Statistik Indonesia* (various years)

Table 6.2 Minerals production in Indonesia, 1940, 1966–80 (volume indexes, 1975 = 100)

	1940	1966	1970	1975	1980
Iron sands	–	–	–	100	17.8
Copper concentrates	–	–	–	100	92.5
Bauxite	27.7	70.6	123.8	100	125.8
Tin concentrates	173.7	50.4	75.3	100	128.4
Tin metal	na	na	29.1	100	170.9
Nickel ore	6.9	14.7	74.9	100	371.0
Manganese ore	84.8	7.2	80.6	100	31.4
Coal	973.6	155.0	83.5	100	147.3
Crude petroleum	12.2	35.8	65.3	100	116.8
Natural gas	22.7	47.7	48.8	100	470.5

na-not available
Source: Department of Mines and Energy *Buku Tahunan Pertambangan Indonesia* (various years)

Table 6.3 Minerals exports, 1966–80 (volume indexes, 1975 = 100)

	1966	1970	1975	1980
Iron sands	–	–	100	7.9
Copper concentrates	–	–	100	100.0
Bauxite	64.8	114.7	100	116.1
Tin concentrates	147.0	185.3	100	16.9
Tin metal	–	28.6	100	207.4
Nickel ore	15.7	67.7	100	151.5
Manganese ore	–	84.4	100	0.1
Crude petroleum	26.1	62.9	100	99.6

Source: See Table 6.2

Table 6.4 Export earnings from minerals, mineral fuels and fertilisers, Indonesia, 1977–83 (US$ million)

Commodity	1977	1978	1979	1980	1981	1982	1983
Minerals				897.8	797.1	636.9	621.4
Iron sands					0.2	0.1	0.1
Copper concentrates					130.3	120.0	126.1
Bauxite					13.4	9.8	11.3
Tin concentrates					21.4	14.1	8.2
Tin metal				499.3	437.4	342.5	313.5
Nickel ore					29.7	23.3	12.8
Ferronickel ingot					30.7	22.5	25.4
Nickel matte				141.0	122.7	97.4	110.3
Manganese ore					–	0.5	–
Other minerals					11.1	6.7	13.6
Mineral fuels					17 846.7	15 289.8	13 925.6
Coal					5.3	6.2	12.2
Crude petroleum and petroleum products	7 222.2	6 940.2	8 705.0	13 827.7	15 328.6	12 642.7	11 407.0
Natural gas					2 512.8	2 640.9	2 506.4
Fertiliser							
Urea (for fiscal year)					10.6	21.4	50.0
Total exports					21 976.2	18 857.2	18 621.5
Percentage of minerals to total exports					84.0	84.0	78.0

Source: Department of Mines and Energy *Buku Tahunan Pertambangan Indonesia* (various years); Bank Indonesia *Laporan Mingguan* (various issues).

Table 6.5 Indonesia's minerals trade balance, 1975, 1980 (US$ million)

	1975			1980		
	Exports	Imports	Balance	Exports	Imports	Balance
Ore and concentrates	144.8	3.1	141.7	245.5	196.1	49.4
Iron	na	0.0	–	0.1	11.9	(11.8)
Copper	67.8	–	67.8	125.8	–	125.8
Bauxite	6.0	2.0	4.0	14.4	181.0	(166.6)
Tin	47.2	–	47.2	66.4	–	66.4
Nickel	21.2	–	21.2	38.7	–	38.7
Manganese	2.6	1.1	1.5	0.1	3.2	(3.1)
Metal and manufactures	103.9	703.6	(599.7)	632.9	1398.9	(766.0)
Iron	0.5	627.6	(627.1)	23.9	1 204.5	(1 180.6)
Copper	10.1	13.5	(3.4)	1.8	31.5	(29.7)
Aluminium	0.3	36.0	(35.7)	1.5	85.4	(83.9)
Tin	92.6	0.3	92.3	443.6	1.1	442.5
Nickel	–	5.3	(5.3)	162.1	23.2	138.9
Other	0.4	20.9	(20.5)	0.0	53.2	(53.2)
Subtotal	248.7	706.7	(458.0)	878.4	1 595.0	(716.6)
Mineral fuels	5 310.9	255.1	5 055.8	15 742.8	1 748.9	13 993.9
Coal	–	2.6	(2.6)	3.1	5.8	(2.7)
Crude petroleum	4 933.1	8.6	4 924.5	11 671.3	963.0	10 708.3
Petroleum products	377.8	243.2	134.6	1 187.2	776.3	410.9
Natural gas	–	0.7	(0.7)	2 881.2	3.8	2 877.4
Total	5 559.6	961.8	4 597.8	16 621.2	3 343.9	13 277.3

Note: Figures in parentheses denote negative trade balances.
Source: Central Bureau of Statistics *Impor* (various years) and *Ekspor* (various years)

private investment. For instance, the Foreign Capital Investment Law stipulates that foreign companies may invest and operate in Indonesia, either independently or in joint ventures with Indonesian partners, for a maximum period of thirty years. In fact all foreign investment undertaken since 1974, other than investment totally for export production, has been in joint ventures with local partners.

Increased petroleum export earnings have significantly improved the country's balance of payments and have assisted efforts to step up its development. Thus, beginning in the financial year 1979/80, petroleum export earnings consistently represented 64 per cent or more of the government's revenues, illustrating the paramount importance of petroleum to the national economy. As a result of the large expansion in foreign exchange earnings under the 'new order' government, the rate of growth of GDP has averaged 7 per cent per year. However, since 1982, when Indonesia suffered its first blows from the world recession,

Table 6.6 Unit values of minerals exports and imports, 1979–81 average (US$/tonne fob)

	Exports	Imports
Iron		
Ore and concentrates	12.4	35.0
Metal	233.3	436.5
Manufactures	2 254.2	1 700.2
Copper		
Ore and concentrates	580.7	387.3
Metal and manufactures	2 204.2	1 923.3
Bauxite		
Ore and concentrates	12.7	194.2
Metal and manufactures	1 811.6	1 904.6
Tin		
Ore and concentrates	10 514.6	–
Metal and manufactures	14 792.2	4 114.8
Nickel		
Ore and concentrates	37.4	–
Metal and manufactures	3 101.9	4 694.0
Manganese		
Ore and concentrates	16.9	271.4
Coal	30.5	149.4
Petroleum		
Crude	208.9	187.6
Products	191.1	236.2
Natural gas	222.6	323.8

Source: See Table 6.5

the rate of growth of GDP has become more modest, that is, less than 5 per cent per year.

The characteristics of the major minerals industries and their development are described below, followed by an assessment of relevant government policies.

Mineral reserves and production

Mineral development in Indonesia is subject to Article 33 of the 1945 Constitution, which stipulates that natural resources are to be exploited under state control for 'the maximum welfare of the people'. Thus the government acts as management, on the people's behalf, in order to extract the greatest possible benefits from mineral resources development.

The law identifies three categories of minerals:

1 strategic minerals, such as oil, tin, nickel and radioactive minerals,

which can be developed only by state agencies or state-owned corporations;

2 vital minerals, which are metallic minerals not designated as strategic, and which can be developed by both state and private enterprises; and

3 other minerals, which are nonmetallic minerals, also to be developed by state or private enterprises.

The provisions contained in the Mining Law and the Foreign Investment Law from 1967 allow foreign companies or joint ventures to act as contractors. The contracts for strategic and vital minerals are granted by the central government, and those for other minerals by the provincial governments. These contractual arrangements are discussed more fully below under 'Major policy issues'.

Non-fuel minerals

Indonesia is well endowed with mineral resources, being an important producer of a number of metallic minerals, including tin, copper, nickel, and to a lesser extent, bauxite. Indonesia is currently the world's second largest producer of tin. Nonferrous minerals are the country's fourth largest export earners after oil and gas, timber and wood products, and rubber.

Under the Fourth Development Plan (1984–89) annual production of tin-in-concentrate is expected to increase from 26 000 tonnes to 38 000 tonnes and output of bauxite is expected to average 0.8 million tonnes. Mine production of copper concentrate is expected to be in the vicinity of 175 000 tonnes per year for the duration of the Plan.

IRON

Iron sands have been produced in the Cilacap area (Central Java) since 1971 and in Pelabuhan Ratu (West Java) since 1978. In the Cilacap area, annual production peaked at 365 000 tonnes in 1974, but decreased to only 27 000 tonnes in 1980. Combined production from both sites amounted to some 63 000 tonnes in 1980 (Table 6.7). The exploitation of the Cilacap deposit, which has a high titanium oxide content, is based on a contract between the state company, P.T. Aneka Tambang, and Japan's Nippon Kokan Kabushiki Kaisha. The contract stipulates the supply of 300 000 tonnes per year to Japan for ten years.

In a third site, Gombong Selatan[1], iron sand deposits have been shown to amount to around 230 million tonnes with an estimated Fe content of 12.5 per cent, which could be upgraded into 28 million tonnes with a 50 per cent Fe content.

Proposals to exploit this deposit have been made in connection with the establishment of an integrated steel plant in the area. However, the feasibility of the plan has been questioned because of the high

Table 6.7 Iron sands: production, consumption and exports, 1974–83 (tonnes)

	1974	1975	1976	1977	1978	1979	1980	1981	1982	1983
Production	365 206	352 991	292 334	311 519	233 341	79 877	62 914	86 626	144 493	132 887
Cilacap	365 206	352 991	292 334	311 519	193 712	35 889	27 315	47 188	–	–
Pelabuhan Ratu	–	–	–	–	39 629	43 988	35 599	39 438	–	–
Domestic consumption[a]	6 500	5 313	12 594	27 619	40 854	62 282	64 660	100 013	107 430	122 154
Exports	372 771	305 650	284 010	257 602	87 648	9 459	24 112	25 520	11 000	10 251

Note: [a] Based on domestic sales data, primarily used by cement factories.
Source: Department of Mines *Buku Tahunan Pertambangan Indonesia* (various years)

electricity consumption involved in the process[2]. An alternative proposal has been to produce 1 million tonnes of pellets per annum, with a 57 per cent Fe content, as an input to the Cilegon steel plant (Krakatau Steel). However, although the volume is adequate, the deposit's mineral composition is reportedly not suitable for use as a basic material for iron pellets.

Copper

Indonesia's only copper mine, in operation since 1973[3], is located at Tembagapura in Irian Jaya. Known as the Ertsberg, this deposit is 3460 m above sea level and has a copper content of 2.5 per cent. The ore is mined by openpit methods, transported and concentrated to a 30 per cent slurry, then piped to a harbour 125 km away. The investment outlays include the mine, cable tramways, the concentrating plant, the slurry pipeline and harbour facilities. In 1967 a contract of work was awarded to Freeport Indonesia Inc., the first non-oil contract granted to a foreign company by the 'new order' government. The contract was of the 'first generation' type, discussed below under 'Major policy issues'. Although no government assistance was provided, air transportation facilities have been improved by the state.

Reserves are estimated at around 32.6 million tonnes, containing 2.5 per cent copper, with magnetite and chalcopyrites with 0.75 grams of gold and 9 grams of silver per tonne. Between 1973 and 1980, production from the mine amounted to about 1.5 million tonnes in total, or an average of 188 000 tonnes of concentrate per annum. Peak production was achieved in 1982 with a total of 223 000 tonnes of concentrate (Table 6.8).

The second deposit which has been mined since 1980 is located one kilometre from the first deposit. Reserves are 45 million tonnes, grading 2.34 per cent Cu. Because the copper grade from the older openpit mine has declined to less than 2 per cent, this new underground mine has helped to maintain overall output. When the openpit mine is exhausted, production will continue from the underground mine. Reserves under the original deposit will be exploited through two huge underground tunnels now being constructed. The tunnels will be completed in 1987 at a total cost of US$20 million.

A third deposit is known to exist in North Sulawesi, explored intensively by P.T. Tropic Endeavour, Indonesia. A sizeable copper area has been found and centres of copper–gold mineralisation have been established[4]. However the project's feasibility is not clear.

Bauxite/aluminium

Bauxite is mined in four locations on Bintan Island by the Bauxite Mining Unit of P.T. Aneka Tambang[5]. Deposits of bauxite in these locations are estimated at around 78 million tonnes, the largest being Wakopek with an estimated deposit of 36 million tonnes[6]. Bauxite

Table 6.8 Copper ore and concentrates: production and trade, 1972–83 (dry tonnes)

	1972	1973	1974	1975	1976	1977	1978	1979	1980	1981	1982	1983
Production												
Copper ore	...	1 249 684	2 486 028	2 630 669	2 972 000	2 530 464	2 694 055	3 037 905	2 988 081	3 374 138	3 462 821	4 349 632
Gunung Bijih	...	1 249 684	2 486 028	2 630 669	2 972 000	2 530 464	2 694 055	3 037 905	2 911 419	–	–	–
Gunung Bijih Timur	–	–	–	–	–	–	–	–	76 662	–	–	–
Copper concentrates	9 750	125 870	212 620	201 273	223 301	189 103	180 933	188 769	186 087	188 472	223 704	205 015
Exports	9 000	114 218	207 245	193 439	216 829	200 582	185 596	168 634	193 513	197 146	228 799	202 823

Source: See Table 6.7

Table 6.9 Bauxite: production and exports, 1966, 1970–83 (tonnes)

	Production	Exports
1966	701 223	668 365
1970	1 229 168	1 182 239
1971	1 237 610	1 237 435
1972	1 275 957	1 162 132
1973	1 229 375	1 302 161
1974	1 290 054	1 274 653
1975	992 556	1 031 158
1976	940 269	939 865
1977	1 301 416	1 226 910
1978	1 007 746	983 324
1979	1 051 905	1 179 144
1980	1 249 123	1 197 228
1981	1 203 380	1 030 752
1982	700 247	731 331
1983	777 869	847 409

Source: See Table 6.7

deposits in West Kalimantan are also believed to be large, although they have not yet been exploited.

Total production of bauxite increased from about 700 000 tonnes per year in the 1960s to over 1 million tonnes a year in the 1970s. Following the conclusion of a ten-year contract in 1969, most of the output has been exported to Japan (see Table 6.9). Production in 1982 and 1983 fell to 700 000 tonnes and 780 000 tonnes, and similarly exports to Japan fell to 731 000 tonnes and 847 000 tonnes respectively.

TIN

The wide tin belt of Southeast Asia passes through much of the region east of Sumatra and west of Kalimantan and the mineral has been mined since 1815[7]. Due to significant production increases over the period 1976–80, Indonesia has become the largest tin producer in the world after Malaysia. Much of the increase has been a result of joint venture activities. Production of tin concentrates has increased steadily from about 13 000 tonnes in 1966 to over 35 000 tonnes in 1981 (see Table 6.10), but then it fell to 26 000 tonnes in 1983.

The government-owned company P.T. Tambang Timah has several production units on Bangka, Belitung, Singkep and Karimun–Kundur Islands, and in Bangkinang on the mainland of Sumatra. In addition, three foreign companies currently operate on a contractual basis with the Indonesian government: P.T. Koba Tin, P.T. Broken Hill and P.T. Riau Tin Mining. Since 1970 several small national companies have operated under contract with P.T. Tambang Timah. During 1983 75–78 per cent of total production came from this company, and the remainder from the foreign companies.

Table 6.10 Tin concentrates and metal: production, consumption and exports, 1972–83 (tonnes)

	1972	1973	1974	1975	1976	1977	1978	1979	1980	1981	1982	1983
Production												
Tin concentrates	21 765.0	22 646.4	25 709.5	25 337.3	23 434.5	25 926.4	27 410.8	29 535.5	32 527.5	35 391.6	33 806.1	26 552.9
P.T. Tambang Timah	21 765.0	22 490.5	25 023.0	24 391.0	22 202.5	24 020.5	24 064.2	25 163.9	26 116.4	27 178.6	26 218.6	19 952.5
Joint ventures	–	155.9	686.5	946.3	1 232.0	1 905.9	3 316.6	4 371.6	6 411.1	8 213.0	7 587.5	6 600.4
Tin metal	12 008.7	14 631.7	15 065.9	17 825.5	23 322.1	24 004.8	25 829.8	27 789.7	30 465.0	32 518.9	29 754.7	28 390.0
Domestic consumption												
Tin metal	313.7	541.4	334.5	382.0	567.0	446.0	430.2	389.9	335.2	378.4	555.1	364.5
Exports												
Tin concentrates	9 006.4	7 266.4	8 106.7	7 295.3	1 093.5	307.0	385.3	464.1	1 231.0	–	–	–
Tin metal	11 190.3	13 954.9	14 948.4	14 890.6	23 018.4	24 914.0	25 549.0	25 737.0	30 885.0	31 829.0	26 825.0	25 332.0

Source: See Table 6.7; and data supplied by P.T. Tambang Timah

P.T. Tambang Timah has increased its production from 24 000 tonnes mined in 1975 to 27 179 tonnes in 1981. The company has invested in new equipment in order to sustain these levels of production, including modern 22 cubic foot bucket dredges. Because newly discovered tin reserves are deepseated, this capacity is necessary. The Belitung, for example, is designed specifically to dredge deepseated tin horizons, up to a depth of 50 metres below sea level. However, production has since decreased to 19 953 tonnes in 1983.

NICKEL

Nickel was first mined in Indonesia in 1936, and since 1960 has been produced in Pomalaa[8] by P.T. Aneka Tambang, in cooperation with Japan's Sunideco. P. T. Aneka also began operations in Gebe Island in 1979[9].

In 1968 a work contract was awarded by the Indonesian government to P.T. Inco for the development of a mine at Soroako, north of Pomalaa. Subsequently a hydroelectric plant was built in Larona by P.T. Inco, with a capacity of 110 MW, to supply power to the Soroako nickel producing plant.

Another work contract was signed in 1969 with P.T. Pacific Nickel Indonesia, for the development of the nickel mine on Gag Island in Irian Jaya. However, because of the bleak prospects facing the nickel market P.T. Pacific decided to withdraw from the project in 1982.

Production of nickel ore by P.T. Aneka Tambang increased rapidly from 117 000 tonnes in 1966 to 1.54 million tonnes in 1980 and 1.64 million tonnes in 1982, of which 44 per cent came from Pomalaa and the remaining 56 per cent from Gebe Island (see Table 6.11). However, the grade of the ore deposits in Pomalaa has declined due to extensive mining since the 1960s. P.T. Inco's ore production also increased, from 285 000 tonnes in 1977 to about 1.4 million tonnes in 1980, then decreased to 1.2 million tonnes in 1983.

MANGANESE

Manganese ore with more than 75 per cent $Mn0_2$ content is mined by local government enterprises at several locations in West and Central Java. Production from these sites amounted to around 18 000 tonnes in 1974, but declined to about 500 tonnes in 1983. Meanwhile production by private companies was initially low, yielding only 280 tonnes in 1980; it then increased to 15 000 tonnes in 1982, but declined again to 8 000 tonnes in 1983 (Table 6.12).

The peaks of production in 1974 and 1982 were associated with relatively large export shipments to Japan. However, since then most production has been consumed domestically. As a consequence, exports declined from a peak of 16 000 tonnes in 1974 to almost nil in 1980, and manganese ore imports—mainly from Africa via Singapore—more than doubled between 1975 and 1980.

Table 6.11 Nickel: production and exports, 1972–83 (tonnes)

	1972	1973	1974	1975	1976	1977	1978	1979	1980	1981	1982	1983
Production												
Nickel ore												
P.T. Aneka Tambang[a]	935 076	867 319	878 855	801 072	1 102 023	1 302 512	1 256 450	1 552 660	1 537 442	1 543 219	1 640 982	1 278 031
–Pomalaa	935 076	867 319	878 855	801 072	1 102 023	1 302 512	1 256 450	879 419	747 114	617 779	724 117	651 071
–P. Gebe	–	–	–	–	–	–	–	673 241	790 328	925 440	916 805	626 960
P.T. Inco[b]	–	–	–	–	16 562	285 168	372 680	689 200	1 434 386	1 155 883	908 650	1 205 400
Ferronickel ingot[c]	–	–	–	–	16 887	21 574	19 734	17 878	18 314	19 884	21 500	20 708
Nickel mattes[d]	–	–	–	–	–	2 077	5 729	8 597	20 533	19 940	13 744	18 570
Exports												
Nickel ore[c]	838 358	724 904	878 855	849 648	877 025	831 964	864 369	1 114 469	1 287 434	1 184 878	1 055 091	737 024
Ferronickel ingot	–	–	–	–	15 639	21 945	19 455	17 900	16 015	21 736	18 648	24 190
Nickel mattes	–	–	–	–	–	–	5 198	7 922	19 571	16 587[e]	15 710[e]	22 221[e]

Notes: [a] In dry tonnes
[b] In wet tonnes
[c] By P.T. Aneka Tambang
[d] By P.T. Inco
[e] For fiscal year

Source: See Table 6.7

Table 6.12 Manganese ore: Production, consumption and exports, 1972–83 (tonnes)

	1972	1973	1974	1975	1976	1977	1978	1979	1980	1981	1982	1983
Production	7 531.7	15 566.5	18 036.8	13 682.1	8 832.6	5 981.2	5 889.1	5 909.4	4 299.5	2 587.1	17 065.1	8 317.7
West Java	7 431.1	11 788.8	13 814.2	9 939.3	5 076.0	4 902.5	4 562.3	4 713.9	3 758.8	2 002.4	1 521.8	525.9
Yogyakarta	100.6	3 777.7	4 222.6	3 742.8	2 727.1	774.6	727.5	554.5	256.3	75.7	–	–
Other[a]	...	...	...	...	1 029.5	304.1	599.3	641.0	284.4	509.0	15 543.3	7 791.8
Domestic Consumption[b]	...	...	...	2 462.1	2 626.4	3 620.4	3 474.1	3 260.8	2 531.9	1 651.9	1 365.1	874.9
Exports[c]	9 381.0	na	15 977.3	7 938.7	8 072.0	2 351.9	1 580.0	1 770.0	10.0	–	–	–

Notes: [a] East Java and those produced by private companies; reported data incomplete
[b] Based on data of domestic sales from West Java and Yogyakarta production sites only
[c] From West Java and Yogyakarta production sites

Source: See Table 6.7

Fuel minerals

Indonesia has been a prominent producer of fuel minerals since well before the second world war, producing, for example, 2 million tonnes of coal in 1940. Oil exploration began as early as 1872 in West Java, the total amount of oil produced since then being close to 12 billion barrels. In 1985 Indonesia celebrated its centennial of oil production. Again, during the early 1980s huge reserves of natural gas were discovered, making Indonesia the world's largest exporter of LNG.

COAL

Indonesia's total coal resources are estimated at 19.3 billion tonnes, over 15 billion tonnes being located in South Sumatra around Bukit Asam and the rest in South Kalimantan (Figure 6.1). However, proven reserves are much lower, about 600 million tonnes.

Although 1960 production stood at 320 000 tonnes, it declined in the first half of the 1970s to less than 200 000 tonnes annually, due to competition from oil. Since then production has increased, and in 1982 and 1983 reached 480 000 tonnes and 485 000 tonnes (Table 6.13). In 1984 production exceeded a million tonnes.

Coal is exploited by the state coal enterprise, P.N. Tambang Batubara, in Ombilin, West Sumatra, which is used to operate a mine in Mahakam, but a separate state-owned company, P.T. Taba Bukit Asam, operates the mine at Bukit Asam[10]. Increased oil prices have encouraged rehabilitation programs, which are currently under way at both Ombilin and Bukit Asam.

At Ombilin the program is designed to increase production from 300 000 tonnes to 750 000 tonnes by 1985 (Tables 6.13 and 6.14), at an estimated cost of US$129 million. Also included in the program is the development of the Bukit Asam mine to supply 2.25 million tonnes of coal to two Surabaya power plants by 1986. This includes infrastructure development, consisting of the rehabilitation of railway tracks, the acquisition of coal wagons, the construction of a coal terminal at Tarahan[11], and the acquisition of barge-boats for transporting the coal to Surabaya. The total estimated project cost is expected to be in excess of $1250 billion.

The program also entails the development of an entirely new underground mine in Waringin, expected to produce an estimated 600 000 tonnes per year by the end of the 1980s[12]. A nearby deposit in Parambahan, apparently suitable for openpit mining, is also being considered for development to an estimated production capacity of 500 000 tonnes per year[13].

The development of the vast steam coal reserves of East and South Kalimantan is now being undertaken by nine foreign consortiums, operating under a specific cooperation agreement contract with P.N. Tambang Batubara[14]. The agreement, signed by foreign contractors with P.N. Tambang Batubara, stipulates the obligations for contractors as follows:

Table 6.13 Coal: production, consumption and exports, 1972–83 (tonnes)

	1972	1973	1974	1975	1976	1977	1978	1979	1980	1981	1982	1983
Production	179 240	148 826	156 153	206 390	182 910	230 627	264 180	278 588	303 989	336 828	480 987	485 631
Ombilin	87 971	81 841	78 804	76 095	60 151	81 020	87 115	92 318	142 829	241 687	302 572	325 663
Bukit Asam	91 269	66 985	77 349	130 295	122 759	149 607	177 065	186 270	161 160	95 141	178 415	159 968
Mahakam	–	–	–	–	–	–	–	–	–	–	–	–
Domestic consumption	190 666	148 566	157 236	197 696	171 394	203 787	198 326	192 378	244 222	261 595	246 356	267 005
Steamcoal	(189 156)	(146 206)	(151 270)	(191 234)	(161 339)	(187 376)	(176 086)	(166 603)	(214 642)	(234 346)	(223 856)	(230 819)
Own use[a]	59 633	61 624	58 502	60 737	58 257	64 086	67 100	61 892	53 776	52 525	45 578	34 976
Railways	48 130	39 493	34 639	51 003	43 257	43 852	30 748	26 125	22 869	19 064	15 162	9 883
Industries	53 593	45 089	45 213	49 044	38 175	55 718	54 238	66 036	116 298	132 670	146 276	166 010
Mining	27 800	–	12 916	30 450	21 650	24 520	24 000	12 550	21 700	30 086	16 900	19 950
Anthracite	(1 510)	(2 360)	(5 966)	(6 462)	(10 055)	(16 411)	(22 240)	(25 775)	(29 579)	(27 579)	(22 500)	(36 186)
Industries	1 510	2 360	5 966	6 462	3 876	4 256	10 043	13 875	14 421	1 725	1 600	3 455
Mining	–	–	–	–	6 179	12 155	12 197	11 900	15 158	25 524	20 900	32 731
Exports	–	–	–	–	7 000	18 650	27 348	51 679	144 639	123 282	121 284	283 772

Note: [a] In coal mines for power generation
Source: See Table 6.7

Table 6.14 Coal: projection of production and domestic consumption, 1985–93 (thousand tons)

	1985	1986	1987	1988	1989	1990	1991	1992	1993
Production	1 750	3 270	6 050	7 550	10 160	12 810	15 010	18 785	22 985
Ombilin	750	1 010	1 110	1 310	1 660	1 810	1 910	2 110	2 160
Greater Bukit Asam									
–Bukit Asam	500	1 260	2 940	3 240	3 000	3 000	3 100	3 500	3 800
–Muara Tiga	–	–	–	–	500	1 000	1 000	1 175	2 025
Kalimantan[a]	500	1 000	2 000	3 000	5 000	7 000	9 000	12 000	15 000
Domestic consumption	2 131	4 180	4 593	5 403	6 718	7 788	10 263	12 058	12 813
Cement factories	691	1 890	2 123	2 203	2 418	2 658	3 208	3 558	3 758
Electricity generation	1 380	2 230	2 410	3 140	4 240	5 070	6 995	8 440	8 995
Other	60	60	60	60	60	60	60	60	60
Balance (deficit)	(631)	(910)	1 457	2 147	3 442	5 022	4 757	6 727	10 172

Note: [a] East and South Kalimantan, to be developed through production-sharing contracts with foreign companies.
Source: Department of Mines and Energy (1985)

1 to establish an Indonesian-based limited company;
2 to provide 13.5 per cent of annual production to P.N. Tambang Batubara;
3 to pay 35 per cent of taxable income as corporate tax for the first ten years of operation and 45 per cent thereafter;
4 regional taxes to be paid at an agreed lump sum of US$100 000;
5 following ten years of operation, 51 per cent of the equity to be offered for sale to Indonesian nationals or entities; and
6 if large quantities of coal are required in Indonesia, the contractors are to sell part of their coal to P.N. Tambang Batubara at the going prices in the Southwest Pacific region.

In return, contractors are entitled to exemption from import or export duties; the carrying over of losses suffered in the first five years of operation into following years; an investment allowance up to 20 per cent of the total value of investment, divided over a period of four operating years; transfer of profit in any currency without limitation, and finally, the exportation of their products, provided consideration is given first to the host country's current need for coal.

PETROLEUM

After a period of stagnation following the second world war, extensive exploration began again in 1967. Between 1967 and 1982 more than 70 fields were located, some with estimated reserves of more than 200 million barrels. Both onshore and offshore fields have been discovered, the overall exploration success ratios being above average at between 30 per cent and 40 per cent[15]. Although prospects for world petroleum demand have been uncertain since 1982 exploration expenditure has continued at a high level, and exceeded US$3.9 billion in 1982 and 1983.

At present oil is produced from twenty fields scattered over the whole archipelago (Table 6.15). Although so far only eight basins have been developed and only one extensively, industry specialists agree on the existence of some 50 basins in Indonesia[16]. The latest estimates of total recoverable oil resources range from 53 billion to 120 billion barrels, while estimated recoverable resources from known prospects range between 15 billion and 17 billion barrels.

The search for oil is, of course, expensive. Between 1967 and 1978 foreign oil contractors spent some US$5.8 million for exploration, development and production of oil and gas in Indonesia[17]. Since 1966 a total of 80 production-sharing contracts have been signed with foreign contractors such as P.T. Caltex, P. T. Stanvac and P. T. Calasiatic–Topco. Production-sharing contractors have made more than oil and gas discoveries, which might lead to additional contractors' areas being declared commercial.

In 1982 total petroleum production was 488.4 million barrels, or 1.34

Table 6.15 Crude petroleum: production, consumption and trade, 1966, 1970–80 (barrels)

	1966	1970	1971	1972	1973	1974
Production	170 524 851	311 551 834	325 672 500	395 552 332	448 536 230	501 837 800
Pertamina/Lemigas	38 503 416	36 000 126	34 833 200	38 066 500	37 047 800	40 505 700
Contract of work	132 021 535	275 551 708	285 797 000	330 998 900	335 331 830	348 492 600
Production-sharing	–	–	4 524 300	26 487 000	76 156 600	112 839 500
Domestic consumption						
Refinery intake[a]	72 390 000	83 674 895	89 978 109	100 543 451	119 310 540	124 920 225
Trade						
Imports	–	–	–	–	–	–
Exports[b]	94 894 000	228 353 179	239 584 627	229 090 802	369 543 130	378 905 462

(cont.)	1975	1976	1977	1978	1979	1980
Production	476 854 951	550 318 649	615 122 777	596 698 420	580 446 658	577 015 900
Pertamina/Lemigas	32 896 413	31 601 963	30 990 182	33 322 597	30 528 589	30 096 417
Contract of work	316 711 636	319 207 343	307 383 584	289 469 376	278 714 777	271 948 838
Production-sharing	127 246 902	199 507 343	276 749 011	275 762 468	271 203 292	274 971 045
Domestic consumption						
Refinery intake[a]	113 379 244	113 697 564	153 777 608	159 478 623	186 018 503	192 850 747
Trade						
Imports	–	10 055 785	29 661 858	31 059 667	30 478 617	32 918 211
Exports[b]	363 069 249	449 470 769	484 236 781	461 917 911	394 379 692	361 713 026

Notes: [a] Since 1973 includes the so-called processing deal
[b] Excludes condensate

Sources: Pertamina *Statistik Perminyakan Indonesia* (1966–69 data); Department of Mines and Energy *Buku Tahunan Pertambangan Indonesia* (1970–80 data); Department of Mines and Energy, Directorate General for Oil and Gas *Statistik Perminyakan Indonesia 1981* (1981 data)

million barrels per day, the lowest level since 1975. In 1983 it was only 490.5 million barrels, adversely affecting not only the petroleum industry but also the Indonesian economy as a whole. The foreign exchange impact for the financial year 1982/83 was a US$2.5 billion decline in net earnings for the oil sector (excluding LNG), to $5.8 billion. Net earnings declined further in 1984/85 to $5.3 billion though this was partly compensated for by increased earnings from LNG.

NATURAL GAS

In 1972 two large gas reserves were discovered by Mobil and Huffco at Arum and Badak, in East Kalimantan. Further discoveries were also made by Conoco, Shell and others. Indonesia suddenly found itself the owner of huge gas reservoirs, but without the means to utilise them. Although these discoveries were far from local consumers, buyers were found, liquefaction trains, terminals and ships were built, and in 1977 Indonesia exported its first shipments of LNG.

In 1982 natural gas production remained static at 1.1 trillion standard cubic feet (scf). Of the total used, 499 billion scf was delivered to the two LNG plants, and 468 billion scf exported, an increase since 1981 of over 20 billion scf. The amount delivered to LNG plants increased to 547 billion scf in 1983 and 830 billion scf in 1984.

At present, studies are under way on how best to utilise the gas resources and the additional oil for export. One solution is to pipe gas from Natuna to Bataan Island near Singapore, sell the gas to Singapore and use the proceeds to pay for the pipeline and its extension to Duri in order to generate steam for secondary recovery of oil by injection[18].

The natural gas which is found in association with oil is lost, unless it is immediately used or stored on the spot. Gas may be used in the production of fertilisers or in an extraction plant. Alternatives include its use in the production of LNG or of carbon black, or for lifting petroleum in oil wells. Associated gas found offshore from Java could be piped to the Krakatau steel plant and the Kujang fertiliser plant. While a large amount of gas has been found, it is only since the development of LNG exports in recent years that the proportion of utilised gas has reached respectable figures.

As to relative values, natural gas resources are believed to be greater than oil resources[19]. During the late 1960s gas production was about 110 billion cubic feet per annum, increasing to 200 billion cubic feet in 1974 (Table 6.17). In 1976 production reached 312 billion cubic feet, when gas was piped to various industries in West Java[20]. In 1978 production jumped again to 820 billion cubic feet, with five trains coming into production, two of them in Badak and the rest in Arun[21]. LNG production increased in 1984 by a further 6.6 million tonnes following the construction of two additional trains in Bontang (East Kalimantan) and another two in Arun. By 1986 a further train is proposed in Arun for the supply of 2 million tonnes of LNG to the Republic of Korea.

Table 6.16 Petroleum products: production, consumption and trade, 1966, 1970–80 (barrels)

	1966	1970	1971	1972	1973	1974
Production						
Refinery output	66 432 000	81 463 801	87 173 079	98 917 930	118 136 302	123 034 225
(of which fuels[a])	(46 057 000)	(49 473 597)	(48 118 857)	(49 800 258)	(52 677 108)	(68 870 499)
Domestic consumption[b]	37 003 000	41 308 000	46 590 000	51 506 000	61 877 000	71 415 000
(of which fuels)	(36 523 000)	(40 474 000)	(45 550 000)	(50 077 000)	(60 293 000)	(69 642 000)
Trade						
Imports	...	...	8 674 000	12 857 000	11 571 000	13 829 000
(of which fuels)	...	(2 146 094)	(5 915 000)	(11 596 000)	(10 568 000)	(12 962 000)
Exports	31 700 000	36 267 000	33 636 000	45 979 001	56 530 542	45 159 276
(of which fuels)	(11 625 000)	(11 283 000)	(10 520 000)	(3 512 671)	(318 269)	(–)

(cont.)	1975	1976	1977	1978	1979	1980
Production						
Refinery output	111 869 535	111 656 330	150 076 949	155 857 071	181 961 869	188 917 688
(of which fuels[a])	(72 006 640)	(68 779 096)	(95 885 523)	(104 063 720)	(108 920 273)	(113 363 266)
Domestic consumption[b]	80 436 000	90 693 000	103 664 000	118 295 000	128 266 000	144 163 000
(of which fuels)	(78 826 000)	(88 872 000)	(101 563 000)	(115 195 000)	(124 929 000)	(140 941 000)
Trade						
Imports	15 506 268	30 649 534	18 360 138	17 660 977	16 164 411	8 329 856
(of which fuels)	(14 965 000)	(30 390 113)	(18 227 000)	(16 936 000)	(15 255 000)	(7 942 000)
Exports	36 652 821	36 254 621	53 580 002	44 605 218	53 614 545	53 430 995
(of which fuels)	(–)	(352 797)	(4 144 367)	(2 256 748)	(291 531)	(1 205 107)

Notes: [a] avgas, avţur, premium, kerosene, automotive diesel oil, industrial diesel oil (HSD), and fuel oil.
[b] Based on data of domestic sales.

Source: See Table 6.7

Table 6.17 Natural gas: production, utilisation and exports, 1966, 1970–80 (million cubic feet)

	1966	1970	1971	1972	1973	1974
Production	106 038 872	108 561 315	121 162 641	150 766 952	186 136 817	202 335 134
Pertamina/Lemigas	57 848 998	67 093 131	69 693 022	86 906 379	69 554 691	80 257 857
Contract of work	48 189 874	41 468 184	50 496 277	62 182 412	59 197 120	43 807 830
Production-sharing	–	–	973 342	7 678 161	57 475 006	78 269 447
Utilisation	52 816 000	44 698 749	44 917 029	42 538 882	53 296 150	72 185 011
of which:						
–manufacture of fertilisers	3 228 000	4 807 000	4 188 000	4 485 000	4 511 598	8 819 195
–manufacture of LPG and carbon black	–	2 383 000	4 314 000	6 688 000	7 201 481	6 898 940
–manufacture of LNG	–	–	–	–	–	–
–other industries[a]	–	–	–	–	–	–
–consumed locally[b]	...	...	...	...	...	...
–fuel in refineries	–	1 668 000	1 212 000	1 024 000	1 695 649	5 800 000
–gas lift						
–pressure maintenance in oilfields	18 115 000	17 042 000	17 369 000	16 622 000	24 871 000	34 082 000
–fuel for own use	15 272 000	18 799 000	27 441 000	14 600 000	...	11 521 000
Exports						
LNG (MMBTU)	–	–	–	–	–	–
LPG (ton)	–	–	–	–	–	–

(cont.)	1975	1976	1977	1978	1979	1980
Production	222 256 233	312 368 144	542 784 183	820 130 292	998 456 677	1 045 748 025
Pertamina/Lemigas	80 423 200	84 858 177	103 679 621	115 562 756	122 826 049	132 068 176
Contract of work	48 200 444	54 484 902	70 327 661	78 011 859	78 055 018	77 050 729
Production-sharing	93 632 589	173 025 070	368 776 901	626 555 677	797 575 610	836 629 120
Utilisation	82 677 501	126 425 594	282 901 103	596 550 012	771 849 728	810 778 568
of which:						
–manufacture of fertilisers	13 108 185	14 263 949	35 748 418	47 436 162	47 840 047	51 070 225
–manufacture of LPG and carbon black	3 766 968	3 195 444	3 308 519	2 167 098	38 091 966[c]	23 192 480[c]
–manufacture of LNG	–	–	57 103 400	224 931 967	383 302 797	472 275 223
–other industries[a]	–	1 326 500	2 952 400	11 071 100	32 054 922	48 202 716
–consumed locally[b]	. . .	35 300	51 600	51 600	173 677	444 721
–fuel in refineries	6 385 000	6 279 000	6 164 043	6 453 800	8 237 874	10 583 675
–gas lift		31 889 000	38 801 800	42 001 400	42 662 405	42 134 142
–pressure maintenance in	53 952 000					
– oilfields		11 645 300	34 701 900	131 738 400	172 006 652	114 864 908
fuel for own use	20 533 000	25 698 700	33 658 700	44 063 953	47 442 991	48 010 478
Exports						
LNG (MMBTU)	–	–	28 134 920	192 325 800	326 151 080	441 666 460
LPG (ton)	. . .	–	. . .	386 006	365 765	520 073

Notes: [a] Steel, cement, and fertiliser manufacturing as well as supply to State Gas Company in Cirebon, Jakarta and Boger
[b] Small-scale industries
[c] Includes Lex plant

Sources: Department of Mines and Energy *Perkiraan Kebutuhan Energi Pelita III (Tahun 1979/1980–1983/1984)* June 1978; Department of Mines and Energy, Directorate General for Oil and Gas *Statistik Perminyakan Indonesia* (various years)

Gas from the vast fields of Arun and Badak is treated somewhat differently, in that gas from Badak, used in nearby refineries or fertiliser plants, is priced low in order to subsidise production of fertiliser plants, is priced low in order to subsidise production of fertiliser and petroleum products. Since mid-1984 the price of gas (in US dollars) has risen to a more realistic level as follows:

to Krakatau Steel, \$0.65 per million Btu for process use and \$2.00 per million Btu for power;
to fertiliser plants, \$1.00 per million Btu;
to all other industrial users, \$3.00 per million Btu.

Processing of minerals

The 1980s have seen an increase in the use of mineral ores in the manufacturing sector of the economy.

Following the increase in intermediate demand, exports of mineral ores have declined from 61 per cent of total output in 1971 to 35 per cent in 1980. However, imports have risen slightly, from 2 per cent to 3 per cent over the same period. With the completion of the aluminium smelter in the Asahan complex, imports of bauxite have also increased. Likewise, following the expansion of the Krakatau Steel complex, iron ore imports have risen, although imports of iron and steel manufactures may have been reduced. The processing of tin into tinplate will be integrated into the Krakatau steel complex and thus will have a significant effect upon trade. Another important development is the completion of new petroleum refineries in 1984, which make Indonesia self-sufficient in refined petroleum products.

ALUMINIUM

Indonesian bauxite has to compete with older, depreciated plants working high-grade deposits in Australia, which produce better grade bauxite. Recognising the vulnerability of this situation, the Indonesian government has decided to construct an alumina plant on Bintan Island. In 1983, P.T. Aneka Tambang, the principal, contracted Kaiser Aluminium Technical Service and Kaiser Engineering International to act as technical suppliers and designers for the project.

The project, which was scheduled to commence production in mid-1986, includes a water reservoir as well as the alumina refinery and a 30 MW power plant, at an estimated cost of US\$726 million. The plant will eventually process 1.75 million tonnes of low-grade Indonesian bauxite annually into 600 000 tonnes of sandy alumina, of which 150 000 tonnes will be exported and 450 000 tonnes will supply the Asahan aluminium smelter in Kuala Tanjung, Sumatra. The project, however, was rephased in 1983.

The smelter in Kuala Tanjung had an initial capacity of 75 000 tonnes per annum in 1982, and this has since been tripled to 225 000 tonnes. Two hydroelectric power stations on the Asahan river provide to the smelter 603 MW installed power[22]. The aluminium plant is seen as a growth centre for further development of the aluminium industry in Indonesia.

The Asahan project, including the hydropower plants, transmission lines and smelter, was constructed as a package by P.T. Inalum. Because direct government assistance has not been provided[23], P.T. Inalum has built its own access roads and a company town in Kuala Tanjung.

Nickel

Ingot production at the P.T. Aneka Tambang's ferronickel plant reached 21 600 tonnes in 1977, but dropped in 1980 to 18 300 tonnes. Wet garnierit is mixed with coal and limestone as flux, and calcined in a rotary kiln at 900°C[24]. The plant produces ferronickel with 16 per cent nickel and cobalt content. This product is then oxidised to remove sulphur, to produce ferronickel containing 20 per cent nickel and cobalt.

P.T. Inco started nickel production in 1977, and by 1980 annual production had reached 20 500 tonnes. The capacity of the plant was at that time 27 400 tonnes per year, and subsequent expansion has led to a current capacity to process ore at 12 000 tonnes per day, aimed at 47 600 tonnes annual production of nickel matte. However, the production of nickel matte subsequently decreased to 19 940 tonnes in 1981, 13 749 tonnes in 1982 and 18 570 tonnes in 1983.

Petroleum

Crude oil is refined in eight locations. Because of the domestic demand structure, substantial amounts of middle distillates have been imported over the years (Table 6.16). Additionally, although total refining capacity is 403 000 barrels per day, actual production was until recently only 275 000 barrels per day due to the age of several plants. Thus some of Indonesia's crude oil was sent to Singapore for refining, under a 'processing deal' agreement.

Accordingly, the Third Five-Year Development Plan gave special attention to the development and expansion of refineries. As a result, additional refining capacity at Cilacap has recently come on stream. This US$1.2 billion project includes a 200 000 barrels per day fuel complex and a 195 000 tonnes lubricant complex. An extension of the Balikpapan refinery in East Kalimantan completed in October 1983[25], and the Dumai hydrocracker plant, at an estimated cost of US$1.45 billion[26] came on stream in 1984. Overall, the completion of these projects has doubled Indonesia's refining capacity to 800 000 barrels per day, a quantity which will enable it to meet its domestic demand and to increase exports of some products. Crude processed in 1984 was 430 000 barrels per day.

The petrochemical industry

Between 1980 and 1983 several projects were planned with a view to developing the petrochemical industry. However, in 1983, due to financial constraints caused by falling oil prices, most of these projects were postponed.[27]

At Bunyu island off East Kalimantan a methanol plant was planned to start production in 1984, under a US$300 million contract awarded in 1981 to a French affiliate of Lurgi (West Germany). Production was anticipated to yield some 330 000 tonnes of chemical-grade methanol per year.

A second project, known as Joint Venture I, was a cooperative effort between Exxon Chemical Co., Pertamina, and Japan's Tonen Sekiyu, Kagaku, and involved the building of an olefin centre at Lhok Semawe for the extraction of ethane and the production of ethylene and polyethylene.

The participants in a third project, called Joint Venture II, had not been decided upon prior to its rephasing. This project, estimated to cost US$1.8 billion, was designed to produce vinyl chloride monomer (VCM) and caustic soda. With completion now scheduled for 1988, the project has been offered to the private sector for development.

An aromatics centre was also planned at Plaju, South Sumatra, using naphtha as feedstock to produce benzene, paraxylene, cyclohexene and pure terephthalic acid (PTA) for synthetic fibre manufacture. Now rescheduled, the first stage will produce terephthalic acid to supply the domestic synthetic fibre industry. The Musi refinery was to produce the naphtha for feeding the aromatic centre, and a contract for this work was signed in 1982 between Pertamina and a consortium of four Japanese companies.[28]

A carbon black plant at Bilegou (West Java), built by a joint venture, P.T. Indocarb, was to have come on stream in 1986 with a capacity of 200 000 tonnes of carbon black per year[29]. A second carbon black plant, to be constructed in Dumai (Sian) at a later date, has also been rescheduled.

Consumption and trade

As mentioned above, much of the mineral industry's output is for export. In 1975 exports took the entire production of bauxite, 96 per cent of copper, 84 per cent of tin, 76 per cent of petroleum and 58 per cent of manganese ore. When processing of nickel into ferronickel began in 1976, output was taken up by the export market. However, by 1982 the worldwide recession was seriously affecting Indonesia's mineral export performance, and the total annual value of mineral exports dropped by about US$2.8 billion.[30]

Exports of all commodities except LNG have declined in volume since the late 1970s. Bauxite exports were drastically reduced, nickel ore exports were cut, and the tin industry was severely affected by the imposition of export quotas by the International Tin Council. Although iron sands production increased, this went mainly to domestic markets. Coal production also increased by 37 per cent, but here too exports declined while domestic consumption increased.

Domestic consumption

In part the decline in mineral ore exports has stemmed from government initiatives to process some ores at home and to export them in metal form. Furthermore, Indonesia has increased the import of iron ore, bauxite, coal and manganese for use in smelters,[31] adversely affecting the trade balance in minerals to some extent.

PETROLEUM

The petroleum industry has been placed under great pressure to fulfil domestic requirements.[32] The rapid increase in consumption has been due partly to low prices, particularly of subsidised fuels, and partly to the deliberate policy to increase petroleum use.

In order to supply the domestic market, crude oil has been imported from Saudi Arabia since 1976. Imports increased from 10 million barrels in 1976 to around 40 million barrels in 1981. Over the same period, imports of petroleum fuel products have fluctuated between 8 million and 43 million barrels per year. However, in 1982 consumption of petroleum fuels was significantly lower than expected, due to a drop in the growth rate to 2.2 per cent from 10 per cent in 1981. As a result the value of oil imports, both products and crude, was US$3.1 billion, well below the budgeted US$5.5 billion.

Internal consumption still accounts for roughly 32 per cent of total production, or about 430 000 barrels per day. However, earlier forecasts that Indonesia would be a net importer of oil by the 1990s now appear to be pessimistic.

COAL

Domestic consumption declined from about 280 000 tonnes in 1966 to around 150 000 tonnes in 1974, largely because of a significant decrease in the railways' use of coal. However, as shown in Table 6.13, consumption of steam coal by industry has been increasing rather rapidly, rising from about 50 000 tonnes in the 1970s to 166 000 tonnes in 1983. The Fourth Plan forecast increases in domestic coal consumption to 5.6 million tonnes in 1988. Most of this is needed to generate electricity and to fuel cement factories. From 1990 onwards, coal required for electricity generation is expected to account for 5–9 million tonnes annually, with an additional 2.6–3.7 million tonnes being needed by cement factories (See Table 6.14).

NON-FUEL ALLUVIAL ORES

Domestic consumption of iron ore increased during the 1970s from about 2 000 tonnes in 1971 to 65 000 tonnes in 1980. The principal supplier is Pelabuhan Ratu, which produces between 35 000 tonnes and 40 000 tonnes per year.[33] Iron sands are used mainly by cement factories.

Domestic tin consumption is low, around 300 tonnes annually. A tinplate factory was inaugurated in 1985 as part of the Krakatau Steel complex. Manganese production is destined almost entirely for the domestic market, with additional needs being met by imports from Africa.[34]

Minerals trade

The contribution of minerals to Indonesia's total exports increased from 39.2 per cent in 1966 to 78 per cent in 1983. Petroleum was largely responsible for the increase, its share of exports rising from 33.4 per cent in 1966 to 61.9 per cent in 1980. LNG exports, which began in 1977, also grew, totalling over US$2.2 billion in 1980, or 10.5 per cent of total exports, and reached US$3.3 billion in 1984.

Despite the development of new mineral commodities (iron sands and copper) for export and increased ore processing capacity of existing minerals, the contribution of mineral ores and metals remained stagnant in 1980, contributing only 4.0 per cent of total export earnings. Before the oil boom, this contribution had been about 15 per cent, declining to an average of 5 per cent after the boom. The dominance of petroleum in Indonesia's mineral exports has been reinforced, although the export structure as a whole has become more unstable.[35]

Japan is the single most important buyer of Indonesian mineral ores. Iron sands and bauxite were once exclusively exported to Japan, as were nickel, manganese, and about 70–80 per cent of copper. Tin concentrates were marketed through the London and Penang markets. Malaysia has become the main buyer of both steam coal and anthracite.[36]

In the 1960s the main buyers of Indonesian crude oil were Japan, Australia, the United States and the Philippines. In the 1970s new markets were developed in Trinidad, the Bahamas, Singapore, Taiwan and Europe. At the beginning of the 1980s South Korea and Brazil were also added to the list of important buyers, although Japan and the United States remained the major importers, both buying some 75 per cent of all exports. Petroleum products and LNG were also mostly exported to Japan and the United States.[37]

Imports of all mineral ores and manufactures gradually increased during the second half of the 1970s, although overall the balance remained favourable to Indonesia, due mainly to trade in petroleum and gas (see Table 6.5). The trade surplus in mineral ores and concentrates decreased from US$142 million in 1975 to US$49 million in 1980, largely because of increased bauxite imports.

Between 1975 and 1980 trade in metals and associated manufactures remained in deficit, due to increased imports of iron metal and manufactures. Processing of tin and nickel into metals had a positive effect upon trade, but the import value for metals and associated manufactures almost doubled over this period. The increase in metal export value (about sixfold) was not sufficient to overcome the deficit.

Although trade in mineral fuels remains in Indonesia's favour, crude petroleum imports have risen largely as a result of the structure of domestic consumption and the country's existing refinery capacity. However, despite increased imports, Indonesia remains an exporter of petroleum products. Moreover, the country's natural gas (LNG) exports have greatly improved its position in the mineral fuels trade and in trade generally (see Table 6.5).

Non-fuel minerals

Between 1979 and 1981, mineral ore imports were higher in unit value than mineral exports (see Table 6.6). Although copper ore exports, tin metal and manufactures, and iron manufactures commanded higher unit values than their imports, the import unit values of iron metal and nickel metal and manufactures were higher than their export unit values. The continued favourable export performance of copper has been due to the generally high copper and precious-metal content of the ore exploited by Freeport Indonesia in Gunung Bijih Timur.

Beginning in 1982, when the aluminium smelter in Kuala Tanjung started production, Indonesia has exported 16 000 tonnes of aluminium ingots to Japan. However, with the closure of aluminium plants in Japan, bauxite exports have been drastically reduced. Similarly, because of below-capacity production from Japan's iron and steel industry, P.T. Aneka Tambang's nickel ore exports have been cut and prices reduced by about 23 per cent. P.T. Inco has reduced its nickel ore production, and is currently operating only one of its three production lines.

The tin industry has been severely affected by the imposition of export quotas by the International Tin Council. A consequent 14 per cent drop in export volume has been accompanied by a 23 per cent drop in the value of those exports.

Increased iron sands production has been taken up predominantly by domestic consumption. Drastic export cuts have been attributed to a lowering in the quality of the remaining iron sands in Cilacap, which no longer meet Japanese buyers' specifications.

Iron

Between 1971 and 1977 total annual iron sands exports fluctuated between 1.9 million tonnes and 285 000 tonnes, declining in 1979 to only 9500 tonnes (see Table 6.7). In 1981 exports amounted to 25 000 tonnes, declining again in 1982 to about 11 000 tonnes.

In 1972 the export price per tonne was US$5.30, increasing in 1977 to US$9.67 per tonne, largely due to stronger Japanese demand.[38] After a period of stagnation in 1978, prices rose again in 1980 to US$10 a tonne.

Copper

Between 70 per cent and 80 per cent of copper concentrates production is taken up by Japanese smelters, and the remaining 20 per cent is exported to the Federal Republic of Germany. Exports reached a peak of 217 000 tonnes in 1976, declining in 1980 to about 194 000 tonnes (see Table 6.8), and declined further to just over 100 000 tonnes in the early 1980s.

Despite weak global demand, the Indonesian copper price remained strong in 1974, but dropped in 1975 due to excess supply internationally. The Inter-governmental Council of Copper Exporting Countries (CIPEC) responded by adopting a short-term policy of reduced production until 1976, when prices improved. However prices fell again in 1977.[39] From 1978 prices rose, continuing to increase until the second half of 1980, when demand weakened during the world recession. Despite this instability, the Freeport mine survived, partly because of the older depreciated plant, and partly due to the contribution of gold and silver also produced from the mine.

Bauxite

During 1971–73 bauxite export prices remained stable at US$5.60 per tonne. In 1974 Caribbean producers within the International Bauxite Association (IBA) increased the basic price, and Indonesia joined them in 1975. The IBA meeting in that year recommended a minimum export price for Indonesian bauxite of US$15 fob per tonne, as from 1976. Prices continued to increase until 1980, when Japanese smelters cut production below normal capacity, causing a drop in Indonesian bauxite exports to Japan. In 1981 the Japanese informed P.T. Aneka Tambang of their decision to import no more than 700 000 tonnes of bauxite in 1982. Accordingly, production was cut by 42 per cent and exports declined by 29 per cent (see Table 6.9).

Tin

Tin exports have been an important source of earnings.[40] In line with government policy of increasing the proportion of processed mineral exports, tin has been exported almost exclusively in metal form since 1977. Tin metal exports increased from 762 tonnes in 1967 to some 31 000 tonnes in 1980, then declined to 23 000 tonnes in 1984. Concomitantly, tin concentrates exports decreased from around 13 000 tonnes in 1967 to only 300 tonnes in 1977 (see Table 6.10).

In the early 1970s exports were directed mainly to the Netherlands, Yugoslavia and Japan. In recent years, however, export destinations have become more diverse, and now include the United States and several European countries.

As a member of the International Tin Council (ITC), Indonesia is bound by that body's decisions, including recently imposed export quotas. In 1982 Indonesia, together with Malaysia, Thailand, Bolivia and Zaire, formed the Association of Tin Producing Countries

(ATPC), whose aim is to arrest any further decline in tin prices. However, because of differences concerning the export control vote and voting systems, agreement is still pending.

Despite low tin prices and tin export restrictions, Indonesia is better placed than other tin producing countries, which have had to close down several mines. Nevertheless, the industry's future remains uncertain, and currently faces the threat of release of 30 000 tonnes of stockpiled tin from the United States' strategic stockpile.

NICKEL

Nickel ore is exported to Japan for smelting, on a toll basis and subsequent sale to third parties. Ferronickel ingots from P.T. Aneka Tambang are also exported, mainly to Japan, although since 1977 some of the ingots have gone to the United States and Hong Kong.[41] Nickel ore exports increased rapidly from 134 000 tonnes in 1966 to about 1.3 million tonnes in 1980 (see Table 6.11).

In 1973 the price of Indonesian nickel exported under contract to Japan was reduced on Japanese initiatives. After rising briefly in 1974, nickel prices dropped again in 1975 with a reduction in Japanese demand.[42] 1976 and 1977 also saw the market oversupplied, largely due to the entry of new producers. A price war started, with newer producers reducing their prices to safeguard cash flow. The price of Indonesian nickel dropped from a 1976 high of US$3.27 per pound to a 1978 low of US$1.98 per pound, and Inco lost price leadership in nickel. However, with 1979's higher oil prices nickel prices jumped again, partly prompted by Inco's announcement of posted prices for its products. Prices continued to increase in the first half of 1980 but dropped in the second half because of world recession, and a dramatic fall in demand, for example, there was a 40 per cent reduction in the United States alone.

By 1982 the full impact of the recession on the Indonesian nickel industry had made itself felt. Demand from Japanese customers weakened and prices dropped to US$1.60 per pound. Overall, nickel ore exports decreased by 11 per cent and those of ferronickel ingots and nickel mattes by 15 per cent and 22 per cent respectively. These developments adversely affected P.T. Aneka Tambang and P.T. Inco, which suffered substantial losses over four successive years, mainly due to high interest charges on their huge investments. To improve its debt position, P.T. Inco made a special offer of 20 per cent of its shares to the Indonesian government and the Industrial Bank of Japan. However, owing to its own financial difficulties, the Indonesian government was unwilling to accept this offer, leading to the temporary closure of P.T. Inco's mining operation.

FUEL MINERALS

The oil boom reinforced the dominance of petroleum in Indonesia's mineral exports. Before this, as noted above, mineral ores and metals contributed around 15 per cent of total earnings from mineral exports, whereas after the boom this declined to some 5 per cent. The unit value of crude petroleum exports has been higher than the unit value of crude imports, reflecting the higher quality of Indonesian crude compared with Arabian light crude. Indonesia also imported middle distillates,[43] and still exports heavy oil and residues (see Table 6.12).

Coal

Coal exports began in a small way in 1976 with some 7000 tonnes of anthracite. By 1980 annual exports of steam coal and anthracite totalled 115 000 tonnes. However, this pattern is unlikely to continue. Projections suggest that 1985–86 domestic consumption will surpass production by some 1 million tonnes, and that by 1988 3 million tonnes might need to be imported. The development of Kalimantan coal is expected to close the gap, and by 1990 there should again be a surplus for export (see Table 6.13).

Petroleum

Indonesian crude oil exports peaked in 1977 at 484 million barrels, with the United States and Japan being the major buyers.[44] The imposition of production and export ceilings imposed by OPEC has meant a levelling off and then a gradual decline in petroleum exports. Petroleum's contribution to total exports was 82 per cent in the financial year 1981/82, 80 per cent in 1982/83, and around 76 per cent in the financial year 1983/84.

Since 1980 efforts to diversify markets have met with some success, although in 1981 the United States and Japan still constituted 75 per cent of the market.[45]

Throughout the 1960s Australia was an important importer of Indonesian oil, taking some 30 per cent of export production in 1966. Exports to Australia then declined, and ceased entirely in 1973, resuming in a small way in 1976 and increasing rapidly to 11.47 million barrels in 1981.[46]

The Philippines was also an important importer of Indonesian crude oil in the 1960s. In 1966 its share was about 12 per cent. In 1972 this declined to 1.2 per cent, ceasing in 1973. Although imports resumed in 1975, the Philippines' share remained a low 1.4 per cent of the total exported.

Singapore is another ASEAN country which has been a steady buyer. Imports began modestly in 1972, and currently Singapore takes 1.7 per cent of the export total.

Within the framework of ASEAN cooperation, the ASEAN Council on Petroleum (ASCOPE) was established in 1976.[47] Since its inception ASCOPE has advocated the concept of a petroleum emergency sharing scheme. The emergency plan involves a formula for triggering the scheme, according to which the supplying member governments will endeavour to contribute crude and products for which they have entitlement. These emergency oil supplies are strictly for domestic consumption, the plan coming into effect only after notification from the ASEAN economic ministers and consultation by the interested parties. This agreement has been tested in practice during the oil crises of 1979 and 1980, during which, at Thailand's request, Indonesia and Malaysia agreed to sell 10 000 and 5000 barrels of oil a day respectively to Thailand at government contracted prices, that is without premium. In 1980, consistent with this scheme, a proposal to establish an ASEAN Petroleum Security Reserve[48] was advanced.

As a member of OPEC Indonesia abides by its decisions, including its production quotas imposed since March 1982. The Indonesian quota of 1.18 million barrels a day remains effective and represents some 71 per cent of the current production capacity of 1.67 million barrels a day. However, despite its oil reserves and resources, Indonesia does not assume an important role in OPEC. It has been called 'OPEC's quiet member', although its moderating role has in recent years been important in solving intra-OPEC production and pricing disputes.

Natural gas

Liquefied natural gas (LNG) exports jumped from 1.1 million m^3 in 1977 to 8.1 million m^3 in 1978, and by 1980 had reached 18.9 million m^3, making Indonesia the largest LNG exporter in the world. At present all LNG exports go to Japan, although South Korea recently concluded a twenty-year contract for LNG imports from Indonesia, beginning in 1986 (Table 6.17).

Gas sold under a 1971 contract is transported to Japan in light 125 000 cubic meter LNG tankers built by General Dynamics and operated by Burmah Shipping. Gas purchased under a 1981 contract will be transported on seven new vessels constructed for and operated by a specially formed Japanese consortium.

In 1983 Indonesia hosted the Seventh Conference on Liquefied Natural Gas. This was the first time the conference, which is held in a different city each year, was hosted by a developing country.

Major policy issues

As noted above, the minerals industry in Indonesia has developed with a predominantly export orientation, currently contributing more than 80 per cent of total export earnings. The industry is dominated by the oil and gas sector, and in 1982, despite a 24 per cent reduction that year in oil exports, crude oil and petroleum products still accounted for some 79 per cent of the mineral industry's export earnings, while LNG exports added another 17 per cent. Mineral ores and metals contributed the remaining 4 per cent.

Dramatic increases in oil prices during the 1970s greatly enhanced the role of oil in the Indonesian economy. In 1978 the oil sector accounted for roughly 15 per cent of per capita GNP, and was also the main source of foreign exchange earnings.[49] Similarly, since 1974 more than 50 per cent of the government's domestic revenues annually have come from oil. This reached a peak of 71 per cent in the financial year 1981/82.

However, while it has brought many benefits, the 'oil boom' has made Indonesia too dependent upon a single commodity. Recent years have shown that the economy lacks the flexibility to compensate for a significant decline in oil revenue. Efforts to promote non-oil exports have failed to effect any significant changes. Other sources of government revenue cannot easily be found to substitute for any significant reduction in oil revenues. Indeed, the short-term inflexibilities of the economy have reinforced the importance of oil as a source of foreign exchange earnings. This has raised a number of major policy issues in the oil and gas sector, which are discussed briefly below.

The oil and gas sector

The first issue arises from the fact that a policy of maximising oil exports creates heavy pressures to continue to increase production and reserve levels. The Third Five-Year Development Plan, ending in the financial year 1983/84, estimated a production level of 1.83 million barrels per day to be reached in the last year of the Plan. However, production stagnated at around 1.3 million barrels per day, lower than the actual production capacity of some 1.67 million barrels per day. The Fourth Five-Year Plan suggests that production could rise to a potential capacity of 1.9 million barrels per day by the end of the decade.

In essence, a policy of maximising oil exports must be less concerned with determining the optimal rate of depletion than with determining the optimal rate of investment. Without significant investments Indonesia will be unable to increase its estimated petroleum reserves and production. In view of Indonesia's lack of capital and technology, major participation by foreign companies remains inevitable.

Two important dimensions have been added to the business of exploring for and developing oil resources in Indonesia. The first has been the acceptance, not only by the oil companies but also by the host government and the public at large, of the need to continue to search for oil in less prospective areas where the discovery of oil would add only marginally to existing proven reserves. This continuous search for oil requires a continuous flow of investment. In 1982, despite the recession and poor prospects in the world oil market, foreign oil companies remained anxious to invest in Indonesia. A record thirteen production-sharing contracts were signed and in all 248 wells were programmed for completion within the year, with an exploration budget of around US$1.1 billion. Such continuity in the search for Indonesian oil may be due partly to the high success rate. Of the 238 exploration wells completed in 1982, including wildcats and appraisal wells, 121 struck oil or gas, a success ratio of 50 per cent.

A second important dimension to the search for Indonesian oil has been an appreciation of the sensitivities in the relationship between foreign companies and the host government. This relationship is described in detail in Appendix 6.1. This can influence the investment climate in general. The host government, in guarding the country's interest, must be aware of the implications of drastic and unilateral changes in policies on foreign investment. The so-called renegotiations of the oil contracts in 1976 led to a slowdown in exploration, and as a consequence the government was obliged to devise new incentives. Similarly, foreign contractors need to appreciate the government's efforts to promote the transfer of technology and expertise, efforts illustrated by the so-called 'Indonesianisation' policy.[50]

The second policy issue for Indonesia flows from the fact that the foreign exchange earning role of Indonesia's oil resources is being eroded through rising domestic consumption of such resources.[51] Between 1966 and 1980 energy consumption increased by about 11.3 per cent per annum (compound rate), while oil consumption increased by 10.7 per cent per annum (compound rate). The annual rate of growth in oil consumption increased from an average of 8.5 per cent between 1966 and 1973 to about 13 per cent from 1973 to 1980. Oil's contribution to total energy consumption was reduced from 87.6 per cent in 1966 to 81.3 per cent in 1980. This was due largely to increased use of natural gas, which accounted for 7.7 per cent of total energy consumption in 1966, rising to 17.2 per cent in 1980. The contributions from hydro power and coal decreased during that period, from 1.2 per cent to 0.9 per cent and from 3.4 per cent to 0.6 per cent respectively.

In view of this imbalance in energy consumption, coupled with oil's importance as a source of foreign exchange earnings, Indonesia's national energy policy has emphasised four programs:

1 intensification of surveys and exploration for energy resources;
2 diversification to other energy sources for domestic uses;
3 conservation and the more efficient use of energy resources; and
4 indexation to ensure the use of the most appropriate energy source in different sectors.

The basic objective of the national energy policy is to shift the economy from mono-energy to poly-energy dependence, by strengthening the position of exportable energy sources (mainly oil and LNG) as foreign exchange earners and developing nonexportable energy sources to fuel the domestic economy.

The targets of the Fourth Five-Year Development Plan (1984–88) reflect this policy. The contribution of oil to total domestic energy consumption is to be reduced to 62.4 per cent, and the contribution of natural gas increased to 18.9 per cent, coal to 9.7 per cent and hydropower to 8.3 per cent, as part of the Plan.

The problem facing Indonesia in implementing this policy in the short and medium terms is one neither of resources nor of technology, but of finance. For instance, estimates of capital requirements for the energy sector during the Third Five-Year Development Plan ranged from US$17.6 billion upwards. Of this amount about 75 per cent was invested in the oil and gas sector by foreign private companies, leaving a modest amount for the development of other energy resources. However, the total amount is large compared with the overall savings of the Third Plan, estimated at about US$68.5 billion, particularly in view of the investment needs of other sectors of the economy. For the Fourth Five-Year Development Plan, the government plans to spend about US$11 billion (of which about 65 per cent will be spent for electricity development) for mining and energy development, out of overall savings estimated at about US$110 billion.

Thus national policy has been geared to mobilise a range of sources of financing and to develop appropriate financial mixes and structures which: will best suit the special requirements of individual projects; will not preempt other priority sectors; will yield a total debt burden which falls within the country's overall repayment capacity; and will be consistent with national investment policies. The following arrangements exemplify these four guidelines.

1 For export-oriented projects such as crude oil production, LNG development, and East Kalimantan coal development, it is thought desirable to continue the present scheme of financing, that is, straight investment under a production-sharing arrangement or joint-venture investment, borrowing from private sources without government guarantee.
2 For projects supplying domestic as well as foreign markets for example, oil refineries, or projects serving only the domestic market but undertaken in conjunction with an export-oriented project such as the Asahan hydropower project, a joint venture could be developed with loan capital from foreign private sources, either without guarantee or with partial government guarantee. This scheme assumes the existence of realistic pricing policies and rational tariff structures.
3 For projects serving only the domestic market, and operating within realistic pricing policies and rational tariff structures, a joint venture between public and domestic private enterprises could be explored, borrowing from semi-concessionary sources abroad, with partial or full government guarantee.
4 Projects serving the domestic market, with subsidised prices or tariffs, are to be undertaken by public corporations borrowing only from concessionary sources with full government guarantee.

As yet, despite policy initiatives, there has been little development of alternative energy sources (non-oil and gas). The exception is the coal mines' rehabilitation programs and the readiness of at least two contractors to develop coal mines in Kalimantan.

The third policy issue arises in connection with the pricing of domestically consumed energy, especially petroleum products such as kerosene and diesel oil. The government's dilemma over the rapid growth of oil consumption and increased oil subsidies remains unresolved, although in 1982 and 1983 serious attempts were made to overcome the problem by raising the domestic prices of petroleum products by about 50 per cent each year. Thus, while in 1981 kerosene was sold at 30 per cent of its production cost, by early 1983 this had increased to about 55 per cent of production cost. Unfortunately, however, this was reduced again by the 27.5 per cent devaluation of the rupiah against the US dollar at the end of March 1983.

Total government subsidies on petroleum products reached about US$2 billion in the financial year 1981/82, and nearly US$2 billion in 1982/83, roughly 12 per cent of total government domestic revenues. It was clear that further increases in domestic prices would be required to decrease the budgetary subsidies, and particularly to decrease the economic subsidies. The question arises whether this is best done through gradual but more frequent increases, or through annual 'shocks', as has been the practice in the past. The government chose the latter by further price increases in 1983, 1984 and again in 1985. The question of subsidies has for the moment been solved.

Effects on other industries are also in question. Subsidies on energy prices (petroleum products and natural gas) were intended to stimulate domestic industries, although there has been much disagreement on this issue. Indeed, the recurring controversy over the price of natural gas supplied to the fertiliser industries continued until mid-1984. There was recognition that the policy of subsidising fuel prices has an adverse effect on the supply side. In June 1984 the question of gas pricing was resolved by the announcement of the new gas prices (see above, 'Mineral reserves and production—fuel minerals').

The fourth policy issue concerns the determination of the optimal allocation of oil and gas resources between exports and domestic uses. It has been proposed that Indonesia should develop energy-intensive industries as well as petrochemical industries, given her comparative advantage in this regard. This will mean diverting more of the oil and gas to domestic uses, rather than limiting their consumption or using more expensive nonexportable substitutes.

Since most resources are located outside Java, the development of resource-based industries is seen as a vehicle for regional development. This change in direction towards industrialisation is also intended to alter the structure of exports. By reducing the share of crude oil exports and replacing it with processed products such as petrochemicals, Indonesia also looks to diversify its markets, moving away from Japan perhaps, and developing new markets in ASEAN countries and elsewhere. Indeed, the large projects such as the aromatic centre, the olefin centre, and the alumina plant described above were developed with this objective in mind, although Indonesia's recent financial difficulties have required their rescheduling. As it happens, the deferral of these projects coincides with world oversupply in petrochemicals.

Other minerals

The development of mineral ores has also been largely for export purposes. As suggested above, this sector has been developed less vigorously than the oil and gas sector, and has depended largely on participation by foreign investors, under work contracts with the government or a state enterprise.

The 'first generation' work contract was introduced in April 1967 for the development of a copper mine by Freeport Indonesia Inc. It was the first and only agreement under such a contract. Freeport Indonesia Inc. was granted a tax holiday, concessions on normal levies, exemption from royalties, freedom in the use of foreign personnel and goods, and exemption from the requirement for Indonesian equity. However, renegotiation in 1976 led to cancellation of the remaining 18 months tax holiday, and the government's purchase of an 8.5 per cent share in the project.

The 'second generation' contract was introduced in 1968. Under the new rules, no tax holidays were granted, there were varying levels of royalties, and provision was made for the acquisition by the Indonesian public of 20 per cent equity at a rate of 2 per cent a year over the first ten years of production. However, there was no mechanism for these transfers. Consequently, a total of sixteen mining contracts were signed under these terms between 1968 and 1972, at which time the regulations were abruptly rescinded.[52] As a result, for a number of years the country lacked the working arrangements for new non-oil mineral development projects.

To encourage greater foreign participation in the mining and processing of non-oil minerals, a 'third generation' working contract was developed in 1976, based on a Presidential decree. The decree reduced the government's earlier demands in the areas of export taxation and provincial government levies, and was also less restrictive in the field of repatriation of foreign currency earnings.

The new regulations allow mining companies to deposit a proportion of foreign exchange earnings in a special account with a foreign bank, subject to central supervision and the use of funds for agreed purposes.[53] The export tax, originally set at 10 per cent, is to be variable, reducing with increases in the degree of local processing of mineral products. The new contract also grants import tax exemptions for capital equipment. Corporation tax is set at 35 per cent for the first ten years of operation and at 45 per cent for the following twenty years.

Prior to this, corporation taxes varied for the three main groups of minerals. Under the new arrangements the tax rate was increased on copper, lead and zinc operations, was unchanged for bauxite and nickel, and was reduced slightly for tin. However, new tin projects were reserved for the state company, P.T. Tambang Timah.

It is not clear whether the evolution from the first-generation type to the third-generation type of contract has encouraged greater foreign participation. Two other factors are worth considering. First, provincial government levies (local taxes) while not severe, have been a source of great irritation. Second, fiscal incentives alone may not be adequate to induce foreign companies to invest in locations with the poor infrastructure typical of most of the resource-rich areas, unless some of the needed infrastructure is to be developed by the government. This seems to be the major issue in the development of the mineral ores industry, although it is also important to other ventures in the manufacturing sector.

In a very real sense, many dimensions of the problem are not specific to the government's management of the minerals industry, but relate to its overall perspectives and policies in implementing a total program of industrial and regional development.

Appendix 6.1

Foreign investment and the Indonesian minerals sector

Foreign oil companies have played an important role in helping Indonesia to develop its oil and gas services. Under the Dutch, exploration and production were conducted by companies granted concessions. In the early 1960s the Indonesian government introduced 'contracts of work', which provide for profit sharing and transfer of title from the contractor to the state.

Freeport Indonesia Inc. was the first foreign company to conclude a contract of work agreement to develop a copper project in Irian Jaya. Production began in 1973, exclusively for exports. In 1968 P.T. International Nickel Indonesia, a subsidiary of the Canada-based INCO Ltd, signed a contract of work giving the company exploration rights in the Soroako area, where a nickel matte plant was to be located. Operations commenced in 1977, and since then all nickel matte production has been exported to Japan. In 1969 P.N. Aneka Tambang, the state company for general mining, concluded a contract with Nippon Kokan Kabushiki Kaisha to export iron sands amounting to 300 000 tonnes per annum for ten years. Production began in 1971, peaked in 1974 and declined subsequently due to a reduction of exports since 1978. All the bauxite produced was for export, since Indonesia lacked treatment facilities. Most of the ore was shipped to Japan under long-term contracts between P.N. Aneka Tambang and Japanese buyers, namely Nippon Light Metal Co. Ltd, Showa Denko Co Ltd, and Sumitomo Chemical Co. Ltd.

In recent years the government has used the production contract to invite foreign participation in coal development. The guaranteed government share is 13.5 per cent of annual coal production, with income tax fixed at 35 per cent for the first ten years of production and 45 per cent thereafter. The most recent development has been the signing of no less than nine contracts of work for gold mining. The form of the contract is the so-called modified third-generation contract.

In 1960 Indonesia terminated the concession system under which the oil industry had operated, and replaced it with the Oil and Gas Law (Law No. 44), in which the state was given exclusive rights to exploit petroleum resources. In 1961 the creation of three national oil companies, Permina, Pertamin and Permigan, paved the way for direct participation by the state in exploration and development. The law stipulated that private and foreign companies conduct their operations as contractors to the national oil companies. Foreign companies which held oil rights (Shell, Stanvac and Caltex) surrendered their rights and signed a contract of work (*kontrak karya*) agreement with the state oil companies. These contracts were later enacted in statutory form by the Indonesian Parliament and ratified by the President in 1963. In 1968 the three state oil companies were merged into a single national company, Pertamina.

Contracts of work are of shorter duration than concession agreements, and the geographic areas covered are not large. The companies obtained twenty-year exploitation contracts for their former concession areas and thirty-year contracts in designated new areas (ten years for exploration and twenty for exploitation).

Foreign companies retained management control but undertook to explore and develop diligently. They accepted mandatory relinquishment of 25 per cent of the area after five years of exploration and of another 25 per cent after ten years. A 60:40 profit split in favour of Indonesia was also agreed, with a signature and production bonus of US$5 million each for the new area. The 'realised price' concept was retained, and provision was made for a value committee in the event of a dispute about price. The foreign contractor was appointed as the exclusive sales agent for the state enterprise's oil, but the state enterprise reserved the right of electing to take 20 per cent of aggregate production in kind.

The contracts also stipulated the sale of refineries operated by Shell and Stanvac, as well as of all three companies' domestic marketing and distribution assets over a period of time. The companies further agreed to supply the Indonesian domestic market with crude oil and refined products at cost plus fixed fees. In 1965 Shell sold its entire Indonesian operation and withdrew from Indonesia.

The contract of work, as critics saw it, was regarded as being no more than a 'dressed-up concession agreement'. In view of this, a production-sharing contract was devised, which has continued in operation since 1966 in the contract between Pertamina and the Independent Indonesian American Petroleum Company. One of the advantages of the production-sharing system was to remove suspicion concerning whether the price of oil used to calculate the profit split was arrived at fairly.

In essence, the production-sharing contract embodies the following features.

1 The contract calls for the sharing of crude oil production, rather than the realised profits accruing from the sale of such oil.
2 The state or the state enterprise, as the owner of the petroleum resource, retains control over the management of operations. In practice, Pertamina is responsible for the management in the operation contemplated in the contract, and approves annual work programs and budgets prepared by the contractors. The contractor executes the work program.
3 The contractor provides all the financing for the operation and bears all preproduction risks. Where oil is discovered and produced, the contractor can recover total costs up to a maximum of 40 per cent of the annual production.
4 The remaining 60 per cent of the annual production is shared on a 65:35 basis in favour of the state enterprise.
5 The contractor has an obligation to deliver a portion of its share (maximum 25 per cent) for the domestic market (pro rata crude), which is valued at a fixed price.
6 The contractor is solely responsible for furnishing all the foreign exchange, materials, equipment and necessary supplies for the operation. Title to all project-related equipment brought in by the contractor passes to the state enterprise upon entry into Indonesia. However, the contractor can recover the cost of this from the 40 per cent of output set aside for cost recovery.

Because of the sharp increases in the price of oil in 1973 and 1974, the production split was changed to 70:30. But the most important development occurred in 1976, when the government entered into negotiations with the contractors to modify existing agreements. The new terms called for a 85:15

split in favour of Pertamina. As before, Pertamina's share was considered to include the contract payment of Indonesian corporate tax liabilities. With pro rata crude still valued at US$0.20 per barrel, the split was effectively close to 88:12. The revision also stipulated a change from straight-line five-year depreciation to the double declining balance method. The depreciation was to run seven years for contracts with proven reserves of less than 300 million barrels and fourteen years for contracts with proven reserves of more than 300 barrels as of 1 January 1976. According to the depreciation formula, contractors were allowed to recover the total costs of production instead of the previous limit of 40 per cent.

The terms of the renegotiated contracts were considered too restrictive of foreign contractors. Exploration and development expenditures began to decline, although production itself continued to increase. The United States Internal Revenue Service (IRS) ruling in May 1976 was also a factor in the downturn in exploration activities. The ruling stipulated that production-sharing payments to Pertamina did not constitute taxes but royalties for United States tax purposes. As such, they did not qualify for write-offs on the companies' United States tax returns. After extensive negotiations and some modifications of the Indonesian tax code, the problem was solved with a new IRS ruling in May 1978.

Due to adverse developments following the revisions, the government offered exploration incentive programs in early 1977, applied retrospectively from 1 January 1977. The price of pro rata crude from new and secondary recovery wells was valued at the realised export price for the first five years, instead of US$0.20 per barrel. An additional depreciation allowance (credit) equivalent to 20 per cent of capital investment (which actually has a net value after taxes of 5.8 per cent of the investment) in new field development and production was granted, for application against the first and/or second year production, provided that the government received at least 49 per cent of gross production. Furthermore, a depreciation period of seven years was applied, independent of the size of proven reserves. Both the 1977 incentives programs and the 1978 IRS ruling had favourable effects on exploration activities, which were also greatly encouraged by the doubling of the oil price and by developments in the Middle East in 1979.

Another development was the establishment of a joint participation plan, a 50:50 joint operating arrangement between Pertamina and foreign contractors for the exploration and development of areas exclusively reserved for Pertamina. This arrangement reduced Pertamina's own exploration costs and provided foreign contractors with lower risk acreage. It covered acreage where some exploration work had already been undertaken by Pertamina, and committed the foreign contractor to spend at least as much on exploration as had already been spent by Pertamina. The contractor was committed to carry all expenses for the first three years, exploration costs thereafter being shared equally by Pertamina and the contractor. Any investment by the contractor during the initial three years which exceeded the earlier investment would be recoverable against future production. Production was divided on an equal-share basis, and the contractor's 50 per cent share came under the renegotiated terms of the standard production-sharing agreement, including the exploration incentive programs.

During the past five years Pertamina has increased its own share of exploration and development activities, both by obtaining loans from Japan and by initiating joint operations with foreign companies. Subsequent government policy has prescribed the manning of operations by Indonesians and appointment of local sub-contractors, especially for 'economically weak' companies. Contracts for new refineries, hydrocrackers and a methanol plant were given on condition that Indonesian companies participated in the design and engineering phases. These policies were in line with general government policy pursued in other sectors, notably the aircraft industry, for 'progressive local content', that is, increasing local content of products for successive units.

While the government encourages mineral exports, these are taxed as is any other export commodity. The government also collects ordinary corporate taxes and royalties on production. Export taxes of 10 per cent are applied to 'strong' and unprocessed mineral commodities. Foreign companies do not pay export taxes, since these are not stipulated in the contracts, instead paying token royalties (*iuran produksi*) of some 3–4 per cent. On the other hand, state mining companies pay export taxes but royalty payments have been eliminated.

Import duties and import sales taxes on mineral ores amount to 2.5 per cent for all ores except iron and coal, which are not taxed. Crude petroleum imports and kerosene are free of duty and sales tax. The import duty and import sales tax on LPG and LNG are 7.5 per cent and 2.5 per cent respectively. Import duties on base metals and articles thereof generally increase in line with the degree of processing.

Notes

1 The Gombong Selatan site is located between Celacap and Yogyakarta in Central Java.
2 The iron ore was to be ground into pellets, reduced through a reduction process into sponge iron, and then refined in an electric furnace.
3 The first copper deposit was discovered in 1936 by Y.Y. Dozy.
4 This deposit has an estimated resource potential of some 300 million tonnes, containing on average 0.57 per cent Cu and 0.47 grams of gold per tonne.
5 Bintan Island is part of the Riau Islands. Each mining site has its own washing installation. The working area extends over 99.462 hectares.
6 The alumina content at Wakopek is 49.3 per cent, with 8.2 per cent reactive Si and less than 1 per cent TiO_2.
7 Tin mining began in 1815 in Bangka, 1851 in Belitung, and 1857 in Singkep.
8 Near Kolaka, Central Sulawesi.
9 P.T. Aneka Tambang's working area covers 8999 hectares, which includes 2157 hectares for exploitation and 4025 hectares for exploration. Some deposits contain 3.0–3.5 per cent nickel. A recent study ascertained a total of 129.4 million tonnes of nickel ore, containing 1.8 per cent nickel and cobalt, of which 350 000 tonnes per year could assure the supply of a ferronickel plant for thirty years.
10 Bukit Asam is located in South Sumatra and Makaham in Kalimantan. The Makaham mine was closed in 1971, having become uneconomic with a production of only 4000 tonnes. Ombilin is an underground mine

producing anthracite, whereas Bukit Asam is an openpit mine producing steam coal.

11 On the southern coast of Sumatra.

12 The seams at this mine are 500–600 metres in depth. The World Bank has shown interest in assisting this project.

13 Production from this mine is intended to supply the cement industries which will be developed in the mid-1980s at Indarung and Madura.

14 The first seven foreign consortiums are–

P.T. Arutmin, a joint venture of Atlantic Richfield, Indarung Coal and Utah Exploration Inc.

P.T. Utah (Utah Exploration Inc.)

A joint venture between Agip (Overseas) and Consolidation Coal of Indonesia Ltd

P.T. Kaltim Prima Coal, a joint venture of CRA and British Petroleum Ltd

A consortium from the Republic of Korea, which formed a local company, P.T. Kideko Jaya Agung

The Spanish State Coal Company, Enadimsa, which formed another local company, P.T. Adaro Indonesia

P.T. Berau Coal, a joint venture of Mobil Oil Corporation (United States) and Nissho Iwai Corporation (Japan).

15 Indonesia's offshore drilling success ratio is 25 per cent.

16 As reported in December 1982 during a meeting of the Indonesian Geologists' Association.

17 In recent years these figures reached US$3 billion per year.

18 This oil amounts to 300 000 barrels per day.

19 The figures presented in the 1982 Indonesian Geologists' Association are between 361 000 trillion cubic feet and 564 000 trillion cubic feet from estimated recoverable resources, and between 100 trillion cubic feet and 159 trillion cubic feet (18–26 billion barrels of oil equivalent).

20 Between 1960 and 1974 gas was used as fuel and feedstock for the Pusri fertiliser plant in Palembang. From 1967 it was used as fuel in refineries, and from 1970 it was used for the manufacture of carbon black and LPG.

21 The two trains from Badak can treat 15 million cubic metres of gas per day, producing 3.3 million tonnes of LNG annually. The three Arun trains can treat 24 million cubic metres per day, producing some 4.8 million tonnes of LNG per annum.

22 These two power plants belong to P.T. Inalum, a consortium of Japanese companies, with a 10 per cent shareholding by the Indonesian government. The government's share has since been increased to 25 per cent.

23 Except for the provision of roads and harbour facilities built for public use.

24 The Elkem process.

25 This project includes a 200 000 barrel per day hydroskimming complex and an 84 500 barrel per day hydrocracker.

26 The plant will have the capacity to crack 85 000 barrels of Low Sulphur Waxy Residual (LSWR) per day and 72 900 barrels of middle and light distilliates per day.

27 This decision affected 47 projects. The refineries' expansion has not been affected.

28 Japan Gasoline Corp., C. Itoh Co., Nissho Iwai Corp., and Far East Oil Trading Co. The cost of the project is estimated at US$1.35 billion.

29 The contract was awarded in 1982 to Tyssen Rheinstall GmbH and Kellogg Overseas Corp.

30 The decrease in petroleum crude and products exports has been close to US$2.8 billion. The drop in exports of mineral ores and metals has been about US$150 million, but this has been offset by a comparable increase in LNG exports, totalling US$150 million.

31 Between 1975 and 1980 the value of Indonesia's mineral ore imports rose from only US$5.7 million to US$202 million. The sources of such imports are Singapore and China (bauxite), Sweden and Brazil (iron ore), and Taiwan, Japan and South Korea (coal).

32 In the 1970s domestic demand for petroleum projects grew by around 13 per cent per annum.

33 Since 1978 the Fe content has been around 55–56 per cent.

34 Between 1975 and 1980 imports of manganese ore more than doubled.

35 This situation is due in part to the fact that mineral ores development has been pursued less vigorously than in the case of the petroleum and gas sector. Direct foreign investment in the mining sector (petroleum and gas excluded) totalled US$1.46 billion between 1967 and 1980, 17.1 per cent of total foreign investment over the period. By contrast, investment in the petroleum and gas sector reached US$1–2 billion per annum between 1979 and 1983.

36 Taiwan has been the second most important buyer.

37 LNG has been exclusively exported to Japan, although South Korea recently concluded a twenty-year contract for the import of Indonesian LNG, beginning in 1986.

38 However, Indonesian exports accounted for only 0.3 per cent of total Japanese iron consumption, and Indonesia was consequently a relatively unimportant source of imports for Japan.

39 CIPEC's twelfth meeting once more failed to agree on production costs.

40 Throughout 1969 and 1980, more than half of non-oil and gas mineral exports were tin exports.

41 Nickel mattes are exported to Japan as well.

42 Trico, Falconbridge and other producers reduced their production in order to maintain prices.

43 Kerosene and ADO; though now, with the refineries expanded, imports have ceased.

44 In 1966, 58 per cent of crude oil exports went to these two countries. By 1975 their share had increased to 85 per cent.

45 If exports to Trinidad and the Bahamas which ultimately also enter the United States are included, the share is 89 per cent.

46 Or 3.2 per cent of total oil exports.

47 This proposal resulted from an ASEAN Concord in 1976, which stipulated that

> Member states shall assist each other by according priority to the supply of the individual country's needs in critical circumstances and priority to the acquisition of exports from member states in respect of basic commodities, particularly food and energy.

48 Similar to the ASEAN Food Reserve.

49 Since 1974 no less than 70 per cent of total exports have come from oil.
50 The production-sharing arrangements have so far proved reasonably successful.
51 This is due to imbalance in the energy mix usage structure, which remains heavily weighted towards oil.
52 Of the sixteen contracts, five have been withdrawn.
53 Originally, the government had insisted that all foreign exchange earnings be repatriated, perhaps so as not to overheat the economy.

Figure 6.1

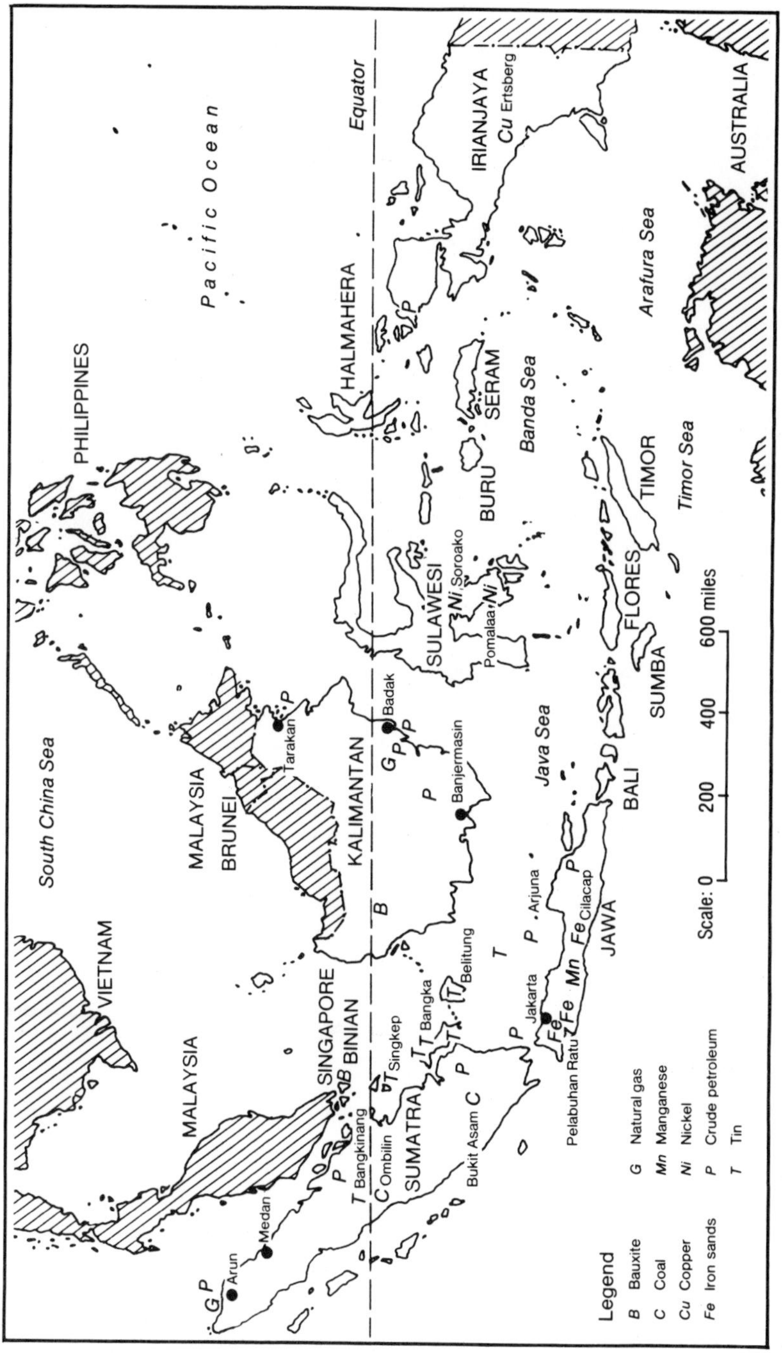
South China Sea
Pacific Ocean
Equator
PHILIPPINES
VIETNAM
MALAYSIA
MALAYSIA
BRUNEI
SINGAPORE
BINIAN
KALIMANTAN
SUMATRA
JAWA
BALI
SUMBA
FLORES
TIMOR
SULAWESI
BURU
SERAM
HALMAHERA
IRIANJAYA
AUSTRALIA
Java Sea
Banda Sea
Timor Sea
Arafura Sea
G P Arun
Medan
T Bangkinang
C Ombilin
T Singkep
T Bangka
T Belitung
Bukit Asam C
Pelabuhan Ratu
Jakarta
Fe Mn Fe Cilacap
Arjuna
Banjermasin
Badak
Tarakan
Ni Soroako
Ni Pomalaa
Cu Ertsberg
Scale: 0 200 400 600 miles
Legend
B Bauxite
C Coal
Cu Copper
Fe Iron sands
G Natural gas
Mn Manganese
Ni Nickel
P Crude petroleum
T Tin

7 Minerals trade and economic development in Malaysia

SAHATHAVAN MEYANATHAN

MALAYSIA has been associated with minerals production for more than a millenium. However, the mining industry developed in a significant way only in the latter part of the nineteenth century, when large tin deposits were discovered in the Larut and Kinta districts of Perak. In the early stages the industry attracted Chinese immigration interested in the exploitation of the mineral. At a later date, the industry was dominated by British capital.

Besides tin, Malaysia has been known also for its gold production. Indeed, in 1511, prior to the conquest of Malacca by the Portuguese, the country was known as 'Aurea Chersonese', or the 'Golden Chersonese'. Coal mining has also been important, and coal was exploited for the first time in 1915 at Batu Arang in Selangor.

Today, tin production still dominates non-fuel mineral production, although other important minerals are now produced, including iron ore, bauxite and copper. Tin played the major role in the country's development, stimulating the early growth of the region and contributing significantly to the development of infrastructure and linkages. Furthermore, the revenues earned enhanced the government's ability to undertake development projects in the early stages. However, over the past decade tin production has begun to turn down due to the depletion of alluvial ores. Paralleling this change has been the discovery and production of oil, which has now assumed the leading role in mineral production. This development has been further strengthened by the production of non-associated gas and liquefied natural gas exports, which began in 1983.

In this chapter, changes in production, reserves, processing, consumption and trade are first explored. This is followed by an analysis of the relevant national policies that impinge on the sector. Finally, an assessment is made of the role of this sector and its probable impact during the 1980s.

Minerals production

Minerals production in Malaysia has been dominated by tin. Changes in its production have an overwhelming influence on the index of mining production, although this influence has declined recently due to the growing importance of petroleum and copper. The mining sector's share of GDP declined from 6.3 per cent in 1970 to 4.6 per cent in 1975, reflecting changes in tin production. However, the recent increase in copper and petroleum production arrested this decline, and in 1980 the share of mining in GDP was maintained at the 1975 level. For the period from 1970 to 1980, output from this sector grew at 4.6 per cent per annum, compared with the 1976–80 growth rate of 8.9 per cent per annum, and with growth of less than 1 per cent per annum between 1971 and 1975 (Table 7.1). The expansion in petroleum output during 1976–80 increased the share of crude petroleum in the sector's output from 29 per cent in 1970 to about 63 per cent in 1980 (Table 7.2). By 1984 the minerals sector accounted for 10 per cent of Malaysia's GDP.

A general feature of the mining industry is its capital intensity. Employment in the mining sector stood at 88 600 in 1970, dropped marginally to 88 300 by 1975, and rose by 1300 persons by 1980. The share of mining employment to total employment fell from 2.6 per cent in 1970 to 1.7 per cent in 1980. This contrasts with 40.6 per cent represented by the agriculture, forestry and fishing sector and 15.8 per cent by the manufacturing sector. The index of labour productivity in the mining sector (based on 1970 = 100) rose by 150 per cent between 1970 and 1979, mainly due to increased use of mechanical equipment and automation in mines, mills, smelters and refineries. This is also reflected in the capital–labour ratio, which doubled between 1970 and 1979. Within the tradeable sectors of the economy, mining had the highest value added per worker, at US$6103 (in 1970 prices) for 1980, compared with US$3462 in 1975. Payments to labour out of total value added in the industry gradually fell to about 23 per cent in 1979. The mining sector is appreciably more capital-intensive than the economy as a whole.

Changes in mining techniques have favoured larger scale mines, and the use of dredging and gravel pumping techniques. Dredging technology began to make inroads around the 1930s. In recent years the number of operating dredges has fallen, to 54 in 1980. However the fall in the number of gravel-pump mines has been greater, declining from 1021 and 1966 to 746 in 1980. Moreover, hydraulic and small working mines have been more severely affected, and the overall outcome has been an increasing proportion of tin coming from more efficient large-scale mines (Thoburn, 1977; Malaysia, *Monthly Statistical Bulletin*).

In the past thirty years, important changes in the ownership of mining companies have occurred in Malaysia. An increasing proportion of shares has been purchased by Malaysian residents, and this has

Table 7.1 Malaysia: Gross Domestic Product (GDP) by sector of origin, 1970–80 (US$ million in 1970 prices)

Sector	1970	1975	1980	Average annual growth rate (%)			Share of GDP (%)		
				1971–75	1976–80	1971–80	1970	1975	1980
Agriculture forestry and fishing	1232.8	1996.6	2668.5	4.8	3.9	4.3	30.8	27.7	22.2
Mining and quarrying	252.6	329.2	557.7	0.4	8.9	4.6	6.3	4.6	4.6
Manufacturing	535.7	1184.5	2468.6	11.6	13.5	12.5	13.4	16.4	20.5
Construction	154.2	271.8	544.8	6.6	12.6	9.6	3.9	3.8	4.5
Electricity, gas and water	74.4	151.7	271.9	9.8	10.2	10.0	1.9	2.1	2.3
Transport, storage and communications	188.6	445.1	779.1	13.0	9.6	11.3	4.7	6.2	6.5
Wholesale and retail trade, hotels and restaurants	530.2	922.2	1513.6	6.3	8.2	7.3	13.3	12.8	12.6
Finance, insurance, real estate and business services	336.4	610.1	989.9	7.2	8.0	7.6	8.4	8.5	8.2
Government services	443.8	918.5	1560.9	10.1	9.0	9.5	11.1	12.7	13.0
Other services	99.4	198.7	301.8	9.3	6.6	7.9	2.5	2.8	2.5
Less: imputed bank service charges	38	87.7	141.5	–	–	–	–	–	–
Plus: import duties	186	276.4	514.5	–	–	–	–	–	–
Equals: Gross domestic product at purchasers' value	3996.1	7217.1	12029.8	7.1	8.6	7.8	–	–	–

Sources: Fourth Malaysia Plan 1981–1985 Kuala Lumpur, 1981; Malaysia, Ministry of Finance *Economic Reports* various issues, Kuala Lumpur.

Table 7.2 Production of minerals, Malaysia, 1965–81 (tonnes unless otherwise stated)

Year	Tin-in-concentrates	Tin metal	Bauxite	Iron ore	Manganese	Wolfran	Oil (US barrels)	Natural gas (cubic metres)	Scheelite	Gold (ounces)	Barite	Antimony	Stibnite	Copper concentrates	Columbite	Silver (ounces)
1965	64 689	73 633	995 190	6 982 674	1 118	9	351 298	–	–	6 652	–	–	–	–	46	–
1966	69 988	72 185	955 494	5 854 639	58 785	4	345 654	–	–	3 980	–	–	–	–	69	–
1967	73 273	77 553	899 555	5 435 376	85 101	13	328 334	–	10	3 811	–	–	–	–	88	–
1968	76 270	89 600	798 619	5 166 707	45 119	100	1 521 423	–	7	4 172	–	–	–	–	53	–
1969	73 322	88 486	1 072 965	5 233 438	10 334	47	3 387 418	–	1	5 422	–	–	–	–	64	–
1970	73 792	91 494	1 139 259	4 490 865	–	134	6 558 749	–	1	5 177	–	–	–	–	61	–
1971	75 441	87 094	977 897	949 942	27 193	237	25 070 892	–	2	5 919	–	–	–	–	24	–
1972	76 830	91 001	1 076 456	520 773	28 639	258	33 866 949	–	2	5 431	–	–	–	–	88	–
1973	72 260	82 468	1 142 762	517 178	28 346	241	33 079 863	90 253 773	5	3 719	–	–	–	–	92	–
1974	68 122	84 395	947 498	480 579	85 262	241	29 540 949	81 767 598	14	4 439	–	456	143	–	83	–
1975	64 364	83 070	703 561	348 200	133 303	194	35 780 152	166 299 470	11	3 676	814	587	–	21 190	50	–
1976	63 401	77 297	660 235	308 184	94 112	109	60 549 019	121 651 653	15	4 538	6 096	600	–	77 617	46	–
1977	58 703	66 304	616 214	329 971	94 732	180	66 983 603	616 130 887	7	4 914	11 074	440	–	99 992	45	–
1978	62 650	71 953	615 060	320 034	42 729	137	79 170 963	2 410 807 086	–	6 776	5 079	439	–	112 510	23	–
1979	62 995	73 068	386 520	360 132	31 605	100	103 286 406	2 990 557 000	3	6 335	1 401	511	–	104 697	40	432 525
1980	61 404	71 319	920 356	371 186	4 003	68	100 912 476	2 676 122 000	–	5 000	–	221	–	114 222	33	437 379
1981	50 938	69 169	700 366	532 455	2 000	67	94 299 640	2 430 040 000	–	6 333	19 365	318	–	120 327	22	518 811

Notes: Gold: from 1979 inclusion of gold contained in copper concentrates would total 61 627 ounces (1979); 65 905 ounces (1980); 83 035 ounces (1981).
Manganese: From 1965–76 includes Ferrolmanganese. For 1981 export figure used.
Silver: figures denote silver contained in copper concentrates.

Source: Malaysia, Department of Mines *Bulletin Relating to the Mining Industry of Malaysia* various issues.

been accompanied by some companies moving their headquarters to Malaysia. Gravel-pump mines which still dominate tin output in Malaysia are mainly Malaysian-owned, while the Malaysian Mining Corporation Berhad (MMC) dominates the dredging sector, operating 38 out of the 54 tin dredges in 1980. In 1979, Malaysian ownership in the industry accounted for 92 per cent of total output (Malaysian Industrial Development Authority, 1981).

Non-fuel minerals

TIN

The mining of tin deposits has been more important to Malaysia than that of any other non-fuel mineral deposits. Malaysia was still the world's largest tin producer in 1980, producing about 28 per cent of the world total (International Tin Council). In 1981, the number of active mining units totalled 710, compared with 852 and 873 at the end of 1980 and 1979 respectively. Of the total tin concentrates produced in 1981 (60 000 tonnes), 76 per cent was mined by gravel pumping, 31 per cent by dredging, 6 per cent by open cast, 4 per cent by underground mining, and the rest by retreatment, dulang washing (panning) and other methods (Department of Statistics).

In terms of producing units, there were 60 dredges, 593 gravel-pump mines, 35 open cast, 20 underground units, and other smaller mining units, comprising a total of 710 units at the end of 1981. These mining units were mainly in the States of Perak, Selangor, Pahang, Johore, Kedah, Perlis, Trengganu, Negri Sembilan and Malacca (in descending order of importance). A total of 35 198 persons were employed in the tin mining industry in 1980, 57 per cent being engaged in the gravel-pump sector, 27.6 per cent in dredging, 5 per cent in open casting, 3.5 per cent in underground operations, and the remainder in other types of operations.

The value of tin production has been in the region of US$1 billion annually for 1980 and 1981, accounting for about 3.6 per cent of Malaysia's GDP for the above period. Production of tin-in-concentrates averaged around 62 000 tonnes for the period 1978–81, and has shown a declining trend since 1972, due mainly to the depletion of alluvial tin deposits[1] and the inability to obtain mining rights to exploit new deposits.

Almost all gravel-pump mines are Malaysian-owned, and the Malaysia Mining Corporation (MMC), a state enterprise, now operates major dredging concessions previously run by the London Tin Corporation. The MMC operates more than 60 per cent of the dredges and controls about 30 per cent of Malaysia's tin output[2].

The most important tin mining technology in Malaysia is gravel-pumping, which is usually employed on small deposits. This and other techniques (with the exception of underground mining) are employed on alluvial or eluvial deposits.[3] This technology involves high-speed water jets directed at an open pit, the liquid mud being sucked up from the bottom of the pit by pumps ('gravel-pumps') on to an elevated sluice for concentration. Dredging is the next most important method. Dredging is capital-intensive, and dredges are ship-like structures which dig the ore from the bottom with a continuous bucket chain. Open casting is dry mining and excavators are used to mine deeper deposits. Underground mining is similar to coal mining to get at 'hard rock' deposits. Small quantities of cassiterite are also recovered by dulang washing and ground sluicing.

IRON ORE

There were eight iron ore mines in operation in 1981, compared with nine and ten in 1980 and 1979 respectively. The largest mine, located in the northern state of Kedah at Gunung Jerain, accounted for 150 500 tonnes in 1980, or about 40 per cent of Malaysia's total output. Smaller mines are to be found in Pahang, Perak and Johore. Since 1965, when production peaked at 7 million tonnes, output has steadily declined (see Table 7.2). In 1970 production stood at 4.5 million tonnes. The same year saw the closure of major mines on the east coast of the Peninsula, and production levels for 1980 and 1981 fell to 371 186 tonnes and 532 455 tonnes respectively. The value of production for these years was US$12.17 and US$12.31 million respectively.

BAUXITE

Bauxite deposits of the lateritic type occur in parts of Peninsular Malaysia, Sabah and Sarawak. However, only one mine, the Johore Mining and Stevedoring Company Sdn. Bhd. (JOMIS), is currently in production. The mine is located in the southeast region of Johore near the town of Pengerang, and has an annual production capacity of 700 000 tonnes. At present the concession area of JOMIS is 1200 hectares. The bulk of the ore mined is of the trihydrate gibbsite type. The ore is of medium export quality, with about 50 per cent available alumina and 4–5 per cent silica. The ore from the above area is derived mainly from volcanic material, and to a lesser extent from sediments and granite–micropegmatite.

In 1955 Alcan established Southeast Asia Bauxites Ltd (Seaba) as a Canadian incorporated company with a branch in Malaysia which then took over the bauxite mining leases. When the Seaba leases came up for renewal in 1977, Seaba agreed to sell 30 per cent of its equity to the Malaysian Bumiputras in compliance with the New Economic Policy. In 1960, when Ramunia Bauxites abandoned its contract mining for Seaba, JOMIS, a wholly owned subsidiary of Seaba incorporated in Malaysia, was established to serve as its mine contractor. The current equity structure of JOMIS is 61.5 per cent Alcan, 8.5 per cent Japanese shareholders of Nippon Light Metal, the Japanese Aluminium Company (NKK) and 30 per cent local Malaysian shareholders. With the change in the equity structure of Seaba in 1977, its mining leases were transferred to JOMIS. Bauxite production in Malaysia peaked in 1973 with an output of 1.1 million tonnes. Production was 920 356 tonnes for 1980 and 700 866 tonnes for 1981. About 80 per cent of the bauxite produced is sold to NKK in Japan and the rest to Taiwan, the Philippines and Singapore (ESCAP, 1983).

Copper

No copper production was reported in Malaysia before its production by the Overseas Mineral Resources Co. Ltd (OMRDC). However, in 1975 some copper had been recovered as a byproduct of tin smelting, mainly at the Sungei Lembing mine in Pahang. Ores containing copper were found on the eastern slope of Mount Kinabalu in Sabah during a United Nations geochemical survey in the early 1960s. The State government subsequently issued a prospecting licence to OMRDC.[4] Drilling commenced in 1968, and was followed by mine development after a lease had been drawn up covering 3600 hectares. The porphyry copper deposits in the area are of low-grade quality, being associated with tertiary adamellite–granodiorite intrusions. The ore at the mine contains an average grade of 0.608 per cent copper and associated gold and silver content. The method of mining is open cast and output in 1981 was at near capacity, at a rate of about 10 000 tonnes per month. The output levels for 1980 and 1981 were 114 222 and 120 327 tonnes respectively of copper concentrates (see Table 7.2), with values of production (inclusive of gold and silver) of US$83.15 million and US$73.96 million respectively.

OMRDC is a joint venture, and was originally owned by Overseas Mineral Resources Development of Japan (51 per cent) and Malaysian private interests (49 per cent). However, in the mid-1980s the Sabah State government acquired majority interest (51 per cent) through the State Economic Development Corporation.

Gold and silver

Gold is recovered as a byproduct of tin mining in the Bidor area of Perak, Kuala Lumpur, and through small-scale mining operations in the Raub area of Pahang and the Bau district of Sarawak. However, the only producing unit is the Mamut copper mine, which yielded 5000 and 6333 troy ounces of gold in 1980 and 1981 respectively, and 437 379 and 518 811 troy ounces respectively of silver (see Table 7.2). Gold and silver contributed some 45 per cent to the total turnover of the Mamut copper mine (*New Straits Times*, 29 November 1982).

Other minerals

Other minerals produced in Malaysia, the 'amang' minerals, include ilmenite, monazite, zircon, xenotime, columbite, struverite, tin slags, haematite and wolfram, recovered as byproducts of alluvial tin operations and the tailings of dredging. The amang re-treatment industry was established in the north of the Peninsula to meet the demand for such minerals from western markets. Of these, ilmenite, tin slags and struverite are the more important in terms of value, yielding between US$2.30 and US$5.97 million in 1981. Other minerals produced in small quantities include manganese and antimony, valued at about US$230 000 in 1981 (Department of Mines).

Fuel minerals

PETROLEUM

The main source of energy in Malaysia is petroleum (94.4 per cent). Other sources include hydroelectricity (3.2 per cent) and traditional energy sources (2.4 per cent). As of 1984, crude oil output is derived from 23 fields (Ministry of Finance). Production averaged 200 000 barrels per day in 1982, compared to 18 000 barrels per day in 1970. The value of production in 1981 amounted to US$3.7 billion.[5] In 1980, production offshore from Trengganu, which commenced in 1978, exceeded that from Sarawak or Sabah. Trengganu accounted for 42.5 per cent, Sabah for 22 per cent and Sarawak for 34 per cent of the total.

Production is carried out by Esso Production Malaysia Incorporated, Sarawak Shell Bhd and Sabah Shell Bhd, who act as contractors under a production-sharing arrangement with Petronas. (For Petronas, see 'National policies' below.) The above companies are wholly foreign-owned. From May 1980, Petronas Carigali, a subsidiary of Petronas, in association with Oceanic Exploration and BP, was engaged as a contractor. However, production has not yet begun.

NATURAL GAS

Production of gas until 1983 was confined to oil-associated gas, which is either reinjected or flared during crude oil production. Production in 1981 was equivalent to 13 350 barrels of oil, of which 65 per cent was flared, 32 per cent used in petroleum operations, and only 2.6 per cent sold. Production of natural gas commenced in Bintulu, Sarawak in 1983, and annual production capacity is projected to reach 6 million tonnes in 1986. Exports, mainly to Japan, were 1.83 million tonnes in 1983, while exports for 1984 are estimated at 3.7 million tonnes, representing a 102.2 per cent increase over the previous year (Ministry of Finance). Sarawak Shell/Petronas sells the gas from Central Luconia to Malaysia Liquefied Natural Gas (MLNG), which in turn liquefies it for export[6]. The ownership structure of MLNG is Petronas (65 per cent), Shell Sarawak (17.5 per cent) and Mitsubishi (17.5 per cent). The US$1 billion project will elevate Malaysia to one of the world's largest producers of LNG when the plant is at full capacity.

Offshore gas from Trengganu, discovered by Esso, was piped to shore in 1984 to meet the power generation and industrial requirements of the east coast states of the Peninsula. In Sabah, gas which was formerly flared has been fed into two energy-intensive processing plants. This was considered preferable to liquefying the gas, as the medium size of the fields did not justify the investment in liquefaction. The plants are described in the next section.

Energy-related materials

FERTILISERS

The four principal nutrients used in Malaysia are nitrogen, phosphorus, potassium and magnesium. Of these, nitrogenous fertiliser in the form of urea is the most widely used by the plantation companies and by semi-government agencies such as FELDA and FELCRA.

Owing to the fall of world fertiliser prices in the early 1980s, current domestic requirements are almost entirely met through imports from East Germany, Japan, Taiwan and Canada. These are usually shipped in bulk container loads to Malaysian ports, where they are bagged, ground and/or mixed. Mixing takes place to raise the nitrogenous–phosphoric–potassic (NPK) content to suit customer requirements.

As of 1983, 27 companies had been approved for the processing of fertilisers, of which thirteen are currently in operation. While approved capacity is 7 million tonnes, existing capacity is about 5.4 million tonnes. Most of these companies, however, are involved only in mixing and bagging processes (Malaysian Industrial Development Authority, pers. comm.).

In 1970, domestic processing of fertilisers reached some 250 400 tonnes valued at US$16.56 million. Since then production has increased steadily at an annual rate of 2.21 per cent over 12 years. By 1982, total domestic production reached 324 190 tonnes (US$64.5 million), peaking at 462 616 tonnes in 1981 (Department of Statistics, Ministry of Finance).

With the ASEAN fertiliser plant at Bintulu, Sarawak (a project under the ASEAN cooperative scheme agreed in 1980) coming into production in 1985, 1000 tonnes of ammonia and 1500 tonnes of urea will be produced per day, utilising 45–50 million standard cubic feet (scf) of Bintulu gas per day as feedstock (Fourth Malaysia Plan and Mid-Term Review). The project is 70 per cent financed by a 48 billion (US$220 million) loan at $5\frac{1}{8}$ per cent provided by the Japanese government.

CEMENT

Three companies currently dominate the five existing producers in the Malaysian cement industry. Associated Pan Malaysian Cement Sdn Bhd (APMC) is an equal partnership between Malayan Cement Bhd and Pan Malaysian Cement Bhd, and has a current capacity of 2.1 million tonnes a year. Another producer, with a current capacity of 1.2 million tonnes, is Tasek Cement Bhd, of which the largest shareholder is the Singapore Cement Industrial Co. (Pte) Ltd[7]. Kedah Cement Sdn Bhd is a Kedah SEDC joint venture, also with a capacity of 1.2 million tonnes. Although large, this company is still relatively new (circa 1983), and has yet to capture a sizeable market share.

Two major additions to industry production capacity are expected over the next two years. Perak Hanjoon Sdn Bhd, a Perak SEDC joint venture, is in the process of completion and should result in a further

1.2 million tonnes by early 1986. Cement Industries of Malaysia Bhd (CIMA) is also planning a 0.6 million tonne plant extension, which should be operational by early 1987. Together, these plants will raise present annual industry capacity from 5.0 million tonnes to 6.8 million tonnes.

During the period 1968–78, production grew at an average rate of 8.9 per cent to reach 2.2 million tonnes in 1978. The average growth rate in cement production for 1981–85 is estimated at 12 per cent compared with 10 per cent for the period 1971–80. The value of production for 1981 was estimated at US$232 million. Owing to massive construction activity undertaken in the early 1980s, however, demand growth over the medium term is expected to be slower. The industry expects a halving of volume sales growth of between 5 per cent and 6 per cent until at least 1989. Significant additions to capacity and lower rates of demand growth should result in lower industry-wide capacity utilisation than that observed in 1985 of 84 per cent. There are, however, potential cost-saving measures available to producers. As energy costs account for between 55 per cent and 60 per cent of total cement production costs, savings of about 30–40 per cent can be effected over old processes, viz. through a switch from oil to coal fired plants and from wet to dry processing. This, together with import duties introduced in 1983, will greatly assist in maintaining long-run profitability in the industry.

METHANOL AND HOT-BRIQUETTED IRON

In Sabah, natural gas is piped ashore at Labuan Island on the southwest coast and fed to two plants, operating since mid-1984. The methanol plant employs the Klockner–Lurgi low-pressure process, and its output of 600 000 tonnes/year is some 5 per cent of the world market. It is reported that the product is competitive with United States and Canadian methanol on the United States west coast.

The hot-briquetted iron plant has a capacity of 730 000 million tonnes per year, and uses the Midrex direct reduction process to produce a metallic briquette of 93 per cent Fe purity, which is more stable than sponge iron in ocean transport. The product is a substitute for steel scrap and pig iron as feed to steel plants. The source of iron ore for the plant is Brazil.

Reserves and resources

Since minerals are nonrenewable, their production is dependent on the resource base. Assessing the quantity of mineral resources is a difficult task, even more so when capital and technology are constraining factors. In Malaysia, the need to gather such information was perceived only in the 1970s. Before this, the existence of alluvial tin reserves and Malaysia's comparative advantage in tin production precluded the intensive search for mineral resources, evidenced by the low levels of financial allocation given to the mining agencies—the Geological Survey Department and the Mines Department. However, a shift has been apparent since 1975; while development expenditure for both departments totalled less than US$0.5 million for the period 1971–75, it increased to US$6.8 million between 1976 and 1980, with a further 205 per cent increase planned for 1981–85.[8]

Offshore exploration and marine seismic activity directed at oil and natural gas gathered momentum from the late 1960s and led to Malaysia's becoming a net exporter of oil in the mid-1970s. Available estimates of Malaysian minerals reserves are given below.

Non-fuel minerals

Fragmentary evidence (Table 7.3) suggests that Malaysia's tin reserves are in the vicinity of 1.2 million tonnes. Copper reserves appear much higher at 77 million tonnes, while bauxite reserves are estimated at 7.5 million tonnes. If the potential reserves for bauxite are also included, the figure may exceed 20 million tonnes. Iron ore reserves are estimated at 8 million tonnes (Majid and Chue, 1977).

The estimates for tin may be biased downwards, as they exclude reserves under existing agricultural and reserved lands in or adjoining known tin deposits. They also exclude inferred offshore deposits.[9] Preliminary data from the Central Belt Project area covering Kelantan and Pahang, begun in 1977 with technical aid from the Canadian International Development Agency, suggest the possibility of resources in that area of copper, molybdenum, lead–zinc, iron, tin and tungsten, gold, and uranium (*Malay Mail*, 11 April 1981).

Fuel minerals

As of 1982, estimated oil reserves offshore from Sabah, Sarawak and Peninsular Malaysia exceeded 2.5 billion barrels (see Table 7.3).

The total recoverable gas reserves of 36 trillion cubic feet (more than 5 billion barrels of oil equivalent) are found in 45 offshore gas fields around the oil producing fields.

Potential coal reserves are estimated at 400–500 million tonnes (about 1.2 billion barrels of oil equivalent). These deposits are found primarily in Sarawak, the main areas being Bintulu, Silantek, Mukah–Balingin and Merit–Pila. Most of the coal is poor-quality lignite.

Processing

Just as the availability of capital and technology and Malaysia's mineral endowment have influenced the production of minerals, so too processing and fabricating have been affected by a range of factors.

The processing of tin into tin metal came about through British investment, while further processing and fabricating of the metal have been dependent upon the availability of domestic capital, technology

Table 7.3 Reserves and resources of minerals and fuels, Malaysia

Non-fuel minerals		
Tin		
K. Langat, Selangor	302 000 tonnes	possible reserves (1980)
Malaysia	1.2 million tonnes	reserves (1978)
Iron ore		
Central Pahang	8 million tonnes	resources (1977)
Bauxite		
Pengerang, Johor	7.5 million tonnes	reserves (1982)
Central Pahang	10–20 million tonnes	potential reserves (1977)
Copper		
Mamut, Sabah	77 million tonnes	reserves (1975)
Fuel Minerals (Non-renewable)		
Oil		
Offshore, Malaysia	2.5 billion boe	reserves (1981)
Gas		
Offshore, Malaysia	5.0 billion boe	reserves (1981)
Coal		
Sarawak	1.2 billion boe	potential reserves (1981)
Fuel minerals (renewable)		
Hydro[a]	200 000 bdoe	
Others[b]	90 000 bdoe	

Notes: [a] Represents indicative technical potential
[b] Includes estimates for rubber, wood, timber wastes, oil palm shell, fibre, bunches, and effluent; domestic and animal wastes; rice wastes, sugarcane wastes, and solid municipal wastes.
boe—barrels of oil equivalent.

Sources: Us Bureau of Mines (1980); Majid and Chue (1977); Ministry of Energy (1982); ESCAP (1983)

acquired through joint ventures, and existing local skills. For iron and steel, the availability of iron ore led to the emergence of a number of small end-user industries catering for the domestic market. In the case of petroleum, proximity to the supply source was the overriding reason for the construction of the first refinery in 1917 in Lutong, Sarawak. Access to local and foreign capital and an expanding domestic market have further stimulated petroleum refining. The access to technology and markets critical to natural gas processing was achieved through joint ventures with foreign partners for the processing and marketing stages of the enterprise. The resultant availability of energy in the form of natural gas also attracted potential bauxite and copper smelters to the country. However, the state's failure to provide preferential energy pricing has led to the abandonment of such projects.

The current state of Malaysian processing for individual minerals is outlined below.

Tin

Tin processing is carried out in a variety of ways, depending on the characteristics of the deposit being worked. Preliminary treatment (concentration) is usually done at the mining site (dredge or mine).

Following this, the tin ore is taken to the dressing sheds (which may serve a number of mines), where the free minerals are separated and impurities removed. Shaking tables, classifiers, magnetic and electromagnetic separators, as well as flotation, are commonly used to separate and concentrate the tin ore from associated minerals such as ilmenite, pyrites, monazite, rutile and zircon. The dressing shed operations produce a high grade concentrate of up to 75 per cent purity, which is dried and sent to the smelters.[10] Further processing at the smelters involves a 25 per cent weight reduction, converting the tin concentrates from the mines into 'Straits Refined Tin' metal of 99.9 per cent purity. The process consists of mixing the tin concentrates with limestone and anthracite and smelting them in large furnaces at high temperatures to produce crude metal. The metal is then poured into refining kettles, where it is treated to remove the last traces of impurities. The metal is then ready for casting into 100 pound ingots of 99.9 per cent purity. In Malaysia, smelting costs have ranged between 1 per cent and 5 per cent of the gross value of metal output (Thoburn, 1977).

Malaysia's entire tin output is smelted in roughly equal proportions by two companies, Straits Trading and Datuk Keramat[11] (formerly Eastern Smelting) in Butterworth; these two companies also conduct the Penang tin market. In 1981 the Malaysia Smelting Corporation (MSC) was formed to acquire Straits Trading. The Malaysian Mining Corporation (MMC) acquired 42 per cent of MSC, and MSC in turn took over all the smelting activities of Straits Trading. MMC also has interests in Datuk Keramat.

A network of ore buying agencies and licensed dealers in the main tin producing regions serve the smelters.[12] However, most dredges deal directly with the smelters. In addition to concentrates from Malaysia, concentrates from Burma, Laos, Tanzania, Indonesia and Australia have been processed at the smelters. The total capacity of the smelters is around 102 000 tonnes per year, and current tin metal output is between 61 000 and 71 000 tonnes per year.

Tin metal in Malaysia is used in the manufacture of pewterware and in tinplating. Pewterware from Selangor Pewter and Penang Pewter is 95 per cent pure tin. An electrolytic tinplate plant, Perstima, started production in March 1982; estimated production capacity for 1983 is 60 000 tonnes (Malaysian Mining Corporation, 1983), directed mainly to the local market of about 25 general-line tin-can factories. The Perstima mill is a joint venture between three government agencies (51 per cent), and investors in Singapore (15 per cent) and Japan (34 per cent). In December 1982, a 20 per cent import duty was imposed on foreign tinplate in order to protect domestic production.

Iron and steel

The processing industry in Malaysia began at about the same time as the mining industry, mainly for the production of cast iron parts to serve the tin mines. Foundries and engineering workshops cluster in the tin bearing areas of Perak and Selangor. With the introduction of rail transport, casting of replacement parts for locomotives and coaches also began. In 1967 commercial production of milled bars from domestic iron ore started at Malayawata Steel Berhad. Before 1967 steel bars produced were based on imported billets.

Malaysia currently produces four main categories of iron and steel products: bar mill products (including rods), secondary wire products, iron and steel castings, and flat products and secondary sheet products.

At the end of 1980 there were eight steel mills in Malaysia, with 377 000 tpa melting capacity and 416 000 tpa rolling capacity. The largest of these are Malayawata Steel Bhd in Prai, with a capacity of 156 000 tpa both for melting and for rolling, and Amalgamated Steel Mills Sdn Bhd in Kelang, with 150 000 tpa melting capacity and 60 000 tpa rolling capacity (Malaysian Industrial Development Authority, pers. comm.).

While fabricated metal products have been manufactured in Malaysia for some time, major involvement in iron and steel products dates from 1967, when crude steel and milled bars were produced at Malayawata Steel Ltd in Prai.[13] Primary iron and steel production contributed significantly to the growth of output of the whole industry, which reached US$276 million in 1974 with an average annual increase of 35.3 per cent from 1968. Growth was somewhat slower (29 per cent per annum) from 1975 to 1979, and in 1979 production of steel totalled 199 000 tonnes, and of bars 251 000 tonnes.[14]

The production of primary iron and steel involves three main processes: the smelting of iron ore into iron, the conversion of iron into steel, and the rolling of steel into various shapes. These processes, as conducted at Malayawata, which consumes most of Malaysia's iron ore, are described below.

Iron ore, charcoal, and limestone are the main raw materials used in the blast furnace. The raw materials, that is, ferrous burdens (sintered ore, iron ore concentrates), fuels (charcoal and coke) and small quantities of flux (sand limestone) are charged into the blast furnaces. Hot air blasted in at the furnace bosh burns the fuels through the production of heat and reducing gas. The heat melts the ferrous burdens, which are reduced to molten iron containing impurities. The flux and impurities combine to form slag, which is separated from the molten (pig) iron.

The next stage, steel making, involves the use of two 15 tonne LD converters, or a $1\frac{1}{4}$ tonne electric-arc furnace. Molten pig iron from the first process, together with steel scrap and oxygen, are used in the LD converter to produce steel. The electric-arc furnace, on the other hand, produces steel from steel scrap.

The third stage involves pouring molten steel from the LD converter or electric furnace into moulds to produce ingots, or feeding molten steel into a two-strand continuous casting machine to create billets. Finally, the steel ingots or billets are reheated in the reheating furnace and passed through a series of rolls to produce round, flat and angle bars. For wire rod production, the billets are passed through additional rolls.

For primary iron and steel establishments, a 1981 survey indicated that material costs were the most significant cost component, accounting for about 43 per cent of the total. About 17 per cent was contributed by fuel and utilities, mainly firewood and charcoal, heavy oil, electricity and water. Materials also represented the largest cost component in secondary iron and steel establishments producing steel pipes, plates and so on, accounting for 90 per cent of the total. Fuel, mainly diesel oil, accounted for about 2 per cent of total production costs. Reduced local availability of charcoal and local iron ore will be critical problems in future production. Production of charcoal declined from 130 457 000 scf to 65 427 000 scf between 1974 and 1978. Firewood also fell from 3 009 000 scf to 902 000 scf over the same period. Increased local demand is reflected in the already expanded capacities of Malayawata and two other mills, and in the approval for five new mills, to increase total melting capacity by 140 000 tpa and rolling capacity by 670 000 tpa.

The State planned to build by 1985 a direct reduction iron processing, electric furnace steelmaking and flat products complex in Teluk Kalong, Trengganu, at a cost of US$227.3 million. Amalgamated Steel Mills has awarded a contract to Danieli (Italy) for the construction of Malaysia's largest electric-arc furnace steelworks. This plant will include a 71.12 tonne electric furnace and a six-strand continuous billet caster. Danieli will build the caster and supply the feed for the existing wire rod plant, while Tagliaferri will supply the furnace (Wu, 1980).

Bauxite/aluminium

The crude ore found in Johore is easily mined because of its accessibility and physical composition. The ore is generally found near the tops of small hills, beneath one or two feet of overburden. As the JOMIS deposits are relatively soft, no explosives are required to loosen the ore. Instead, bulldozers are used to clear the trees and overburden and to break up the ore, which is then loaded into trucks and hauled to the central washing plant, some one to three miles away. About ten mining faces are prepared at any one time, coordinated at different rates, so that the mixing of ores of various grades maintains the average grade of the ore reserves.

After the crude ore from the mining faces has been crushed, screened, washed and cycloned in the washing plant, fine particles of bauxite are recovered and the slurry of clay-like residue which remains is discharged later into tailing ponds. This residue amounts to 45 per cent of the feed into the washing plant. The washed ore containing 13 per cent moisture is reloaded onto trucks and transported to open stockpiles for shipment overseas (ESCAP, 1983).

No further processing of domestic bauxite is undertaken in Malaysia. Downstream operations depend on imported ingot. The Aluminium Company of Malaysia (ALCOM), in which ALCAN has interests, operates a hot- and cold-rolling mill and extrusion plant producing plate, flat sheet, coiled sheet, strip, roofing sheet, cladding, circles, slugs and extrusions in mill, anodised and analok finishes. Raw materials (mainly imported) constitute 60 per cent of total production costs, and the products of the company are consumed domestically. The plant has a capacity of 3050 tonnes per annum. A new sheet and foil plant in Klang with sheet-rolling and foil mills is currently under construction. Other companies with extrusion plants include Fujisash, Pony Metal and Jennik Metalworks. A large number of small fabricators produce aluminium architectural products, packaging products, hollow-ware, electric cables and estate–industrial equipment.

A US$27 million aluminium smelter complex projected for Labuan Island, Sabah, involving Péchiney (France) and Hyundai (South Korea) was shelved in 1980 due to the lack of enough natural gas to generate the required electric power. In Sarawak, a smelter project proposed by Reynolds of the United States was rejected after a request for special gas pricing failed (Wu, 1980; *Far Eastern Economic Review*, February 1982).[15]

Copper

The copper ore mined in Sabah is crushed, ground and concentrated to separate the mineral from waste material. All the concentrates produced are shipped to Japan for smelting and refining. A US$136 million copper smelter proposed for Sabah, with an estimated annual capacity of 61 000 to 102 000 tonnes per year, was scrapped in 1981, again due to energy-related problems (*Far Eastern Economic Review*, April 1983).

Petroleum products

About 80 per cent of Malaysian crude is exported, and imports from the Middle East are necessary to feed the local refineries. Since Malaysian crude is light, heavier crude must be imported to meet the mixture of end products required, with emphasis on heavier products. In 1981, about 70 per cent of the petroleum products for internal consumption were produced by the three local refineries, one in Lutong (Sarawak Shell), and two in Port Dickson in Peninsular Malaysia (Shell and Esso). Additional product requirements were met through imports, mainly from Singapore. Part of this quantity was accounted for by processing arrangements between Petronas and the Singapore Petroleum Company. About 12 million barrels of oil products were brought back into the country in 1981 under this arrangement. From 1983 onwards the quantity should be reduced, with a fourth refinery built by Petronas coming into operation. This refinery, located at Trengganu, has a capacity of 30 000 barrels per stream day (bsd). A fifth refinery to be located at Malacca, also by Petronas and with a larger capacity of 120 000 bsd, is expected to come on stream in 1987.

Esso started production at its Port Dickson refinery (65 per cent owned by Exxon, and with a 42 000 bsd capacity) in 1960. Shell's two refineries have a combined capacity of 137 200 bsd. Its refinery at Lutong was established in 1917, and its Port Dickson refinery in 1963. The refinery in Port Dickson is 75 per cent owned by Shell[16], while its Lutong refinery is wholly owned. Local ownership accounts for the remaining shareholding in these refineries.

Crude oil from Malaysia is mixed with West Asian crude in the ratio of 30:70 and distilled to produce the required end products for the Malaysian market at the two Port Dickson refineries. However, this ratio is expected to change in favour of Malaysian crude as the demand for petrol in the overall consumption mixture increases.[17]

Malaysia LNG Sdn Bhd has built a three-train gas liquefaction plant near Bintulu, Sarawak. The plant uses an advanced cryogenic technology which achieves liquefaction by chilling the gas to minus 161°C. The plant will process some 1 million cubic feet of gas per day, or 6 million tonnes of LNG per year at full capacity. The LNG is sold to Tokyo Electric Power Inc. and Tokyo Gas Ltd. These two companies have contracted to buy 4 million tonnes and 2 million tonnes per annum respectively over a period of twenty years. The gas is transported by the Malaysian International Shipping Corporation Bhd, using five specialised LNG carriers owned by the corporation. Each carrier has a capacity of 130 000 cubic metres.[18]

A contract to build a methanol plant at Labuan in Sabah was agreed upon in 1981 between Petronas (Sabah Gas Industries), Klockner and Lurgi of West Germany, Borden of the United States, and Sulser of Switzerland. The plant was commissioned in July/August 1984. Sabah Gas Industries is also expected to establish a hot briquette iron plant and a power plant using gas from Petronas/SSPC Pecten. This gas would be sold first to Sabah Energy Corporation, which has an offshore gas terminal, and subsequently to Sabah Gas Industries.

The petrochemicals industry in Malaysia is still in its infancy, and is characterised by a handful of petrochemical products manufacturers, selling to a large base of finished product fabricators. There are two PVC manufacturers, one polystyrene manufacturer and one polyester fibre manufacturer. At the downstream end, plastics and synthetic fibre converters estimated to number around 300, produce a host of products ranging from plastic household wares and plastic components, parts and pipes, to textiles.

Consumption and trade

The exploitation of the minerals discussed above has mainly been directed to export markets, with the exception of petroleum and iron ore. Domestic demand for other minerals has been small, since industries using these minerals are few. In the case of tin, for example, domestic consumption accounts for less than 1 per cent of production. Demand factors have therefore not affected Malaysia's revealed comparative advantage in the production of tin.[19] Production of other minerals is small by world standards. In the following sections, the estimated consumption of fuel and non-fuel minerals, metals and manufactures, and their trade implications, are reviewed.[20]

Consumption of non-fuel minerals

Estimates of 'apparent' consumption, taking into account stock changes, indicate that iron ore consumption is most significant, rising sharply from about 300 000 tonnes in 1976 to more than 500 000 tonnes in 1980. Tin metal consumption has risen slowly but has remained low (400 tonnes for 1979–80). Net imports of lead and nickel ores have been minimal at about 5 tonnes or less, while net imports of zinc ore rose to 50 tonnes in 1979 but declined to 6 tonnes in 1980.

Consumption of metals and manufactures

Apparent consumption of iron and steel has risen from 490 000 tonnes in 1968 to about 1.3 million tonnes in 1979[21], generally reflecting GDP growth. Net imports of manufactured steel products have been in the vicinity of 21 000 tonnes, valued for 1976–79 at US$27 million.

Next in importance has been the consumption of aluminium metal and manufactures. Net imports in this category have increased rapidly, rising from about 14 000 tonnes in 1975 to 36 000 tonnes in 1980. Copper metal and manufactures have followed a similar pattern in the same period, with net imports rising to 24 600 tonnes from about half that amount in 1978. Between 1978 and 1980, 12 000 tonnes of lead metal and manufactures were imported. Although nickel metal and manufactured imports rose to about 8000 tonnes in 1979, these dropped by half in 1980. A steady increase has been noted in the consumption of zinc metal and manufactures, amounting to 16 500 tonnes for 1980.

Fuel minerals

About 95 per cent of Malaysia's total energy requirements are met by petroleum products. Between 1970 and 1980, energy consumption grew at an average annual rate of 9 per cent, about one percentage point higher than the average annual growth rate of GDP. The country's consumption of petroleum products has also increased by 9 per cent per year, a rate substantially higher than that of the developed countries.

Of the petroleum products consumed in 1980, the largest proportion was contributed by fuel oil at 33.9 per cent, followed by diesel at 33.5 per cent, petrol at 18.3 per cent, kerosene at 4.9 per cent, LPG at 2.3 per cent, and others at 7.1 per cent. Fuel oil is consumed in industry and in electricity generation, while diesel is consumed mainly by industries (Ministry of Energy, 1982). Estimated sectoral energy consumption for 1980 reveals the electricity sector as the largest user, accounting for about 29 per cent of total energy consumption. The transportation sector was next in importance, accounting for about 20 per cent. Manufacturing represented 19 per cent, and agriculture 16 per cent. However, if the vehicles used in mining, quarrying and other sectors are included in transportation, the share taken up by that sector rises to 37 per cent, exceeding that of the electricity sector.

Energy-related materials

Apparent consumption of all types of fertilisers has shown a dramatic increase. The bulk of production is used in the rice industry, where an increase in double-cropping acreage has meant increased demand for fertilisers. Between 1970 and 1975, consumption almost doubled from a base of 550 000 tonnes, and it increased further to 1.6 million tonnes in 1979.

Expansion in the construction industry since the mid-1970s, and increased government expenditure on infrastructure projects, have led to a growth in cement demand of over 10 per cent a year. Cement production in Peninsular Malaysia doubled from about 1.4 million tonnes in 1975 to about 2.8 million tonnes in 1981, mainly to meet domestic demand. Between 1978 and 1980, per capita consumption rose from 180 kg to 250 kg, a higher increase than that in other ASEAN countries. The elasticity of cement consumption to GNP was 2.13 in 1978.

While production of minerals reflects resource endowments, their domestic consumption is determined by manufacturing production. With the acceleration of manufacturing production during the 1970s, increasing consumption has several significant trade implications, briefly reviewed below.

Trade implications

During the period between 1975 and 1980, trade in mineral fuels was more important to Malaysia than that in non-fuel minerals. A positive trade balance was evident in the case of mineral fuels, tin, copper and bauxite. For iron ore, a positive balance was noticed until 1979. However, in 1980 the exportable surplus was eliminated, due to increased domestic consumption and declining production, resulting in net imports of iron ore valued at US$2 million. For lead, nickel and zinc, imports were needed, and an increasingly negative trade balance was seen during the decade 1969–79 (Department of Statistics, *Annual Statistics of External Trade*).

These trends are also apparent from the trade ratios presented for selected products in Table 7.3. The total ratio of exports to imports indicates a strong export dominance for tin, bauxite and copper. For petroleum and petroleum products this dominance is less marked. In the case of iron ore, the trade ratios switched from being strongly positive to negative in 1980. Trade ratios for iron and steel and the fertiliser industries reveal the substantial shortfall in output to meet domestic demand, and the consequent need for imports. These ratios have remained strongly negative.

Table 7.5 summarises Malaysia's trade patterns during 1975–79 for fuel, non-fuel minerals and metals. The focus of Table 7.5 is on Malaysia's trade with ASEAN and the Pacific region. It is clear that Malaysia is highly dependent on the ASEAN–Pacific region for its trade in both energy and mineral resources. The only exception is crude petroleum, which is imported mainly from Saudi Arabia, Iran and Kuwait. Most other fuel and non-fuel minerals are imported from the Pacific region. Japan is an important source of imports of iron ore (indirectly),[22] iron and steel, copper, coal and coke. The United States features prominently in nickel ore and nickel imports. Lead ore, lead, zinc ore, zinc and tin ore are imported mainly from Australia, although tin ore is used for refining and re-export, rather than for domestic consumption. From the ASEAN region, Malaysia imports mainly zircon, petroleum products and gas.[23]

ASEAN is also a significant market for Malaysia's mineral exports, although nickel, bauxite, tin and zircon are exported predominantly to the Pacific region. Singapore is important for the export of iron ore, iron and steel, copper, petroleum products, and the re-export of nickel, aluminium, lead ore, lead and zinc. Malaysia also re-exports coal and coke to Thailand. Japan is an important buyer of copper, petroleum and bauxite, while the United States is a major market for tin and crude petroleum.[24]

National policies

The determination and execution of minerals policy affect materially the incentive to produce, through such matters as the terms and conditions, institutional support and administrative practices, and the fiscal regime. The current state of policy on non-fuel minerals is largely a result of historical accretion. The colonial administration, recognising the significance of the large tin deposits, established the following policies: mining given priority over other forms of land use; tin mining encouraged to increase revenue;[25] conversion of royalty paid to rulers to export duty; and the imposition of legislation concerning mining.

Two important features may be distinguished in national policy. The first is the dichotomy of responsibilities at State and Federal levels, and the second is the different treatment accorded to various specific minerals. National policy on mineral fuels (petroleum and natural gas), found offshore is a relatively recent development. Much of this legislation came into being in the late 1960s and the early 1970s as the sector grew in importance. The ensuing sections describe policies of state support for mineral development, land and leasing policies, and fiscal and other policies as they relate to fuel and non-fuel minerals.

Table 7.4 Trade ratios for major minerals and mineral products, Malaysia

	Tin metal			Iron ore			Bauxite			Copper			Iron and steel			Petroleum and petroleum products	Fertiliser		
Ratios[a]	1976	1979	1980	1976	1979	1980	1976	1979	1980	1976	1979	1980	1970	1974	1979	1981	1976	1979	1980
Imports/output	0	0	0	0.00	0.00	0.18	0.00	0.00	0.00	0.00	0	0.00	49.24	65.85	287.30	0.445	1.075	1.887	1.460
													1.05	1.09	0.80	0.737	0.346	0.377	0.598
Imports/domestic availability[b]	0	0	0	0.00	0.00	0.13	0.01	0.00	0.00	0.00	0	0.00	−0.88	−0.82	−0.81	0.303	−0.999	−0.998	−0.976
Trade ratio[c]	1	1	1	0.99	0.99	−0.70	0.99	0.99	0.99	0.99	1	0.99							

Notes: [a] Explanation of ratios:

Imports/domestic availability: domestic demand is represented by apparent demand (that is, domestic availability), obtained by adding imports to production and subtracting exports, assuming that the level of stocks remains constant. Thus the ratio of imports to domestic availability indicates the proportion of domestic demand that is satisfied by imports.

Trade ratio: the value can vary from minus 1 (no exports) to plus 1 (no imports), the degrees of import dominance being reflected in negative ratios and of export dominance in positive ratios. A zero value indicates a point of balance between exports and imports in the industry.

Imports/output: this ratio indicates the extent of import dependence. The value can vary from zero to infinity. A ratio of one implies that imports are equal to domestic production, a zero value indicates that none of the commodity is being imported, while a high import dependence relative to domestic production would be indexed by values greater than one. If a country is totally dependent on imports, the ratio will tend towards infinity.

[b] Domestic availability = apparent consumption = production + imports − exports

[c] Trade ratio defined as $\left[\frac{\text{exports} - \text{imports}}{\text{exports} + \text{imports}}\right]$

Source: Compiled by the author

Table 7.5 Nominal protection coefficients, minerals, Malaysia, 1979–81

	Copper	Gold	Bauxite	Iron[a] ore	Amang[b] metals	Tin	Crude petroleum	Average[c]
1979	0.84	0.95	0.90	0.90	0.90	0.76	0.29	0.48
1980	0.77	0.95	0.90	0.90	0.90	0.76	0.24	0.45
1981[d]	0.87	0.95	0.90	0.90	0.90	0.83	0.30	0.50

Notes: [a] The following minerals would also fall under the same duty structure: bauxite, manganese, wolfram, and columbite.
[b] Includes struverite, mozanite, tin slag and zircon.
[c] Weighted average. Weights based on value of production at border prices.
[d] Weights for 1980 used for 1981.

Source: Calculated by the author, where the nominal protection coefficient is defined as:

$$NPC = P_i^d / P_i^b,$$

where P_i^d = domestic price of good i;

P_i^b = border price of good i.

It indicates the differential between domestic prices and world prices that exists as a result of protective measures (not necessarily tariffs): ie, how far the domestic price departs from that which would prevail if the commodity concerned were allowed to be fairly traded between countries.

Non-fuel minerals

Mineral development is related to the resource support in terms of institutional and administrative machinery provided by the State. Different administrative policies apply to the fuel and non-fuel mineral sectors. The Mines Department and Geological Survey, established during colonial times, have as their main function the enforcement of relevant legislation. Both departments are part of the Ministry of Primary Industries, which is responsible for policy on non-fuel minerals. Administration and policy formulation on petroleum and natural gas are under the purview of Petronas and the Petroleum Development Unit of the Prime Minister's Department. While the growing importance of the petroleum sector elicited appropriate administrative responses, support for non-fuel minerals was not forthcoming until recently—largely a reflection of early complacency due to reliance on alluvial tin mining.

RESOURCE SUPPORT

Since the decline in the production of tin, iron ore and bauxite the government has placed emphasis on this sector, with an increased allocation to the Geological and Mines Department under the Fourth Malaysia Plan (1980–85). The total sum allocated was US$22 million, compared with less than US$460 000 for 1971–75 (Fourth Malaysia Plan, 1981). In 1981 alone, the allocation to the Mines Department was increased to US$16 million for research, prospecting and conducting airborne geophysical surveys (Fourth Malaysia Plan, 1981).

In 1977, the Geological Department initiated the Central Belt Project to prospect for tin and other metals as part of the effort to increase tin output and to pursue a minerals diversification policy. The project area covered about 31 000 square kilometres in the States of Kelantan, Pahang, Perak and Trengganu. An aerogeophysical survey of the area was completed in 1980, and the geochemical survey of the remaining areas of the project was scheduled for completion by 1985. Recently the Mines Department, through its Mineral Investigation Drilling Unit, carried out scout prospecting works in thirteen new areas in Malay Reservations, covering an area of 20 300 hectares. The Mines Research Institute and the geochemical laboratory of the Geological Department at Ipoh have only lately been strengthened.

Land and leasing policies

The State can exert a strong influence on the development of the minerals sector through its land policies. While the existence of adequate mineral resources is not in doubt, their exploitation depends entirely on such policies. Decisions concerning mining in all States and Territories are at present on an ad hoc basis. Although a National Mineral Policy was initiated in 1980, to encourage diversification beyond alluvial tin mining and to ensure orderly exploitation and development of non-hydrocarbon minerals, a national mining code still does not exist. Legislation at present consists of a collection of old State laws on mining and mineral ores. The issue of prospecting rights and mining titles is unsystematic; there are difficulties in renewals of mining leases; and security of tenure is poor (Majid and Chue, 1977). The Federal Constitution ensures that all land matters in Malaysia come under the purview of individual States, which hold the right to grant land leases. State policies determine the duration and area of leases, prospecting licences, lease renewal procedures, and equity requirements to obtain leases.

To streamline Federal and State laws on mining, formulation of a National Mining Code was begun in 1977 through the promulgation of a Draft National Mining Act and a Draft Mines Control Act. The rationale for the Mining Code was the establishment of the mining industry on a firmer basis through the standardisation of all mining enactments concerning prospecting, alienation, issue and renewal of leases, and conversion of mining land in the various States. However, the joint Federal–State National Land Council rejected the draft code in early 1982. State governments, working within a Federal fiscal system which to them seemed inequitable, were anxious to preserve their rights over mining, and thus opposed the National Mining Code.

Meanwhile, the Selangor State government's formulation of a new mining policy in 1978 was the first model to systematically affect the development of the minerals industry in any State. Under this policy, the State government no longer renewed the expired mining leases of foreign-owned or majority-foreign-owned mining companies. New leases are issued to the State-owned company, Kumpulan Perangsang

Selangor Berhad (KPS), which in turn may sublease to the original owners or other miners working on the land. Such subleases are granted subject to a payment of tribute to be agreed upon by all parties.

Furthermore, where applications for new land are made by foreign or majority-foreign-controlled companies, such applications must be made on a joint-venture basis, with KPS holding an equity of 70 per cent and the foreign company holding the remaining 30 per cent. Applications for new land leases by local companies for areas in excess of 202 hectares must be made on a joint-venture basis, with KPS holding 51 per cent equity. The policy implies that leases will no longer be issued to the private sector in order to ensure for the State government a greater share of the benefits arising from a depleting resource. It can also be noted that some other State governments, Johore and Perak, have emulated the Selangor model, whereby leases are granted provided the State has interests in the company.[26]

Fiscal policies

'A basic premise of all mineral policy is that the returns from mining activities should benefit the country under various forms' (Bosson and Varon, 1977: 167). The fiscal regime governing the minerals sector is an important mechanism for deriving such benefits. Included under fiscal measures in Malaysia are ore and gold licences, dulang passes, premium payments, rent for leases, permit fees, and miscellaneous charges that accrue to the Department of Mines and the land offices of the various States. Royalties also apply to gold, barite and copper concentrates. For gold, a 5 per cent ad valorem royalty is imposed, while for barite it is 10 per cent ad valorem. In the case of copper concentrates, the royalty rate is determined through the application of the following formula:

$$\left(\frac{P - 36}{3.5}\right) \text{ per cent ad valorem,}$$

where P is the copper metal price in US cents per pound. Figure 7.3 shows the royalty rates payable during 1979–81, indicating a range from about 13 per cent to 23 per cent.

Other fiscal measures relate mainly to tin, and include export duty[27] and tin profits tax. The tin export duty until the 1973–74 commodity price boom was essentially a specific ad valorem tax, as the rate schedule varied in such a way that export charges averaged about 15.5 per cent before 1961 and 16 per cent of prices thereafter (Lam, 1978). Only in 1974, with the imposition of the surcharge, was some element of progressivity introduced. The surcharge which was incorporated into the export duty in 1977 was part of the government's anti-inflationary program, intended to absorb excess profits accruing to the industry.

In an attempt to encourage increased tin production, the government introduced a two-part export tax structure in the 1980 Budget. Under this cost-plus concept, the cost of production of the commodity is taken into account and the appropriate duty is imposed at prices above the prevailing cost of production. The average cost of production[28] was assessed at US$8.37 per kilogram, and the Budget proposed a duty only if prices exceeded US$9.05 per kilogram. A graduated scale of duty was announced, relating to the prevailing price of tin, up to a price of US$11.31 per kilogram, beyond which a maximum marginal rate of 50 per cent applied. The restructuring of the export duty was intended to benefit the more profitable mines. Until 1978, tin profit tax was levied at a flat rate of 10 per cent on tin profits. In 1978 the tax was changed to a scaled rate, at 5 per cent on the first US$86 500 of taxable tin profits, at 10 per cent on the next US$86 500, and at 12.5 per cent on any amount exceeding US$173 000. The marginal tax rate on tin profits was raised from 12½ per cent to 15 per cent on taxable tin profits in excess of US$184 000.

The Malaysian tin industry is therefore subject to two major specific taxes: export duty and the tin profits tax[29]. The changes announced in the 1980 budget involved an apparent shift in emphasis from the former to the latter. Figure 7.4 illustrates the net impact of the change in export duty structure as announced in that budget. It can be seen that for prices prevailing in 1979 the new rate reduced taxes by about 4 per cent ad valorem, while it did not alter the tax rate at the higher 1980 price. The new rates were 16.7 per cent and 19.8 per cent of external prices for 1979 and 1980 respectively.

For the purpose of export taxes, the two smelters are considered to be outside the country and the value of tin concentrates delivered to the smelters depends on the daily tin price at the Penang exchange. The net value of tin concentrates is derived from the posted price at the exchange, after deductions are made for locational and insurance charges, smelting costs, and the appropriate amount of export duty. As all tin concentrates are assumed to assay at 75 per cent tin metal for duty imposition, low assay concentrates are penalised while higher assay concentrates avoid payment of a higher rate of duty[30] (Majid and Chue, 1977).

The smuggling of tin concentrates out of Malaysia to avoid export taxes is a perennial problem. It has been estimated that up to 7 per cent of production is sometimes smuggled out, the proportion tending to increase in those years when export controls are imposed by the International Tin Council (Majid and Chue, 1977).

The tin industry is also subject to the normal 40 per cent corporate income tax and 5 per cent development tax. In the case of corporate income tax, capital expenditures can be written off and, if unabsorbed, can be carried forward. In the 1978 Budget, an accelerated depreciation allowance comprising an initial 80 per cent for capital expenditure on plant and machinery used directly in the working of a mine was extended to all industries for assessment years 1979 and 1980. The period of this offer was extended in the 1980 budget for a further three years, ending with assessment year 1983. The existing Malaysian taxation structure also permits exploration expenditures to be claimed under certain provisions as an allowable deduction. For any year in which exploration expenditures prove to be abortive, such expenditures can be claimed as a deduction against aggregate income. Exploration expenditures which prove successful are capitalised and written off over the life of the mine.[31]

It is pertinent to note that, of all the major fiscal measures imposed on the tin industry, only low-value payments such as land premiums and royalties go directly to the State governments. One-tenth of the tin export duty collected is assigned to the producing States in proportion to the amount of tin they produce, as provided in the Federal Constitution. The Federal government receives all other forms of payments (Majid and Chue, 1977). This provision has thus been an important consideration in the Federal–State land problems discussed earlier.

Table 7.5 indicates the nominal protection coefficients for the major non-fuel minerals, which for the years 1979–81 were less than one. That is, minerals were net taxed. Tin had the lowest nominal protection (that is, it was taxed most), followed by copper and other minor minerals.

OTHER POLICIES

The production measures on non-fuel minerals relate only to tin. Malaysia has been a member of all International Tin Agreements (ITA) since the postwar period, and abides by the export quotas imposed on it. From 1981 Malaysia decided to reduce its tin production unilaterally by 25 per cent in order to arrest declining world prices. This quantity, equivalent to 15 000 tonnes, was pro-rated among the larger producers. Export controls are based on the historical shares of production of mining units, while quotas cannot be transferred.[32]

Contributions to the ITA are currently borne by the Federal government. When the future of the Sixth ITA was in doubt, Malaysia took the lead in setting up a tin producers' association as a fallback strategy. In June 1982, draft articles of the Association of Tin Producing Countries (ATPC) were adopted by Malaysia, Indonesia and Thailand. Agreement on the formation of the ATPC by the above countries was reached in March 1983.[33]

The encouragement of resource-based industrialisation in the late 1970s led to the granting of incentives to the iron and steel industries, and more recently to the tinplating industry. The metal products processing industry is accorded tax relief by an accelerated depreciation allowance and duty-free imports[34], while the tinplating company, Perstima, operates under an investment tax credit scheme.

Fuel minerals

Policies relating to mineral fuels in Malaysia have been more consistent than those for non-mineral fuels, due mainly to the powers exercised by the Federal government. Legislation pertaining to mineral fuels, pricing policies, production policies and the nation's energy policy are reviewed in the following sections.

LEGISLATION

In recognition of the growing importance of the oil sector, the consultancy firm of Walter J. Levy was commissioned in the early 1960s by the Malaysian government to prepare a report outlining appropriate measures to ensure the optimal development of the country's petroleum resources. The report, completed in 1965, resulted in the passing of three acts pertaining to petroleum: the Petroleum Mining Act (PMA), 1966; the Petroleum (Income Tax) Act (PITA), 1967; and the Petroleum Mining Rules Act (PMRA), 1968. Of these, the PMA and the PMRA established a concession system (an acreage-leasing system), under which oil companies operated. The companies were licensed to explore, produce and market petroleum with a free hand, in return for royalty payments of between 8 per cent and 11 per cent of the value of the output, payable both to the State and Federal government in approximately equal shares. The companies were further subject to an income tax of between 55 per cent and 60 per cent of their profits. By the early 1970s, offshore oil production from Sarawak had gained momentum, setting the conditions for the formulation of the Petroleum Development Act (PDA), passed in 1974. The PDA established Petronas, the National Oil Corporation, and vested in it ownership of and exclusive rights to explore for and exploit this resource. With the PDA, the National Petroleum Advisory Council was formed, representing various Ministries advising the Prime Minister on the affairs of the Corporation and other general policies concerning the industry. Petronas had first to negotiate oil agreements with the Federal and State governments, and then with the companies. After about two years of negotiations, production-sharing contracts were signed with Shell and Esso in late 1976.[35]

Petronas also licenses any business or service related to the exploration and exploitation of petroleum, including the supply and use of rigs, derricks, ocean tankers and barges, and the supply of equipment, facilities and services. The Ministry of Trade and Industry also grants licences for the processing or refining of petroleum, the manufacture of petrochemical products, and the marketing and distribution of petroleum products.[36]

Thus the implementation of the PDA and the formation of Petronas have been the instruments through which the Malaysian government controls the development of its oil and gas resources. Under the Act, the government reserves full control over the exploration and development areas, and requires ownership of all equipment which the companies introduce. It also requires the companies to employ and train Malaysian nationals in all aspects of their operations, and retains the ownership of the petroleum and natural gas up to the wellhead, where it assigns a certain proportion to the company in order to compensate it for capital investment and production costs. The assignment of a share of the output to the investing company is governed by a production-sharing contract. Petroleum exploration and production operations in Malaysia all take place offshore, and are carried out by Petronas' contractors (normally foreign oil companies), with or without the participation of Petronas' fully-owned subsidiary, Petronas–Carigali.

The main features of the production-sharing contract are as follows:

1 a contract period of 24 years for oil and 27 years for gas;
2 relinquishment of the contract area after five years if no commercial discovery is made;
3 the purchase of Malaysian goods and services where possible;
4 the training of Malaysian personnel;
5 Petronas–Carigali to have a participating interest;
6 each area contracted out to be subject to bidding with a minimum of US$100 000 in *signature bonus*;
7 an *information bonus*, to be fixed in line with the value of the data already collected for the area concerned;
8 a *discovery bonus* of US$0.98 million, paid for each discovery of a commercially viable petroleum field;
9 a *production bonus* of US$1.97 million, paid when production reaches 725 000 kilolitres per quarter;
10 a *research cess*, paid at the rate of 0.005 per cent of the proceeds from the sale of the contractor's cost and profit oil; and
11 apportionment of the petroleum and gas produced as follows:
—10 per cent of production per annum for crude oil and 10 per cent of gross sale of natural gas as royalty payments to the Federal and State governments, each receiving 5 per cent;
—a maximum of 20 per cent of gross production per annum of crude oil and 25 per cent of gross sales of natural gas to the contractor for the recovery of costs:
—the balance (profit oil), that is, 70 per cent of gross production of crude oil and 65 per cent of gross sales of natural gas, to be shared between Petronas and the contractor in the ratio 70:30;[37]
12 a supplementary payment on the difference of profit oil, valued at the current price per barrel and its value at the 'basic price' (US$12.71 per barrel), escalated at a cumulative 5 per cent a year since April 1975;

13 the contractor to submit an annual work program and budget and procurement proposals, for consideration and approval by Petronas.

In the case of the Elf Aquitaine contract of November 1982, the percentage for the recovery cost was increased to 30 per cent for crude oil and 35 per cent for natural gas. New contracts are also likely to specify a Petronas–Carigali participating interest.

In addition to the provisions contained in the production-sharing contract, the contractors must pay the usual corporate income taxes,[38] and a 25 per cent export duty on the volume of profit oil exported. At present, five contractors are working for Petronas under six production-sharing contracts.[39] Table 7.6 shows Federal tax revenue from petroleum (1970–80), and Figure 7.5 indicates the estimated total government fiscal take from one barrel of oil. The proportion retained by the government, excluding corporate taxes and research cess, was about 70 per cent of the gross value per barrel in 1981, compared with 60 per cent in 1979. This difference can be attributed to the export duty and the higher supplementary payments caused by higher world prices in 1981.

Pricing policies

Malaysian crude oil prices from 1971 to 1981 are shown in Figure 7.6. As can readily be seen, in nominal terms they rose from US$2.36 per barrel in 1971 to over US$40 per barrel in 1981, having been set with reference to OPEC prices and spot prices in Rotterdam. They have almost always commanded a premium compared with both Arabian and Kuwaiti prices. However, the price differential in part reflects a quality differential, since Malaysian crudes are light, and low in sulphur.

Energy pricing policy at the retail level in Malaysia can be considered something of a policy variable. Two significant developments substantially altered the price structure of petroleum products during the 1970s. First, there were the massive price increases in the Arabian, Kuwaiti and domestic crudes that are used in Malaysian refineries. These price increases were not unnaturally reflected in higher retail prices for petroleum products, which have more than doubled in nominal terms over the past decade. Second, these price hikes, coupled with the licensing since 1974 of petrol, diesel and kerosene as controlled items and the institution of subsidies for diesel and kerosene, led to a widening in price differentials. No subsidies are afforded to premium petrol, regular petrol or fuel oil, whereas diesel and kerosene received subsidies between 1977–80 and 1976–80, respectively (Meyanathan and Wells, 1982). The subsidies for petroleum products in 1981 amounted to US$324.4 million, derived from product taxes and the export duty on crude oil.

Table 7.6 Federal tax revenue, Malaysia, 1970–80 (US$ million and percentages)

		Direct taxes				Indirect taxes					
							Export duties				
Year	Total revenue	Total	Petroleum income tax	Other Income tax	Petroleum Royalties/ cash payments	Total	Petroleum	Rubber	Tin	Palm oil	Others
1970	746.8	213.3	—	213.3	na	403.5	—	26.0	42.2	5.8	329.5
1971	759.8	233.6	1.3	232.3	—	416.1	—	16.4	41.6	9.2	348.9
1972	992.7	284.1	—	284.1	8.2	522	—	17.4	45	11.3	448.3
1973	1331.8	363.1	11.1	352	10.2	822.7	—	95.4	53.2	20.5	653.6
1974	1891.1	542.1	59.8	482.3	19.1	1151.2	—	159.1	112.6	94.7	784.8
1975	2046.5	800.4	133.8	666.6	32.4	1021.1	—	50.3	81	117.2	772.6
1976	2336.7	812.9	126.7	686.2	33.1	1261.8	—	204.2	114.5	65.3	877.8
1977	3032.1	1134	315.3	818.7	45.1	1618.3	—	226.3	179.2	140.6	1072.2
1978	3690.5	1368.4	333.8	1034.6	50.2	1961.0	—	309.9	216.5	89.6	1345.0
1979	4618.4	1678.8	378.8	1300	73.1	2484.4	—	510.9	249	107.8	1616.7
1980[a]	6056.3	2415.4	797.5	1617.9	158.5	3163.3	332.6	503.5	264.1	76.3	1986.8

	Total tax revenue		Total non-tax revenue		Total non-oil tax revenue				Total petroleum tax revenue		
	Value	% of total revenue	Value	% of total revenue	Value	% of total revenue	% of total tax revenue	% of GDP[b]	Value	% of total revenue	% of total tax revenue
1970	616.9	82.61	129.9	17.39	616.9	82.61	100.00	17.94	—	—	—
1971	649.7	85.51	110.1	14.49	648.4	85.34	99.80	15.20	1.3	0.17	0.20
1972	806.1	81.21	186.5	18.79	806.1	81.21	100.00	15.96	8.2	0.82	1.01
1973	1185.7	89.03	146.1	10.97	1174.6	88.20	99.07	18.05	21.3	1.60	1.79
1974	1693.3	89.54	197.7	10.46	1633.5	86.38	96.47	22.82	78.9	4.17	4.66
1975	1821.6	89.01	224.8	10.99	1687.8	82.47	92.65	23.39	166.2	8.12	9.13
1976	2074.7	88.79	262	11.21	1947.9	83.36	93.89	25.56	159.7	6.84	7.70
1977	2752.2	90.76	279.9	9.23	2436.9	80.36	88.54	28.73	360.4	11.89	13.09
1978	3329.4	90.22	361.0	9.78	2995.7	81.17	89.98	31.08	384	10.40	11.53
1979	4163.3	90.15	455.1	9.85	3784.5	81.94	90.90	34.26	451.9	9.79	10.86
1980	5578.6	92.11	477.7	7.89	4448.5	73.45	79.74	37.08	1288.5	21.28	23.10

Notes: [a] Estimates
[b] Gross domestic product at 1970 constant market prices
na—not available
Source: Malaysia, Ministry of Finance *Economic Report* various issues

Table 7.7 Malaysia: trade with other ASEAN countries and the Pacific region, 1975–79 (per cent)

	Imports			Exports		
	Other ASEAN	Pacific region	AP region[a]	Other ASEAN	Pacific region	AP region[a]
Iron ore	—	99.97 (Japan)	99.97	67.70 (Singapore)	32.30	100.00
Iron and steel	5.17	78.82 (Japan)	83.99	64.48 (Singapore)	24.88	89.36
Copper ore	1.16	0.19	1.87	0.001	99.992 (Japan)	99.993
Copper	2.57	88.31 (Japan)	90.88	63.67 (Singapore)	11.06	74.73
Nickel ore	—	93.86 (US)	93.86	—	—	—
Nickel	3.33	88.82 (US)	93.15	45.78 (Singapore)	50.30	96.08
Bauxite	—	3.25	3.25	5.68	92.53 (Japan)	99.21
Aluminium	—	55.66 (Canada)	55.66	52.30 (Singapore)	23.84	76.14
Lead ore	0.04	92.69 (Australia)	92.73	86.96 (Singapore)	13.04	100.00
Lead	20.63	64.50 (Australia)	85.13	77.94 (Singapore)	19.03	96.97
Zinc ore	7.28	88.13 (Australia)	95.41	—	—	—
Zinc	10.17	86.22 (Australia)	96.39	50.65 (Singapore)	40.26	90.91
Tin ore	16.83	57.02 (Australia)	73.85	—	—	—
Tin	8.93	71.74 (US)	80.67	9.18	46.08 (US)	55.26
Zircon	47.29 (Indonesia)	51.25	98.54	1.45	37.60	39.05
Energy products						
Coal and coke	25.63	50.39 (Japan)	76.00	99.15 (Thailand)	0.007	99.157
Petroleum crude[b]	—	—	—	22.09	74.14 (US)	96.23
Petroleum products	87.69 (Singapore)	1.86	89.55	61.67 (Singapore)	34.44	96.11
Gas	96.66 (Singapore)	0.85	97.51	5.86	76.49 (Taiwan)	82.38

Notes: Countries in parentheses indicate the most important trading partner.
[a] ASEAN–Pacific region.
[b] Imports from the ASEAN–Pacific region are negligible, and are mainly from Kuwait, Saudi Arabia and Iran.

Source: Compiled from the *Annual Statistics of External Trade* Department of Statistics, Malaysia, various issues.

National oil depletion policy

Malaysia's crude oil production increased during the 1970s. However, in 1980 the National Oil Depletion Policy (NDP) was adopted, and as a result production fell to 276 000 bpd, from 283 000 bpd in 1979. The policy was aimed at prolonging the lifetime of Malaysia's oil output. Its key features are as follows: of the fields with oil initially in place, only nine fields of above 400 million barrels each would be controlled;[40] production from each of these fields would not be allowed to exceed 1.75 per cent of the oil initially in place in any one year; production from Esso's Guntong and Palas fields would be postponed from 1981 to 1985; Petronas would be given the power to postpone for three to five years production from any major field which might be found; and in cases of such postponement Petronas will, if necessary, extend the contractor's production period. However, there seems to be conflict between the NDP and a statement in the Fourth Malaysia Plan (1981–85), which envisages an increase in production to reach 362 900 bpd by 1985.[41] Also in apparent conflict with the NDP, production increased to reach 446 800 bpd in 1984.

Energy policy

The relatively low cost of petroleum products until 1973, ease of transportation, and divisibility were the three factors making petroleum products the most suitable and economical source of commercial energy. Almost all the needs of Malaysia's transportation sector are currently met by petroleum products. Petroleum products also constitute the basic fuel for electric power plants, with coal, natural gas and hydro resources being used where they are available. Kerosene and LPG are the favoured cooking fuels, and the former is the main lighting fuel in areas still without electricity.

One of the objectives of Malaysia's planners is to provide the nation with adequate and secure energy supplies by developing and utilising alternative sources of energy so as to reduce dependence on oil. A four-fuel energy strategy based on oil, hydropower, gas and coal has been formulated, with the aim of diversifying the energy base as well as assuring supplies for continued economic growth (Ministry of Energy, 1982). The nuclear option has been deferred, and the development of other indigenous energy sources is currently being given priority.

Recognising the importance of energy as a major input to economic growth, the sum of US$993.3 million was allocated for the energy sector in the Fourth Malaysia Plan (1981–85), a thirteenfold increase over that allocated in the previous Plan. Most of this is earmarked for hydropower, rural electrification and transmission.

In keeping with the objective of diversifying the country's energy base, priority is currently being accorded to the use of Malaysia's gas reserves for power generation. The National Electricity Board will have commissioned a 900 MW gas-fired combined-cycle power station in

Paka, Trengganu by 1984/85. Provision has been made to increase the installed capacity of this station to 1800 MW at a later date (Ministry of Energy, 1982).

Plans are also being made to utilise associated gas produced from oil fields throughout the country. Associated gas will be used for LPG production, power generation, and in heavy industries such as the iron reduction plant and the methanol plant. In keeping with the government's intentions to reduce Malaysia's dependence on crude oil, LPG extracted from natural gas will be used more extensively in its traditional markets, that is, the residential and commercial sectors, and will also be introduced in new applications such as transportation.

Malaysia is relatively rich in hydropower resources, although these resources are not evenly distributed throughout the country, with Peninsular Malaysia contributing 13 per cent, Sarawak 53 per cent and Sabah 34 per cent. Nevertheless hydropower development has been given priority, because of the large potential for development available in Sarawak. A feasibility study is being undertaken to examine the hydropower potential of Sarawak and the possibility of transmitting the excess through submarine cables to Peninsular Malaysia. It is estimated that some 5000 MW can be developed by the year 2000, with 1500 MW transmitted to Peninsular Malaysia (mid-term review of the Fourth Malaysia Plan, 1981–85).

Mini hydropower schemes to augment the nation's rural electrification program are also being developed. These schemes will be sited at remote and isolated villages which are not connected to the national grid, and will be simple run-of-the-river designs with capacities ranging from 50 kW to 500 kW each.

Solar energy development in Malaysia is still at the experimental stage. The contribution of solar radiation to Malaysia's energy sources is likely to be negligible, at least before the year 2000. The government will implement several pilot solar installations, as part of its rural electrification program throughout the country, during the period of the Fourth Malaysia Plan.

Preliminary forecasts set coal demand at about 600 000 tonnes per annum by 1985, the main user being the cement industry. Depending on the extent of coal utilisation, demand from the power generation sector will range from 1.5 million tonnes per year in 1986 to about 6 million tonnes by the year 2000. Affecting this demand, the government has recently directed the Electricity Board to establish the nation's first coal-fired power plant at Port Klang.

Concluding remarks and assessment of the industry

During the 1980s, Malaysia is likely to remain a major exporter of tin, despite the current decline in production. In other non-fuel mineral production, with the exception of copper, a decline in growth may be expected. It is noteworthy that demand factors have not reduced Malaysia's comparative advantage in tin production, but have rather eliminated its exportable surplus of iron ore. Imports of iron ore will probably rise in the 1980s, given Malaysia's emphasis on heavy industrialisation over this period with its implications for current iron ore exporting countries. Increasing domestic demand for petroleum products may also reduce the surplus in exports of crude oil over this decade, despite attempts to diversify the nation's energy sources. It is appropriate to note that gas, and to a lesser extent coal, would gain in prominence in the nation's energy mixture. Considering the availability of low-quality coal and its planned usage, imports from Australia could play an important role in this context.

Given the present reserves, it is conceivable that Malaysia will also have a 'revealed' comparative advantage in the production of non-associated gas over the 1980s. However, due to the current state of the world LNG market, export expansion may be limited in the short term, unless new markets are secured. Production of natural gas, therefore, would be directed more towards the domestic market, and possibly towards Singapore. The contribution of natural gas to total energy consumption is expected to increase from its present 1 per cent to 38 per cent by 1990 (Energy Work Group, 1982). Because of its abundance, this energy resource will also be used in processing sponge iron in Labuan and Trengganu. With the anticipated increase in the absolute size of the industrial sector, lead, nickel and zinc imports will also continue to rise, but will not be as important as imports of iron ore and other metals and manufactures. This in turn will have implications for exporting Pacific Basin countries, with whom Malaysia has high trade intensities.

In the above review of national policies, it was observed that the state placed emphasis on the extraction of a significant rent from the minerals industry. The nominal protection coefficient for the minerals produced (weighted by production valued at border prices) averaged 0.55 for the year 1980 (see Table 7.5). This contrasts with coefficients of 1.49 and 1.33 in that year for the grain and livestock sectors respectively. The minerals sector is thus taxed more highly than the food sector, but at a rate comparable to the export-oriented agricultural sector, with a nominal protection coefficient of 0.66 for that year. It appears that Malaysia's price and trade policies have been to tax minerals and export agriculture while assisting the food and manufacturing sectors (Meyanathan and Sivalingam, 1983).

The role played by the minerals sector in Malaysia's economy is reflected not only in the provision of rents to the state, but also in the country's overall development. It is generally accepted that exploitation of tin resources provided the basis for the early growth of Malaysia. The industry has generated backward linkages, such as the light engineering industry, construction of tin dredges, dredge casting and so on, and forward linkages as well, especially through the smelting of tin. Substantial final-demand linkages, technological and employment externalities, have also been adduced. The retained value, the proportion of export income paid locally, which provides a summary of an export development contribution, was more than 70 per cent for tin in the early 1970s (Thoburn, 1977). The relatively higher level of development of the west coast states in the Peninsula has been attributed to the discovery and exploitation of tin. Roads and railways followed mine development, and settlements grew into the present-day towns. Small workshops and trading houses, set up to serve the needs of pioneer miners, were the forerunners of secondary industries and commercial enterprises. The minerals industry also contributed to the development of ports and power: 'The tin industry has been the most important factor in determining the historical growth trend and the growth pattern of the electricity supply industry' (Lim, 1967: 291).

To this date the tin industry continues to provide housing, education, electricity and water supply for its employees. The township of Sungei Lembing, located west of Kuantan, owes much of its development to the tin mining operations conducted since 1887 by the Pahang Consolidated Company Ltd. The company developed the road network and funded the settlement of the Sungei Lembing township with its population of 15 000, most of whom are directly or indirectly dependent upon the company's mining operations. The company built the township's housing, hospital and school. A more recent example is that of the Mamut copper mine in Sabah. The local township of Ranau was built around facilities provided by the Overseas Mineral Resources Development Company. The road between the mine and port facilities at Usukan was developed partly with company funds.

This pattern, particularly as regards the significant role of the early tin industry, has been repeated in other parts of Malaysia, development now being predicated on the discovery and exploitation of oil and gas. Bintulu and Miri in Sarawak, Kuala, Trengganu, Paka, Kemaman and Kerteh in Trengganu, and Labuan in Sabah, have all experienced rapid development, and will continue to do so in the near future. The development of ports such as those in Kota Kinabalu, Bintulu, Kuching and Tanjung Berhala (Trengganu), also stems from resource-based development in the surrounding regions. Forward linkages are represented by the refinery gas plant, steel mill, LPG terminal, and other industrial developments in Kuala Trengganu. However, backward linkages are few at present, capital and intermediate products being imported mainly from Japan and the United States. There has been some employment creation, but due to capital intensity and the high skill levels needed, this has been relatively low. Nevertheless, through training, these skills have been upgraded in the areas mentioned.

The role of the state vis-à-vis the industry has gradually changed.

With the implementation of the New Economic Policy, ownership structures have altered and state corporations have begun to play an important role in mineral exploitation and development. Such a trend is apparent in the operations of the MMC, Pernas–Mining, Petronas–Carigali, and State-level agencies.

At the national level, export earnings from minerals have been substantial. During the 1970s petroleum became the largest export earner, contributing significantly to Malaysia's rapid overall growth. Short-term to medium-term management problems ensued, with effects on the structure of the economy (see Tables 7.1 and 7.6). A decline in the relative role of the agricultural sector in particular has been evident, while the mining sector registered an 8.9 per cent growth for the 1976–80 period, arresting the earlier decline due to falling tin production. Also worthy of observation is the growth in non-tradeables due to the export boom, such structural changes generating real exchange-rate movements (Meyanathan, 1984). However, the weakening of the external sector over recent years has affected the domestic budget, the growth rate and the external debt. While the role of tin has declined, its place has been taken over to some extent by LNG since 1983, and minerals-based development can be expected to continue at least over the 1980s.

Notes

1 About 90 per cent of the gravel-pump mines in Perak and 65 per cent of those in Selangor are now working in mine tailings or areas which have been previously mined (Majid and Chue, 1977).

2 MMC and Kumpulan Perangsang Selangor (KPS), the mining arm of the Selangor State government, reached a joint-venture agreement to develop the 304 815 tonnes of possible deep alluvial reserves in Kuala Langat, Selangor, in 1980. The new mining company, Kuala Langat Mining Sendirian Berhad, will be established with KPS holding a 65 per cent interest in the US$91.87 million venture. Production is expected to begin in 1985 with over 2000 tonnes of concentrates in that year, reaching 6000–7000 tonnes per year in 1990 (Wu, 1980). A new tin company was formed in early 1980. The Modal Seri Pandan Sdn Bhd (MSPS), a US$7.3 million joint venture owned 30 per cent by Conzinc Rio–Tinto Malaysia and 70 per cent by the State government of Pahang, was expected to begin operations in 1982 on 3000 acres of alluvial plain at Sungei Pandan, Kuantan.

3 On alluvial deposits, tin ore is broken down by the action of water, and is found fairly close to the surface. On eluvial deposits, the tin deposit is decomposed but not transported by the action of water (Thoburn, 1977).

4 The mine was started with a capital investment of US$107 million.

5 Malaysia ranks 22nd among oil producing countries.

6 The LNG price is believed to be closely related to the average selling price of crude oil landed in Japan.

7 The company obtained a licence to expand production capacity by

another 1.5 million tonnes, but this project was shelved indefinitely in 1982.

8 A similar trend is noticeable in the case of large mining companies. MMC's total exploration expenditure for 1982 was US$1.8 million, and a further expenditure of US$21.7 million for 1983–87 was anticipated for the completion of existing exploration projects (MMC, 1983).

9 In the early 1970s approval was given to three companies, Conzinc Rio–Tinto Ltd, N.V. Billiton, and Ocean Mining Co. Ltd, to explore for tin in areas offshore from the west coast of Peninsular Malaysia (UNIDO, 1972).

10 The concentrate from underground mines has a lower purity level (about 70 per cent), and the tin ore is reduced to the necessary size by conventional crushing and grinding before the other stages, which are described in Figure 7.2.

11 Datuk Keramat also operates a small secondary lead smelting and refining plant for the supply of soft and antimonial lead to automobile battery manufacturers in Malaysia, and for the production of various grades of solder and other lead–tin alloys.

12 The smelters also have interests in tin mining companies.

13 This company was incorporated in 1961. It is a joint-venture company, with 51 per cent Malaysian and 49 per cent foreign investment. Of the foreign component, 39 per cent was accounted for by a consortium of Japanese companies, mainly Yawata Iron and Steel Ltd, Tokyo, which also designed and set up the fully integrated plant and provided technical assistance.

14 The balance of the output consisted of steel pipes and fittings, wire rods, galvanised sheets, iron castings, crude iron, and wire rods.

15 Comalco Ltd of Australia submitted a plan for the establishment of an alumina plant in Malaysia in 1974. Since then, however, no further developments have taken place.

16 A lubricant blending plant was added in 1971, and in 1974 a second crude distiller was installed. The latest project is to install a second platformer, which will increase the output of gasoline.

17 The demand for petroleum is estimated to increase by 10 per cent per annum between 1980 and 1990.

18 LPG would also be extracted from the plant when it is in full operation in the mid-1980s. Estimated supply is in the region of 300 000 cubic metres per year.

19 The revealed comparative advantage index for tin in 1979 was 1.28. See Anderson (1982) for calculation and discussion of the index.

20 The data base for consumption analysis is weak for minerals, and even weaker for metals and manufactures. Stock figures exist only for tin, iron and bauxite. Consumption data have been published only for tin metal.

21 This takes into account production, imports and exports.

22 This presumably constitutes pre-exports, since Japan has no iron ore resources.

23 Malaysia also imported substantial quantities of phosphate rock, urea, ammonium nitrate, ammonium sulphate and superphosphate from the Pacific region and Indonesia to meet its fertiliser needs.

24 Malaysia's trade intensity index with the ASEAN and Pacific Basin

countries was close to two in the case of fuels, minerals and metals for 1977 (Anderson, 1982).

25 Although other minerals, such as tungsten, coal, gold, manganese and iron ore, were mined in the early nineteenth century, it was tin that occupied the attention of the British (see Fermor, 1943; Li Dun Jen, 1982).

26 Mining leases in urban areas are generally limited to a period of about three to five years. Leases granted to areas in the hinterland may extend for periods of fifteen to twenty years. In July 1983 Rahman Hydraulic Tin Berhad obtained a fifteen-year lease from the Perak State. However, this was made possible through a 20 per cent holding in the company by the State Economic Development Corporation. State governments may also be favourably disposed when dealing with the Federal-government-owned state enterprises, MMC and Pernas Mining.

27 The export duty for iron ore and all other metals and metalliferous ores stands at 10 per cent ad valorem (Department of Mines, 1978). A research cess of 78 US cents per kilogram of tin is also imposed.

28 This cost of production has subsequently been raised twice, first to US$10.04 per kilogram, and more recently to US$11.48 per kilogram.

29 This tax was first levied in 1965 to help to finance increased defence expenditures due to 'confrontation' with Indonesia, and has continued ever since. Losses in any one year are not deductible against profits in another year. Certain costs, such as financing charges, may be deducted for corporate tax purposes, and are project/mine-specific.

30 However, a rebate on the export duty is provided in the case of Pahang Consolidated Co. Ltd, due to the lower assay content of its concentrates. The actual assay content of this, the only underground lode tin mine in Malaysia, is about 67 per cent, rather than the 75 per cent generally applicable to the tin industry (States of Malaya, Chamber of Mines, 1981). The imposition of export duty is of long standing. As early as 1877 a duty was levied on the export of tin in Selangor, a practice followed not long afterwards by the neighbouring States. In 1903 a prohibitive duty on tin ore was imposed, mainly to prevent the Americans from building smelting factories outside Malaya (Li Dun Jen, 1982).

31 The industry has called for a depletion allowance which, as practised in other countries, would reflect the depletion of ore reserves. The industry also holds the view that there is a strong case for levying tax on net rather than gross mining income. It would prefer a system of taxation which dispenses with export duty, and which is based entirely on profitability (States of Malaya, Chamber of Mines, 1981).

32 The machinery for implementing production cutbacks exists in the form of the 1954 Tin Control Ordinance, under which the Department of Mines reviews individual operators' export quotas every three months.

33 The agreement includes economic provisions, the details of which have not been spelt out, and voting rights and financial contributions, both of which will be proportional to each member's production (MMC, 1983).

34 Due to shortages of raw materials, especially steel bars and steel scrap for conversion into steel ingots, a ban on the exportation of steel scrap was enforced in 1981. In 1974 the effective rate of protection for basic iron and steel products was 139 per cent, while for structural and metal products in that year it was 166 per cent and 140 per cent, respectively (Fong et. al., 1982).

35 Six foreign companies, including the French state-owned company Elf Aquitaine, pulled out of Malaysia after failing to reach agreement on production-sharing terms. However, Elf returned in 1982. Two Amendments were made to the PDA, in 1975 and in 1977: the first ensured more explicit Bumiputra control of the industry; and the second made provision for the payment of compensation to any oil company whose rights under the former concession system had been abrogated by the introduction of the 1974 Act.

36 The following guidelines govern the issue of licences: incorporation of a Malaysian company with capital of not less than US$43 478; local participation of up to 70 per cent, with Bumiputra holdings of not less than 30 per cent; and observation of government policy with regard to management, employment, and utilisation of local resources.

37 In the case of the Elf contract, the following changes were made: for the first 16 000 kilolitres of oil production and the first 20 million cubic metres of gas sales, the split is 70:30; for the subsequent 16 000 kilolitres and 20 million cubic metres, the split is 75:25; and for volumes in excess of 32 000 kilolitres and 40 million cubic metres, the split is 80:20.

38 Petronas is also subject to corporate taxes, and to export duty on any crude oil exported.

39 These include Sarawak Shell Bhd, Sabah Shell Bhd, Esso Production Malaysia, Petronas–Carigali/BP/Oceanic, and Elf Aquitaine.

40 These nine fields accounted for 72 per cent of the oil initially in place in 1980.

41 From the statement, it appears that the rate of oil production might emerge as a policy variable, given the trend export earnings from other commodities.

References

Anderson, K. (1982) 'Intensity of Trade Between Pacific Basin Countries' Paper presented at the Second Western Pacific Food Trade Workshop, Jakarta (August)

Bosson, R. and B. Varon (1977) *The Mining Industry and the Developing Countries* New York: Oxford University Press

Datuk Keramat Holdings Berhad (1981) *Annual Report 1980* Kuala Lumpur

Department of Mines (Malaysia) (1978) *Bulletin Relating to the Mining Industry of Malaysia* Kuala Lumpur

Energy Work Group (1982) *Malaysia's Energy Outlook* Kuala Lumpur: Malaysian International Chamber of Commerce and Industry

ESCAP (1983) *Transnational Corporations and the Distribution of Gains in the Bauxite/Alumina Industry of Malaysia* Bangkok (April)

Fermor, L.L. (1943) *Report Upon the Mining Industry of Malaya* (3rd impression) Kuala Lumpur

Fong, C.O. et. al. (1982) *Comparative Advantage of Manufacturing Industries in Malaysia* Tokyo: Institute of Developing Economies

Lam, N.V. (1978) 'Incidence of Tin Export Taxation in West Malaysia' *The Developing Economies* XVI, 4 (December)

Li Dun Jen (1982) *British Malaya, An Economic Analysis* (2nd edn) Kuala Lumpur: Institute for Social Analysis

Lim Chong Yah (1967) *Economic Development of Modern Malaya* Kuala Lumpur: Oxford University Press

Majid, S.A. and H.C. Chue (1977) 'A Review on the Development and Problems of the Mining Industry in Malaysia' Paper presented at the National Seminar on the Mining Industry, Kuala Lumpur

Malaysian Industrial Development Authority (1981) *Annual Report 1980* Kuala Lumpur

Meyanathan, S. (1984) 'Energy and Adjustment: Indonesia and Malaysia' in R.M. Bautista and S. Naya (eds) *Energy and Structural Change in the Asia–Pacific Region* Manila: Asian Development Bank

Meyanathan, S. and G. Sivalingam (1983) 'Malaysian Economic Development and Food Trade' Paper prepared for the ASEAN–Australia Food project, Kuala Lumpur (August)

Meyanathan, S. and R.J.G. Wells (1982) 'Petroleum Products Pricing in Malaysia' *Kajian Ekonomi Malaysia* XIX, 1 (June)

Ministry of Energy, Telecoms and Post (Malaysia) (1982) *An Introduction to Malaysia's Energy Sector* Kuala Lumpur

MMC (Malaysia Mining Corporation) (various years) *Annual Report*

National Printing Department (Malaysia) (1981) *Fourth Malaysia Plan, 1981–1985* Kuala Lumpur

—— (1984) *Mid-Term Review of the Fourth Malaysia Plan* Kuala Lumpur

States of Malaya, Chamber of Mines (1982) *Annual Report, 1981* Kuala Lumpur

Thoburn, J.T. (1977) *Primary Commodity Exports and Economic Development: Theory, Evidence and a Study of Malaysia* London: John Wiley & Sons

UNIDO (1972) *Non-Ferrous Metals: A Survey of Their Production and Potential in the Developing Countries* New York: UNIDO

Wu, J.C. (1981) 'The Mineral Industry of Malaysia' in *Mineral Yearbook, 1980* (Vol. III) US Bureau of Mines, US Department of the Interior

Figure 7.1

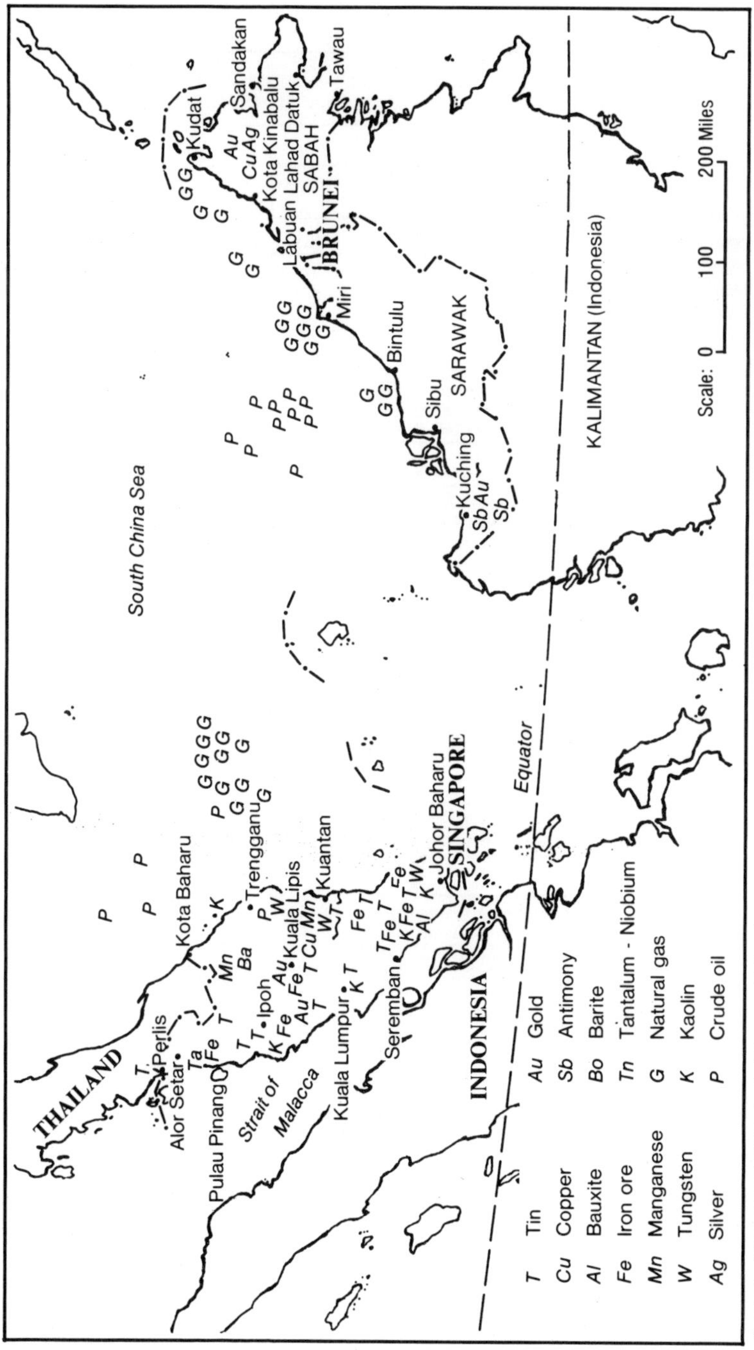

Source: Department of Geological Survey, Malaysia.

Figure 7.2

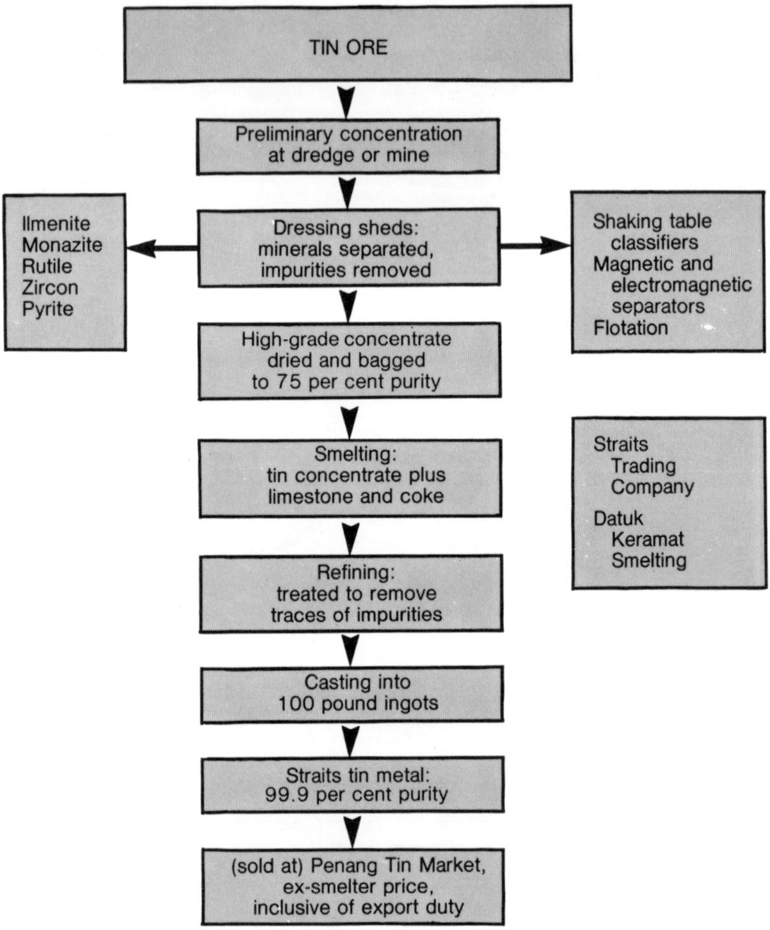

Source: Author's compilation.

Figure 7.3

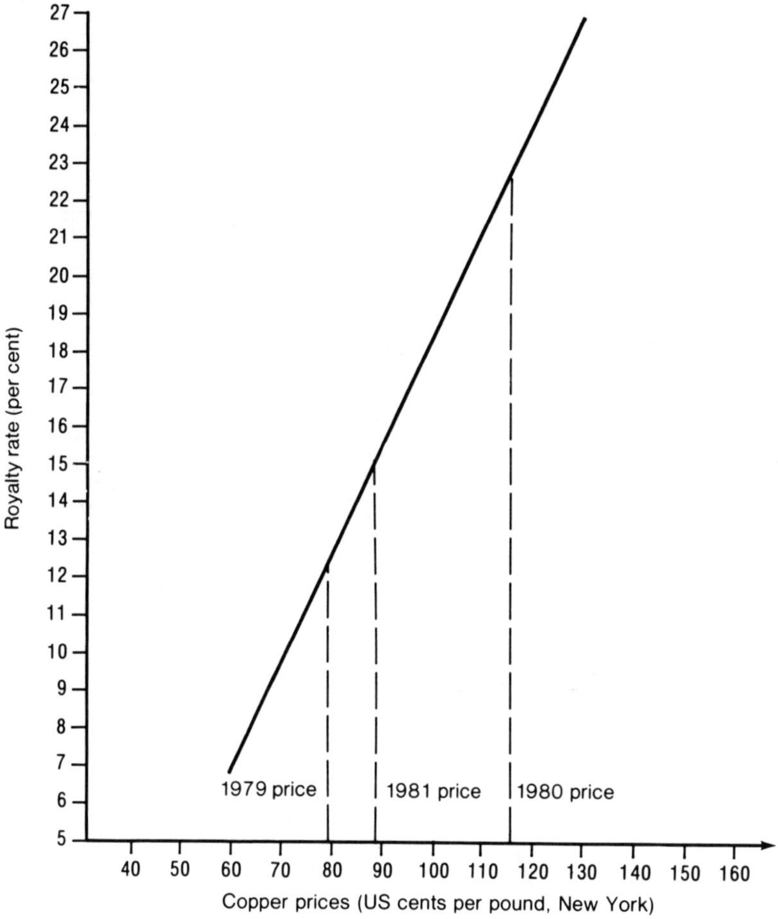

Source: Derived by the author from the formula [(P−36) divided by 3.5)] ad valorem, where P is the copper metal price in US cents per pound.

Figure 7.4

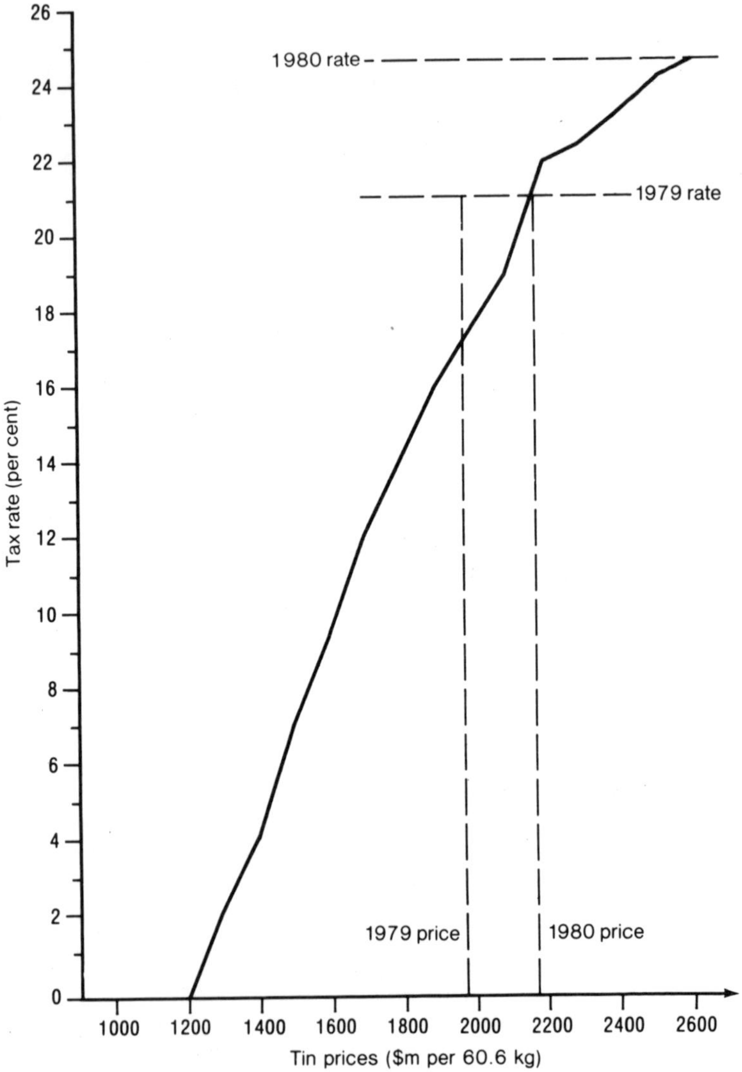

Source: Derived from Ministry of Information (Malaysia), *The 1980 Budget* Kuala Lumpur, October 1979.

Figure 7.5

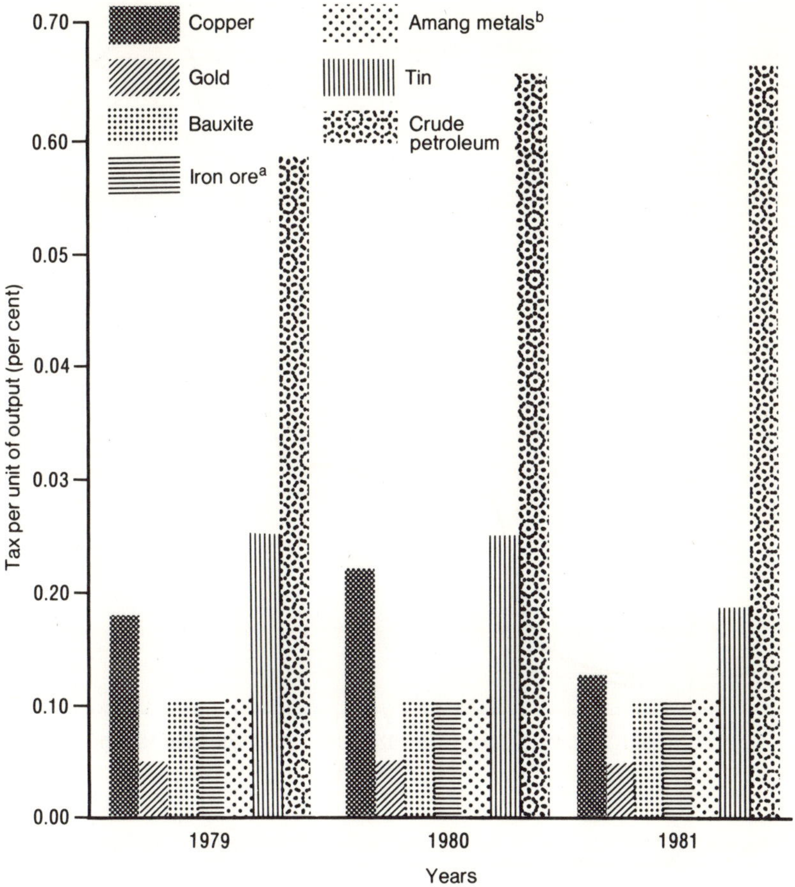

Notes: Taxes exclude research cess and corporate taxes.
a includes bauxite, manganese, wolfram and columbite
b includes struverite, mozanite, tin slag and zircon

Source: Author's estimates

Figure 7.6

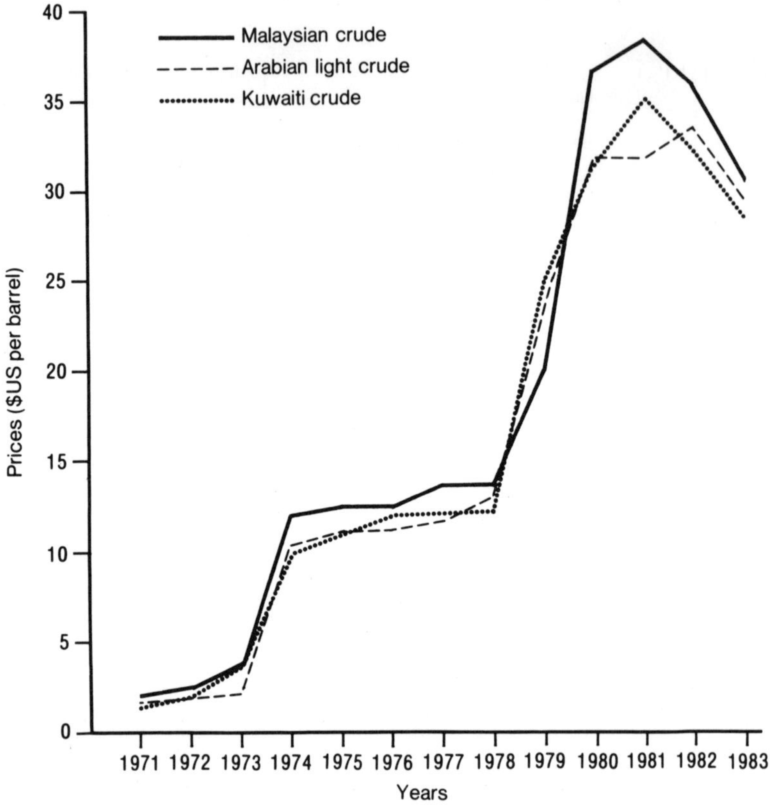

Sources: *Petroleum Economist* (various issues); Ministry of Finance (Malaysia) *Econcomic Reports* (various issues)

8 The Philippine minerals industry, development and trade

TEODORO M. SANTOS

THE possession of abundant natural resources might be expected to give a country a decided advantage in the pursuit of its social and economic development goals. Such a proposition is anchored in two important economic principles: first, that a country can produce only those goods which can be manufactured from its own resources (the principle of absolute advantage); second, that a country which can produce a good relatively more cheaply than others will specialise in producing such a good (the principle of comparative advantage), so that the possession of abundant natural resources will lead a country to concentrate on producing natural-resource-based commodities. It follows that one might expect a natural-resource-rich country, other things being equal, to develop more rapidly than one with few or no natural resources.

However, in the light of the rapid development over the past two decades in resource-poor countries such as Japan, Korea, Taiwan and Singapore, and the relatively slower growth rate of neighbouring countries such as Indonesia, Malaysia, the Philippines and Thailand, which are relatively rich in natural resources, the importance of natural resources in development has been called into question (Lim and Wong, chapter 9 of this volume). Moreover, the experience of a minerals-poor country like Singapore, which has derived much of its economic growth from activities related to minerals, such as tin smelting and petroleum refining, casts doubt on the comparative advantage conferred by the possession of mineral resources. Nevertheless, the importance of natural resources, particularly mineral resources, to the development of mineral-rich countries like the United States, Canada and Australia must be acknowledged.

In attempting to explain the growth of the resource-poor countries mentioned, two additional assumptions may be necessary. First, the role of natural resources in the development of a country depends on that country's present state of development. Early in development, when the manufacturing and service sectors are nonexistent and the dominant agricultural sector is marginal, providing no surplus for growth, mining and related natural-resource industries can provide the key to the surplus earnings necessary to finance development. With more intensive development, however, the country's manufacturing and service sectors become larger and more important than the primary sector, and thus more significant as revenue earners. In this case the importance of natural resources, the basis of the primary sector, may be reduced. In the context of the theory of comparative advantage, as a country develops, man-made resources become relatively more abundant (and hence cheaper) than natural resources, and more important to its further development.

Second, the transformation of a mineral resource into useful products involves a series of steps, each requiring a different balance of resource inputs, capital, labour and technology. For instance, before copper wire can be produced the following steps are necessary: exploration for the copper deposit; development of the deposit into reserves; mining the ore; milling and beneficiating the ore into a form ready for extracting the metal; smelting the ore to obtain crude metal such as blister or matte; refining the crude metal into ingots, slabs and related forms; fabricating the refined metal into forms ready for use or for making consumer goods such as wire, tubes and sheets; producing consumer goods from the semifabricated products; and finally, transporting and marketing the product (Santos, 1982a). Only the first three steps require mineral resources endowment. The remainder can be undertaken by any country with the requisite capital, technology, skilled labour, and ability to secure ores and concentrates from elsewhere.

Nevertheless, it remains true that even if a mineral-poor country can engage in minerals processing a mineral-rich country has greater freedom to select the stages in which it wishes to participate. Particularly given this freedom, but even without it, it is essential for any country set on the path of economic development to understand the potential role of minerals in that development.

This chapter has a dual focus. First, it describes the Philippine minerals industry in terms of resources, production, exports and imports, with emphasis on the potential for trade with Australia and other ASEAN countries. Second, it examines important policies or issues affecting minerals industries. The two-part structure of the paper reflects these twin objectives. The review of the minerals industry in the first part centres on the years 1970 to 1980, although other years are considered where relevant.

Development of the minerals industry

History and significance

Mining in the Philippines dates from long before European settlement. Gold was the main metal mined before the arrival of the Spaniards, and was used mainly for making earrings, bracelets and other ornaments. It was also used as currency in trade with Chinese and other foreign merchants. After settlement, one of the Spaniards' first actions was to explore for the sources of gold then in currency. Their search brought them to the premier gold districts of Benguet and Paracale, worked by the natives, then as now, using crude technologies.

Systematic mining began with the American regime in the early 1900s, when gold mills and cyanidisation were introduced, along with American capital. The Chinese and Spanish revolutions of the 1930s caused an inflow of surplus capital to the industry, which, together with the devaluation of the dollar from US$20.67 per ounce to US$35 per ounce in 1933[1], ushered in a gold mining boom in the late 1930s and early 1940s which has never been surpassed or equalled.

Before the Pacific War in 1942, iron, chromite and copper were also mined in modest amounts, with extraction increasing substantially immediately before 1942 due to growth in demand engendered by war preparations.

Most mines were devastated during the Pacific War, and production was minimal during the period of Japanese occupation. Massive reconstruction was needed after 1945[2] to restore even a part of prewar production.

The Vietnam War and the rapid growth of the Japanese economy in the 1960s and 1970s provided the main stimuli for the development of the copper industry, with increased demand and high prices in these years. These events established the Philippines as a leading copper producer. After the Pacific War the Philippines also became a leading producer of chromite ores, particularly refractory chromite. In 1975, the Philippines also became a major producer of refined nickel. However, at about this time the once flourishing iron ore industry finally died due to a government ban on beach sand mining.

By 1980 the Philippines was a leading producer of copper, nickel, gold and chromite. Among world producers, it ranked fourth in chromite, sixth in nickel, eighth in gold, and ninth in copper.

Although some level of mineral processing has existed for many years, it is quite small in comparison with crude mineral production. Only nickel and primary gold were produced, smelted and refined domestically as of 1983. Metals like aluminium, zinc, lead, tin and iron and steel are transformed into semifinished and finished goods, mainly from imported scrap, ingots and semifinished products.

The minerals industry is seen by the government as playing an important role in the country's socioeconomic development.

Traditionally, crude mineral production has been one of the most important sources of foreign exchange, vital for financing the country's development programs. In 1983 crude minerals accounted for 6.5 per cent of the value of the country's total exports.[3] In addition, domestically produced gold is bought by the Central Bank to form part of the government's gold reserves.

However, since the Philippines produces little crude oil, the costs of petroleum imports have negated the foreign exchange earnings from crude minerals.[4] In 1972 crude mineral exports accounted for 21 per cent of total exports, while petroleum imports appropriated 13 per cent of export earnings. However, by 1981 crude minerals contributed only 16 per cent to total export earnings, whereas the payment for petroleum imports took 43 per cent of the country's total export earnings. The country's oil bill, particularly since 1974, has eaten up a large chunk of its foreign exchange earnings, a heavy burden indeed to its development thrust.

The minerals industry is also seen as important in the country's industrialisation program. To provide the needed infrastructure, the government has proposed twelve major industrial projects. Of these twelve, six will require crude and semiprocessed minerals as inputs (a copper smelter, a phosphatic fertiliser, the cement industry, an integrated steel mill, and petrochemical and aluminium smelters), two will need processed minerals as inputs (diesel engine, heavy engineering), and the rest can be thought of as dealing in mineral substitutes (alcogas, pulp and paper, and the coconut industry). These mineral-related industries are seen as likely to generate numerous smaller scale industries, particularly secondary and tertiary, in the countryside.[5]

Mineral resources[6] may be classified as identified reserves, hypothetical resources, and speculative resources. These are briefly defined as follows.

Identified reserves: estimates of known mineral deposits, evaluated as to extent and grade, and whose contained minerals may or may not be economic to extract under prevailing technology and economic conditions.

Hypothetical resources: mineral deposits, whether economic or uneconomic to extract, whose existence is predicted on the basis of geological knowledge, in known mineral districts.

Speculative resources: both economic and uneconomic material deposits whose existence is predicted on the basis of geological knowledge, in unknown mineral districts or in unconventional form.

The data presented in this chapter are derived mainly from the Philippine Bureau of Mines and Geosciences (Bacuta, 1980). Data on identified reserves come from the files of private companies and the Philippine Bureau of Mines. Data on hypothetical and speculative resources are derived from the work of a team of government and

Table 8.1 Distribution of copper reserves in the Philippines (million tonnes)

Location	Tonnage
Negros Island	920
Cebu Island	938
Baguio District, Luzon Island	591
Marinduque Island	360
Davao del Norte, Mindanao	289
Batangas, Luzon Island	112
Kalinga Apayao, Luzon	107
Total	3317

industry analysts and academics—that is, they are intelligent guesses rather than empirical measures.[7]

Copper resources

Copper deposits occur mainly as low-grade disseminated bodies with some smaller massive sulphide deposits. Appendix Table 8.1.2 shows reserves in 1978 at 3657 million tonnes, with a weighted average of 0.48 per cent copper. Hypothetical and speculative resources have been estimated at 30 000 million tonnes. More than 90 per cent of known copper reserves are concentrated in seven mineral districts (Table 8.1).

Gold resources

Gold resources are of two kinds: those which are mined for gold as the major product, called primary gold, and those which are extracted as copper byproduct, classified as secondary or byproduct gold. Reserves of primary gold are estimated at 169 million tonnes, with an average gold content of 2.1 grams per tonne, including the placer deposits containing 0.24 grams per tonne. The lode deposits alone, of which there are 66 million tonnes, have an average grade of 4.97 grams per tonne. Hypothetical and speculative resources are estimated at 475 million tonnes. Byproduct gold reserves are estimated at 835 million tonnes with an average gold content of 0.74 grams per tonne. Hypothetical and speculative resources are estimated at 4050 million tonnes (Appendix Table 8.1.2). Since byproduct gold occurs with copper, its distribution is similar to that of copper.

The distribution of gold reserves in the five major gold districts of the country is shown in Table 8.2.

Chromite resources

Chromite may be classified, according to use, into refractory, metallurgical and chemical reserves. These deposits are associated with ultramafic rock sites. Of known chromite reserves, 59.4 per cent are metallurgical and 40.6 per cent are refractory. Hypothetical and speculative resources are estimated at 121.5 million tonnes, of which 45 per cent are refractory and the rest metallurgical. Although there are

Table 8.2 Distribution of hypothetical/speculative primary gold reserves in the Philippines (million tonnes)

Location	Tonnage
North and Central Luzon	174
Camarines Region, Luzon	84
Visayas/Mindoro	110
Mindanao	107
Total	475

Table 8.3 Distribution of chromite reserves in the Philippines (million tonnes)

Region	Tonnage
Zambales	10.95
Palawan	5.43
Surigao del Norte	0.24
Misamis Oriental	0.25
Camarines, Mindoro, Samar	0.25
Total	17.12

known chemical-grade deposits, they are not yet officially reported. Chromite reserves are distributed among six mineral districts, as shown in Table 8.3.

Other metallic resources

Nickel resources are of two kinds, nickel sulphides and nickel laterites. However, only small amounts of nickel sulphide resources (4.05 million tonnes containing 0.29 per cent nickel) are known. Most common are the laterites. Identified nickel reserves in laterites are estimated at 1366 million tonnes with an average grade of 1.08 per cent nickel. Hypothetical and speculative nickel resources are estimated at 4500 million tonnes containing 0.23–2.4 per cent nickel. Major nickel resources are located at Surigao del Norte, Palawan, Davao Oriental, Samar, and Zambales.

Other mineral resources, like lead, zinc, cadmium and manganese, are too insignificant in tonnage or grade to warrant special discussion. Although iron ore reserves and resources appear large, about 35 per cent of nickel laterite resources being iron, they are currently not economic. Only the lump iron ores show some economic promise in the near future. There are also more than 100 million tonnes of bauxite deposits, probably economic, which are not reported in Appendix Table 8.1.2.

Fuel minerals

Known geothermal and hydroelectric resources, as well as coal and petroleum, are discussed here.

In 1981 proven and potential (hypothetical and speculative) coal reserves were estimated at 220 million tonnes and 1750 million tonnes respectively.[8] However, only very limited petroleum reserves—6 million tonnes[9]—were estimated in 1980. As for geothermal energy, by 1981 installed capacity already stood at 446 megawatts, with 1700 megawatts of proven geothermal power capacity, and resources are still being developed. Installed hydroelectric capacity in 1981 was 944 megawatts.

Of the 4666 megawatts of power capacity in 1981, 20.2 per cent was accounted for by hydropower, 9.6 per cent by geothermal, 2.1 per cent by coal, 45.1 per cent by oil thermal and 23 per cent by diesel. It is clear that petroleum products remain the main source of electricity in the country, although geothermal, hydroelectric and coal resources are gradually playing a greater role.

Crude mineral production

Mining in the Philippines is generally associated with the production of hard mineral ores and concentrates, although recently some metal smelting and refining has begun. Copper, gold, nickel and chromite have been the major mineral products, with iron ore, lead and zinc concentrates also important. From 1980 crude oil and coal also became significant to the minerals sector.

Appendix Table 8.1.3 summarises the volume and value of crude mineral production for 1970, 1975, 1980 and 1983. In 1980 the total value of production, excluding nonmetallics, was US$1658 million, contributed mainly by copper (35 per cent), gold (22.1 per cent), crude oil (22.1 per cent), nickel (11.4 per cent), and chromite (2.2 per cent).

Copper is generally produced in the form of concentrates containing 26–30 per cent copper. Gold, silver, and in some cases molybdenum, are associated with the copper, enhancing the per unit value of its concentrates. Most of the important mines are operated as open pits, associated with block caving, except for Lepanto, which is an underground mine.

Gold is the Philippines' second most important metal product. Primary gold is smelted domestically, refined into bullions in the National Gold Refinery, and then sold to the Central Bank. Secondary gold, associated intimately with copper, can be recovered only during copper refining, which is carried out abroad. The major mines are underground, although there are two open-pit mines. In addition, there are many small gold mining operations, generally unregistered, which are difficult to monitor.

Nickel products are of two types, direct smelting ores or concentrates. The concentrates, which account for a small fraction of nickel output, consist mainly of sulphides. The bulk of the nickel ore mined occurs as laterites, and open-pit methods are employed. A small percentage of laterite is exported, but more than 90 per cent is smelted and refined locally by Marinduque Mining. Nickel ores mined and smelted by Marinduque Mining are reported as refined nickel.

Most chromite products are classed as refractory or metallurgical. Metallurgical chromites are those that are suitable for extracting chromium metal or for making alloys. Refractory chromites are used for making high-temperature bricks for furnace lining. The third category, chemical-grade chromite, which is suitable for making chemical products, may form part of the product mixture in the near future.

Two other metals which are currently important to the mining industry are cobalt and silver. Cobalt is a byproduct of nickel smelting and refining, and silver is a byproduct of copper and gold production, although only silver associated with primary gold is recovered by local refining.

Zinc, lead and iron ore contribute relatively little to the total value of mining production, although iron ore was a relatively important mineral product until 1975, when the government imposed a ban on beach sand mining.

Petroleum production started in 1979. Although its value appears impressive, there is considerable doubt about whether these levels can be sustained, since production is limited to small reef reservoirs offshore. On the other hand, although both the value and volume of coal output in 1980 were low, there is a very sound basis for future expansion. The main problem is that most Philippine coal has low calorific value, of 4500–6100 Kcal/kg, being lignitic to sub-bituminous.[10]

Processed mineral production

Although it is not generally realised, minerals processing constitutes a large segment of the economy. The minerals processed may be classified as nonmetallics, metallics and fuels.

Cement and fertiliser are the most important nonmetallic products. Cement is used mainly in the construction industry, and some 4777 thousand tonnes were produced in 1980. About 300 thousand tonnes of fertilisers were produced in the same year, consisting of nitrogen, phosphorus, potassium and related products. Fertiliser production was designed to complement the country's food production program.[11]

The metallic processing industry includes both iron and steel and nonferrous metal processing, and involves those processes and activities that lead to the transformation of metals into useful end products after smelting and refining. Thus this chapter omits smelted and refined gold and silver, which remain as crude mineral products.

The iron and steel industry began in 1949 under the government-run National Development Corporation. It has since become a large and dynamic industry, consisting in 1979 of sixty firms with an aggregate annual capacity of 3.9 million tonnes. In that year it employed 12 400 persons, and produced 900 000 tonnes of products valued at US$500 million. The industry produces billets and ingots, as well as flat and nonflat products used in construction, shipbuilding, automotive manufacture, appliances, packaging, and fabrication. Demand has grown steadily, at a rate of 8 per cent per year.

The nonferrous metal industry involves all activities associated with the production of useful goods, from refined nonferrous metals to semifinished and finished goods. In 1975, the industry employed 6477 people in 111 firms with total assets worth US$121.3 million and sales of US$130.7 million. Roughly 42 per cent of these firms are jobbing operations, producing on the basis of specific job orders. The remaining 58 per cent are independent, buying their own raw materials and deciding independently on their product ranges.

Aluminium, copper, lead, zinc, nickel and tin are the principal metals in the industry. Zinc is used mainly as a protective coating for steel in galvanised iron sheets, with secondary uses in die casting, brass manufacture and paint making. Tin is used principally as a protective coating for steel in tinplates, and secondarily in the manufacture of soft solders (Pb–Sn alloys), brass and bronze. Lead is used in making storage batteries, and in solders and babbitt. Nickel is used to produce stainless steel, as well as in electroplating. No local production of semifinished or finished nickel products existed until 1975. Copper and aluminium are used in the manufacture of a large array of semifabricated and fabricated products, including wires, plates, sheets, foil, tube bronze, brass, bars, rods, wrappings, automotive bodies, cigarette packing and collapsible tubes.

The raw materials for the metallic processing industries are imported bars, ingots, and semifinished products, supplemented by domestically produced scraps. The products of metallic processing are used in the metalworking industry, which transforms semifinished metal products into end products such as machinery and equipment. In 1978 there were 1621 metalworking firms, 71 per cent small in scale, 12 per cent medium-sized, and 17 per cent relatively large (MIRDC, 1979).

The value of production of processed or fabricated metals in 1978 amounted to US$1.4 billion, augmented by imports valued at US$1.8 billion to fill the total consumption requirements.

The processing of fuel and energy minerals includes crude oil refining and electricity generation from various sources, which represent perhaps the largest processing industries in the Philippines. Total refining capacity amounts to around 200 000 barrels per day, distributed among four plants. Together, these plants transform annually some 13 million tonnes of crude oil, worth more than US$2.0 billion, into usable petroleum products,[12] the most important being fuel oil, motor gas, kerosene and liquefied petroleum gas (LPG).

From 1981 the Philippines has an installed power-generating capacity of 4666 megawatts (MW), which in that year generated 19 824 gigawatt hours (GWh), equivalent to 34 million barrels of oil equivalent.

The sources of energy for electricity generation were as follows: oil/diesel (46.6 per cent); hydropower (18.9 per cent); geothermal power (14.0 per cent); and coal (2.1 per cent); the shares of coal and geothermal energy are expected to increase in the near future.

Energy use, which amounted to 86.1 million barrels of oil equivalent in 1981, was distributed to four sectors of the economy as follows: industrial use (49.7 per cent); residential use (12.22 per cent); commercial use (11.23 per cent); and transportation (26.85 per cent).

Expansion plans

Projects designed to effect a major change in the country's industrial structure are expected to come into being in the 1980s, the government's attention and resources having been previously channelled into the development of agriculture and vital infrastructures such as irrigation schemes, roads, bridges, power networks and schools. The program of industrial development will involve seven main areas of activity, varying from accelerated implementation of the twelve major industrial projects, through industry dispersal and promotion of small and medium-scale industries, to export promotion.[13] The focus here is on the major industrial projects because of their importance to the industrial infrastructure and their likely impact on the Philippines' mineral production and trade.

Appendix Table 8.1.5 summarises these twelve projects, including the copper fabrication project which is the Philippines' contribution to the ASEAN joint industrial projects. Another project of some significance to the minerals sector (but not included in the table) is a 620 MW nuclear power plant, planned to be operational in 1984–85, for which uranium (yellowcake) will be imported, very likely from Australia. Finally, it is worth noting that a domestic firm is expected to produce ferrochrome metal, while another has planned to manufacture refractory bricks from chromite for lining cement, steel and other furnaces. A careful examination of these projects gives rise to several observations on their likely influence on the minerals sector.

1 Of the twelve projects, seven will require either crude or semiprocessed minerals as direct inputs, two will require semifabricated or fabricated metals, and three will deal with the production of mineral product substitutes.
2 Substantial amounts of crude minerals in the forms of copper concentrates, bauxite, iron ores and concentrates, coal, fertiliser and cement raw materials are needed for smelting and refining, and for the fabrication of semifinished and finished products. These raw materials must be either produced locally or imported.
3 These projects are designed either to produce raw materials for further manufacturing into end products, or to produce final products such as machinery and equipment.

4 Coal demand by 1985 is projected to reach 5.6 million tonnes, used in the following ways: power (20 per cent); steel plant (19 per cent); mining, cement and fertiliser (55 per cent); and others (6 per cent).
5 Considerable energy will be generated from indigenous alternative energy sources such as alcohol (343 000 barrels of oil equivalent—boe), coconut oil (35 000 boe) and biogas (27 000 boe).[14]
6 The industrialisation thrust is essentially mineral-resource-based, and in the future the Philippines are likely to produce and export more processed minerals and import greater amounts of crude minerals than previously.
7 Crude minerals likely to be imported in larger amounts are iron ore, crude oil, crude fertiliser, uranium, bauxite and coal.
8 Minerals to be exported in more processed forms are copper, aluminium, fertiliser and chromite.

Minerals trade

In this section we consider the following market-related topics: exports, imports and potential for trade.

MINERAL EXPORTS

Crude minerals are generally exported to industrial countries, where smelting and refining facilities are located. As shown in Appendix Table 8.1.6, crude mineral exports have grown from US$214 million in 1970 to US$1142 million in 1980—an 18 per cent annual increase. During this period, the contribution of crude minerals to total export earnings fluctuated between 15 per cent and 20 per cent.

Changes in the pattern of iron ore exports require some explanation. Before 1975, iron ore was exported in the form of lump ores, pellets, and concentrates derived from beach sands. The closure of the only lump ore and pellet producer (the firm decided to shift to copper production) and the subsequent ban on beach sand mining in 1976 dramatically reduced domestic iron ore production. The high volume of exports in 1980 and subsequently is attributable to export of sintered products[15] made from imported iron ores.

Appendix Table 8.1.7 shows that the pattern of exports of processed minerals is dominated by refined nickel, iron and steel, cement and petroleum products. The value rose from US$38.36 million in 1970 to US$264 million in 1980, a 21 per cent annual increase. The share of processed mineral products in total exports fluctuated between 2.5 per cent and 5.0 per cent.

MINERAL IMPORTS

Appendix Table 8.1.8 shows the pattern of crude mineral imports, the value of which increased from US$106 million in 1970 to US$2724 million in 1980, a 37.9 per cent annual increase. The share of total imports likewise increased consistently, from 9.2 per cent in 1970 to 35.3 per cent in 1980.

The pattern of processed mineral imports is shown in Appendix Table 8.1.9. The value grew from US$172 million in 1970 to US$1035 million in 1980, an annual growth rate of 19.7 per cent. The bulk of metallic mineral imports is made up of refined metals, scrap, and semifinished metals used in metal fabricating and metalworking industries. A considerable volume of manufactured metal goods enters the country as equipment and machinery.

The potential for trade with ASEAN and Australia

An assessment of the Philippine potential for mineral trade with Australia and other ASEAN countries depends upon the answers to two questions: Does the country export mineral commodities which Australia and ASEAN import, and does the country import mineral related goods which Australian and other ASEAN countries export? Appendix Tables 8.1.10 and 8.1.11 answer these two questions, and the following inferences can be drawn.

First, as to exports, despite the Philippines' strong exports of crude mineral products such as the ores and concentrates of copper, chromite, lead and zinc, none has been exported to Australia or other ASEAN countries. Australia produces most of these minerals in large quantities, except for chromite, while other ASEAN countries do not have the requisite smelting and refining capacities, a highly capital-and technology-intensive stage in minerals production. However, the Philippines does export refined, semiprocessed and processed mineral products which Australia or other ASEAN countries import. In particular, the data indicate that the Philippines has good prospects for exporting nickel, copper and aluminium products to Australia and other ASEAN countries. This observation is particularly important, given the existence of large copper and nickel smelters and refineries in the Philippines and the expected expansion of the demand for refined and semifinished metals as the ASEAN countries further industrialise.

As for imports, the Philippines can import substantial amounts of crude oil and crude fertiliser from other ASEAN countries, and considerable amounts of iron ore from Australia. The shift to the use of coal for the cement and mining industries in the Philippines is expected to enlarge coal imports rather than reduce imports of other energy sources. As regards processed minerals, the Philippines can import from ASEAN substantial amounts of petroleum products, tin, lead, iron and steel, and from Australia copper, iron and steel, and aluminium.

In short, it is clear that the Philippines cannot hope to export crude mineral products to ASEAN or Australia in the near future, but that it has good prospects in exporting refined, semifabricated and fabricated metals such as copper, aluminium and nickel. On the other hand, the Philippines may well import from ASEAN and Australia crude oil, crude fertiliser and iron ore, as well as processed mineral products such as petroleum products, tin, aluminium, iron and steel, lead, copper and zinc.

Mineral issues and policies

Three main issues require discussion: the role of mineral resources in the country's industrialisation, growth or development; energy development policy; and mineral ownership, investment and taxes.

The role of minerals in Philippine development

Does the minerals industry contribute to the country's development? If so, in what ways does it help in development?

The main contributions of the minerals industry may be seen to derive from the production of crude minerals, mainly ores and concentrates, and the processing of crude minerals into more valuable products, such as refined metals, semifinished products and fabricated end products. Critics of the minerals industry assert that mining does not contribute to development because mines operate as isolated enclaves and develop no linkages with other sectors, and that the minerals industry is highly capital-intensive and consequently benefits only a few—particularly the rich, and the large multinational companies.

In order to examine these arguments, this chapter will consider separately the impacts of minerals production and processing on the Philippine economy.

CRUDE MINERAL PRODUCTION

The mining industry's most universally acknowledged contribution is foreign exchange earnings. This contribution becomes particularly important when one considers the increasing perennial balance-of-trade deficit (Appendix Table 8.1.1). While exports of hard mineral products contribute substantially to export revenues, the importation of crude oil necessitated by limited domestic resources causes a drain on foreign exchange, which in turn impedes the course of economic development. Certain critics have proposed that the outflow of foreign exchange in the course of developing the mining sector perhaps exceeds the inflow. However, independent studies (for example, Radke et. al., 1980) indicate a more substantial net dollar inflow from mining activity.

Other commonly recognised contributions of mining to the economy include tax revenues and employment opportunities. Numbers employed in mining increased from 53 000 in 1970 to 95 000 in 1981, and annual tax revenues derived from mining grew from US$15.6 million in 1970 to US$149.6 million in 1981 (NEDA, 1982).

Although the number of persons directly employed in mining is not very impressive compared with total investment, it must be recognised that for every person so employed, three to five jobs are created in other sectors (Toombs and Andrews, 1976; Smith, 1978). More significantly, these jobs are generally created in rural areas where marginal agricultural employment or hunting and fishing were previously the only economic activities possible.

A little-appreciated contribution of mining to development is the associated construction and maintenance of infrastructures such as roads, bridges, schools and hospitals, in communities where these amenities were unknown. Such facilities undoubtedly promote economic and social intercourse between members of the community, and with other communities. Indeed, as a result, the residents of mining areas that were previously undeveloped and rural seem to enjoy higher incomes, better education, better health facilities and better housing than the residents of similarly situated non-mining areas (Santos et. al., 1982c).

Mineral processing

On the premise that national development can be better served by processing natural resources into higher valued products, natural resource processing has been adopted as a national policy. The government sees industrialisation occurring through the creation of satellite or supplementary industries following the initiation of large, capital-intensive processing projects. These satellite industries will not only raise the level of national production, but will also create numerous employment opportunities, thereby improving the wellbeing of the people. The twelve major industrial projects described earlier are at the core of this policy, but there has been some criticism of their value. It is therefore important to examine their potential contribution more carefully.

As a basis for this examination Figure 8.1 summarises several minerals steps in processing currently being undertaken. For instance, the Figure clearly shows that in 1981, despite intensive copper mining, the Philippines had no copper smelter or refinery[16] to produce ingots and slabs. Thus it had to import copper ingots and slabs to feed into the domestic copper fabricating sector. Subject to issues of scale economies and input costs, further industrialisation should fill in these gaps, making important the construction of the copper smelter (now completed), the aluminium smelter, and the iron and steel making facilities. These projects represent attempts to structure a stable industrialisation program by providing a steady flow of raw materials to the growing manufacturing sector. They will also provide a mechanism for increasing the value to be realised from exhaustible resources.

Finally, the data on the domestic metal processing industry illustrate the explosive multiplication in firms and employment from a few capital-intensive activities. In 1979, sixty firms employed 12 500 persons in iron and steel processing, and 111 nonferrous fabricating firms employed more than 6477 persons. In 1978, however, more than 2500 metalworking firms employed some 50 650 persons, 71 per cent of these being small-scale (fewer than twenty employees), 17 per cent large-scale (50 or more employees), and 12 per cent medium-scale (20–49 employees). These data indicate very clearly the importance of minerals processing to national industrialisation, particularly in generating the growth of small firms and wider employment opportunities (MIRDC, 1979). This is consistent with the experience of other countries (Roemer, 1979).

These observations suggest that the highly capital-intensive smelting and refining of metals such as nickel and copper can be justified in a capital-poor country only if the outputs are used as raw-material inputs to the production of various semifinished and finished products. In this case, they become a medium for realising more benefits in terms of earning more value added, creating employment and industrial linkages. However, to the extent that the bulk of smelted or refined products are exported without further processing, then the capital-poor exporter merely subsidises the usually capital-rich importer. The adverse effect is even worse if the investments in such facilities come from domestic sources or from foreign loans.

Mineral pricing

Pricing in crude minerals differs considerably from pricing in processed minerals. Since crude minerals are sold in the world market, pricing conforms mainly with international practices. Copper concentrates, for instance, are mainly sold under long-term contracts in which the price is tied to the London Metal Exchange (LME) price. Nickel and chromite are similarly priced. Prices for both the exported gold and that sold to the local Central Bank are determined with respect to the LME price.

Processed minerals that are widely used locally, such as fertiliser and cement, undergo a complicated pricing process. 'Fair market prices' are determined by regulatory bodies which 'protect' the interests of the consumers, apparently on political rather than economic considerations. If, for instance, it is perceived that fertiliser must be sold at a price lower than the production cost, the government subsidises the difference. In this way the government prevents widespread social unrest arising from commodity pricing. The system ensures that at best the producers break even in the long term. This therefore results in disinvestment and general deterioration in the industry.

Energy development policy

The 1970s were characterised by grave oil shortages and a dizzying oil price spiral which threatened the stability, and even the existence, of many countries. The policy of the Philippine government in response to this crisis, and some of the attendant problems, are discussed below.

Energy development programs

The problems caused by the 1973–74 oil crisis triggered the adoption of an energy policy designed to liberate the country from excessive dependence on imported oil, so as to avoid balance-of-payment deficits due to oil price spirals and energy supply disruption threatening national security. The government attempted to solve the probiem by:

1 diversification of the sources of petroleum imports;
2 establishment of adequate stockpiles (about 150 days supply) as a protection against possible supply disruption;
3 collaborative efforts with other nations, particularly ASEAN, in emergency petroleum supply sharing and exchange of technical expertise; and
4 development of indigenous energy resources.

The first three were short-term solutions, and easy to implement. The fourth is a long-term goal, and is likely to be fraught with numerous difficulties (Ministry of Energy, 1982). Central to the program is the projected reduction of dependence on imported energy from 95 per cent of the total in 1970 to about 50 per cent in 1987;[17] the recognition of potential domestic energy resources from hydropower, geothermal power, coal, petroleum, and biomass;[18] and a projected cost for energy resource development between 1982 and 1987 of some US$8 billion (1982 prices).

To bring demand in line with policy objectives the government employs two powerful instruments: taxation 'to reflect real economic and social costs'; and direct intervention to create a market for new energy forms such as coal, cocodiesel and alcogas. Thus the entire cement industry has been required to use coal as fuel, instead of bunker oil.

Although the Philippines has won international acclaim as the first third world nation to develop a comprehensive energy policy, several questions remain. Should national resources be channelled to energy development to the detriment of other sectors, perhaps equally important socially, such as the generation of employment or the provision of education? Should the government aspire to energy self-sufficiency instead of supply stability and tolerable import dependence?

However, because of the limitations on the present chapter, we focus here on two important and more specific issues, energy pricing and non-price market intervention. Many economists argue that government interference with the market to solve one problem leads to the emergence of other difficulties. The Philippine experience is consistent with this argument.

Energy pricing

As a mechanism to reduce oil consumption and to raise additional revenues and redistribute wealth more equitably, the Philippine government regulated prices and profits and imposed different rates of tax on various petroleum products.[19] From 1979, taxes as a percentage of prices were 42–45 per cent for regular and premium gasoline, 19 per cent for diesel, and 8.4 per cent for kerosene. Higher taxes in gasoline were justified because gasoline users were believed to be mainly affluent car owners. Low tax rates were levied on diesel and kerosene because they were used by the less affluent. As might be expected, these tax differentials led to radically different price ratios for the various petroleum products, and caused a shift in consumption in favour of the cheaper product (Table 8.4).

As the price of gasoline increased relative to that of diesel, a massive shift to diesel took place. Consequently there has been a shortage of diesel and an oversupply of gasoline. As a result, the government must import diesel at prices higher than those prevailing locally, and must export gasoline at a loss. The problem may be solved by investing in oil refineries to process new types of crude oils so as to yield the desired product mixture. However, this raises the further problem of obtaining the finance necessary for such a capital-intensive project.

Certain sectors of the coal industry claim that government pricing policy stifles the development of that industry (Philippine Chamber of Coal Miners, 1982). It is said that, while the government guarantees the end users a ceiling price of 65 per cent of the delivered price of oil (based on energy content), a minimum price has not been set, so that producers cannot estimate their potential revenues. To make matters worse, the government, through the Philippine National Oil Corporation, is thought to have bought Australian coal at 48 per cent of oil price, while it buys local coal at 32 per cent of oil price.

Again, since only the National Power Corporation can buy geothermal steam for power generation, and since there is no clearly defined pricing policy, the private sector is reluctant to participate in

Table 8.4 Price and consumption ratios for selected petroleum products

	1965	1970	1975	1981
Consumption ratios				
Regular gas/diesel	1.22	1.06	0.76	0.22
Premium gas/diesel	0.24	0.34	0.39	0.32
(Regular gas/premium gas)				
Diesel	1.46	1.40	1.15	0.54
Kerosene/diesel	0.34	0.30	0.24	0.24
Price ratios				
Premium gas/diesel	–	1.39[a]	1.36	1.69[b]
Regular gas/diesel	–	1.18[a]	1.29	1.62[b]

Notes: [a] for 1973;
[b] for 1982.

Sources: Ministry of Energy press releases

geothermal development because of uncertainties which make profit estimations impossible.

One can conclude that the price-intervention techniques described above are not entirely consistent with the policy objective of solving the balance-of-payments problem and stabilising energy supply by reducing import dependence. Understandably, however, there are numerous social, political and informational constraints operating to prevent perfect intervention.

NON-PRICE INTERVENTION

In order to hasten the development of certain energy resources so as to effect the partial substitution of imported energy by domestic sources, the government has formally stipulated for certain sectors the use of some domestic energy resources, and has invested directly in energy projects.

The conversion of the cement industry from bunker oil firing to coal firing was a consequence of government directives to the industry to substitute local coal for imported oil. The government has also ordered the large-scale use of cocodiesel and alcohol.[20] Numerous technical problems were encountered in forcibly introducing cocodiesel and alcogas on a large scale, quite apart from the vigorous protests of the sectors affected.

Government participation in the energy industry also causes apprehension among private firms. For instance, the Philippine National Oil Corporation, which is the largest petroleum refiner and seller in the country, possesses a tremendous advantage over rival firms, not only because of its size and access to government funds, but because its officers are also officials of the Ministry of Energy and therefore well placed to influence in its favour policies or regulations that affect a firm's viability; for example, pricing and contract conditions. These advantages have tended to discourage full private participation in the development of geothermal and coal resources.

It is obvious that, while the national energy policy is correct in general terms, it is flawed in the detail. For instance, previous price intervention had made it very difficult to align current prices with market conditions. While it is obvious that the price of diesel must be raised substantially to avoid shortages in diesel and gluts in gasoline, the government is powerless to act because of the serious social and political implications. Apart from any other considerations, the people would certainly blame the government as the visible cause of the price increase. Without government intervention in the first place, prices would have adjusted according to supply and demand, and only the multinational would be blamed. Moreover, the gasoline glut and the diesel shortage would probably not have occurred.

Again, the forced introduction of cocodiesel and alcogas on a large scale, without preliminary testing for potential problems, is unwise. In the first place, both alcohol and coconut oil are uncompetitive in price terms with gasoline and diesel. Second, the world prices of sugar (the source of alcohol) and coconut oil are highly volatile, and not conducive to the stated goal of supply and price stability. Third, both sugar and coconut oil are already important exports. Their domestic consumption for fuel will not necessarily improve the balance-of-payments problem.

Despite these difficulties, the development of potential domestic energy sources must continue if the Philippines is to avoid the threat, or indeed, occurrence of an oil embargo. The most logical program of development would be first to perfect the required technology so that alcohol and coconut oil can be used as fuel in times of need. No restructuring of the market is needed immediately. Given that the appropriate technology and supply have been put in place, the market, rather than direct government intervention, would be a more efficient means of allocating alcohol and coconut oil, among other sources, to alternative fuel users at appropriate prices.

Mineral ownership and investments

Continuous and intensive investment is needed to ensure the growth of the minerals industry. The major issues affecting minerals investment are ownership, taxes and subsidies. These are discussed below, particularly in the context of minerals production.

Ownership of mines

The Philippine constitution restricts the exploitation of mineral resources to its citizens and to corporations whose capital is owned 60 per cent or more by Filipino citizens. However, in terms of the exploitation of natural resources, American citizens enjoyed the same privileges as nationals until the expiration of the Laurel–Langley Act in 1974. Consequently, before that year most Philippine mines were controlled by Americans, ownership of the mines passing to Filipinos after 1974.

The recent responsibility of the Filipinos for financing minerals projects has had two important effects. The first has been the growth in foreign and domestic loans, and the second has been heavy government lending for several large mineral projects.

Because of high interest charges and escalating costs, many highly geared firms incurred tremendous losses or defaulted in loan repayments in the late 1970s and early 1980s. Since foreign loans were guaranteed by the government, they became the obligation of the government when the debtor defaulted. Consequently, in April 1983 we estimated the government exposure to be about US$3 billion. In some cases, the government reluctantly converted loans into equity in order to preserve employment and foreign exchange earnings, thus becoming the biggest single investor in the Philippine minerals industry.

Since the Constitution discourages foreign investors, a concept called the 'service contract' has been adapted to the local scene. This was tailored to comply with the constitutional constraints on mineral ownership, and to encourage foreign participation in mineral development. Under the scheme a foreign firm could act as a contractor to the government or to a qualified entity, receiving a certain percentage of the net proceeds in payment for its services. This is legally acceptable, since contracting does not convey ownership.

To date the concept has found favour in petroleum exploration, but has been poorly accepted in hard minerals. It appears that service contract terms vary from one resource to the other. It is clear that if this concept is to be successful the government must guarantee reasonably attractive terms to the service contractors.

Taxes and incentives

Tax rates in the Philippines are among the lowest in the world. In 1978, taxation as a percentage of the value of production was 14.1 for copper, 12.7 for chromite, 2.2 for gold, and 0.03 for nickel. Corporate income tax is between 25–35 per cent of profit. Despite these low rates, mining firms object to the imposition of certain tax items in some years but not in others—for example, premium tax and export tax. A higher tax rate, constant over the life of the project, is preferred, since this permits more accurate calculations of profitability. In the same vein, a profit-based tax would be preferred to the existing tax regime.

Criticism of its taxation policies notwithstanding, the government tends to assist the mineral industry whenever it is facing difficulty. Thus the Gold Subsidy Act, first passed in 1954, has undergone some eight revisions designed to prop up the constantly embattled gold industry. More recently,[21] with most copper mines about to shut down because of low metal prices, the government devised the Copper Stabilization Fund, which absorbed the losses corresponding to the difference between the average industry cost (or a reference price) and the prevailing world copper price. As an alternative, it provided loans to firms which decided to stockpile until prices improved. Similarly, without infusions of government money the nickel industry could not survive. Such government assistance indicates recognition of the industry's value to the national economy.

Summary and conclusions

This chapter has focused on three general questions: the minerals industry and trade patterns; the role of mineral resources in development, and policies affecting the minerals sector.

The minerals industry of the Philippines can be divided conveniently into two segments, the crude and the processed sectors, covering, respectively, exploration through the production of marketable ores, concentrates, sinters or pellets, and smelting through the production of finished goods. Since the Philippines remains relatively undeveloped but possesses considerable mineral resource endowment, its crude minerals industry is relatively more important than the processed sector. Crude minerals of importance to the Philippines are copper, gold, chromite, nickel, zinc, iron and lead ores and concentrates. Important processed minerals include iron and steel, nickel, aluminium, tin, lead, copper and zinc.

Mineral exports consist mainly of crude copper, chromite, iron, lead and zinc, complemented by processed nickel and moderate amounts of iron and steel, aluminium, copper and petroleum. Crude minerals imports are dominated by petroleum, and to a lesser degree by fertilisers. Important processed mineral imports are iron and steel, aluminium, tin, zinc, lead and petroleum.

There are good opportunities for trade with the other ASEAN countries and Australia in processed minerals. However, the Philippines cannot hope to export crude minerals to these countries, but rather will remain dependent on imported crude oil and coal.

As for the importance of minerals in development, it is realised that minerals have much to contribute, apart from the traditional foreign exchange earnings, employment creation and tax revenues. However, the precise role of the minerals sector must be planned in relation to national mineral resource endowment, the country's own stage of development, and the already existing components of the minerals industry structure.

Despite flaws in the Philippine's mineral policies, their general thrust appears to be conducive to the country's development.

Appendix 8.1

Philippine mineral resources, production and trade

Appendix Table 8.1.1 Some Philippine development indicators

	1972	1980	1983
GNP (US$ million)	8 322	35 276	26 955
Exchange rate (P/US$)	6.671	7.5114	11.105
Population (million)	38.9	48.3	52
Density (per km²)	129.7	161.1	173.4
Total exports (fob, US$ million)	1 106	5 788	5 005
Total imports (fob, US$ million)	1 230	7 727	7 487
Total energy consumption	62.7	87.7	98.5
Per capita GNP	214	730	518
Per capita energy consumption (boe)	1.72	1.82	1.89
Oil share in energy consumption (million boe)	62.7	77.3	68.2
Share of imported oil			1 763
In total imports (cif, US$ million)	137	1 857	23.5
Per cent of total imports	12.1	20.5	35.2
Per cent of total exports	11.8	32.1	
Share of crude minerals			
In total exports (US$ million)	213.7	1 142	9.8
Per cent of total exports	18.7	19.7	6.5

Sources: NEDA (1984); Ministry of Energy (1983)

Appendix Table 8.1.2 Philippine metallic ore resources, December 1978

	Identified reserves quantity (million tonnes)	Grade	Hypothetical/ speculative resource (million tonnes)
Gold			
Primary	169.17	0.194–19.18 GmAu/Tonne	475.00
Byproduct	834.70	0.150–20.57 GmAu/Tonne	4 050.00
Copper	3 657.29	0.150–32.62% Cu	30 000.00
Chromite			
Refractory	7.01	27.14–38.00% Cr_2O_3	54.70
Metallurgical	10.25	10.00–52.30% Cr_2O_3	66.80
Iron (lump ore)	71.10	10.00–71.00% Fe	52.10
Laterite iron ore	3 513.32	20.00–44.98% Fe	
Aluminum laterite ore	292.01	40.65–41.61% Fe	1 408.00
Magnetite sand	155.82	17.29–65.38% Fe	37.50
Lead	6.49	0.26–2.5% Pb	–
Molybdenum	71.00	0.01–1.52% Mo	–
Mercury	15.90	0.05–2.89 Hg/tonne	–
Zinc	10.52	0.40–7.28% Zn	–
Nickel	1 366.12	0.23–2.40% Ni	4 500.00

Source: Forum on Estimates of Philippine Mineral Wealth Natural Resources Management Center, Ministry of Natural Resources, 1978 (Also in Bacuta Jr., George, 1980); *Development Prospects of Mineral Resources in the Philippines* Manila: Bureau of Mines and Geo-Sciences, 1980, p. 18.

Appendix Table 8.1.3 Summary of Philippine crude mineral production (quantity in thousand units; value in US$ million)

	Unit	1970		1975		1980		1983	
		Q	V	Q	V	Q	V	Q	V
Metallics									
Gold	kg	18.74	20.24	15.61	76.68	20.02	366.43	25.40	344.16
Silver	kg	52.93	2.41	50.37	6.67	60.71	35.30	56.70	19.55
Cobalt (metal)	Tonne	nil	nil	nil	nil	1.33	72.44	0.16	2.06
Copper (metal)	DTonne[c]	160.30	172.84	255.77	218.67	304.5	580.17	1009.28[a] (conc)	364.39
Lead (metal)	Tonne	0.01	3.82	5.49	0.87	1.81	1.64	na	na
Nickel (metal)	DTonne	0.10	0.26	9.36	35.02	25.38	189.11	6.10	29.07
Zinc (metal)	DTonne	3.19	0.65	19.11	4.30	6.84	2.93	4.51[a] (conc)	0.78
Chromite									
Met conc	DTonne	100.00	1.32	97.00	2.81	137.43	9.36	111.82	6.70
Met ore	DTonne	na	–	1.99	–	25.53	1.99	24.53	1.57
Ref ore	DTonne	466.5	8.55	421.05	12.75	355.10	25.09	130.56	14.59
Iron (lump ore)	DTonne	1 869.9	12.9	1 351.45	12.1	nil	nil	2.64	0.05
Subtotal			222.99		369.87		1 284.46		784.94
Non-metallics									
Cement	Tonne	2 555.3	27.5	4 273.0	119.97	4 481.3	263.39	4383.0	237.96
Salt	Tonne	63.6	4.12	202.1	9.05	346.4	26.86	381.9	28.65
Gravel and sand	CuM	4 619.1	3.80	5 264.9	13.79	13 251.2	80.7	15 093.3	92.06
Subtotal			35.42		142.81		370.95		358.67
Energy									
Crude oil	bbl	nil	nil	nil	nil	1 644	366.0	5 147	na
Coal	Tonne	42.4	0.18	105.1	1.75	328.8	7.7	1 019.6	35.35
Subtotal			0.18		1.75		373.7		
Total			258.59		514.43		2 029.11		1 176.96[b]

Notes: [a] Concentrate form
[b] Excludes value of crude oil production.
[c] DTonne: dry tonne

Sources: BMGS *Mineral News Services* various issues; Ministry of Energy Annual Reports, various years

Appendix Table 8.1.4 Summary of Philippine processed mineral production (quantity in thousand units)

Mineral product	1970	1975	1980
Metallics (tonnes)			
Iron and steel	452.00	663.70	900.0
Copper	1.99	5.85	
Lead	2.00	3.00	
Zinc	18.89	10.10	
Aluminum	8.33	14.03	
Tin	0.03	0.07	
Non-metallics (tonnes)			
Cement	2 550.0	4 351.0	4 777.0
Fertiliser			
Nitrogen		57.0	33.7
Phosphorus		34.3	36.7
Potassium		8.7	19.5
Fertiliser Products		291.7	230.0
Fuel			
Petroleum products (barrels)			
Fuel oils, distillate	13 925	13 499	17 220
Motor gas	15 521	16 037	11 781
Kerosene	3 365	3 288	3 629
LPG	1 007	2 318	1 844
Electricity			
Generating capacity (mboe)			
Hydro	3.12	3.75	6.42
Oil/diesel	15.04	16.58	21.66
Coal	0.02	0.04	0.42
Geothermal	nil	0.02	1.21

Sources: Metals Industry Research Development Centre (MIRDC) (1976, 1982); NEDA (1982); Ministry of Energy (1980)

Appendix Table 8.1.5 Major industrial projects

Project	Cost (US$ million)	Remarks (status)[a]
1 Copper smelter	343	Annual capacity: 138 000 tonnes of refined copper; 440 000 tonnes of sulphuric acid. Target date, mid-1983. (Exported 99 371 tonnes of copper cathodes in 1984 valued at $140 million.)
2 Phosphatic	370	Annual capacity: 510 000 tonnes of diammonium phosphate; 745 000 tonnes of other fertiliser products; 495 000 tonnes of sulphuric acid; 384 000 tonnes of phosphoric acid. (Production started in August, 1984. The domestic market will consume between 20% and 25% of production, while the remainder will go to the Asian region.)

Appendix Table 8.1.5 Major industrial projects

Project	Cost (US$ million)	Remarks (status)[a]
3 Integrated iron and steel mill	765	Annual capacity: 140 000 tonnes of steel, use direct reduction for producing iron; may need 1 332 tonnes coal, 3.0 million tonnes iron-ore per year. (Deferred.)
4 Aluminum smelter	580	Annual capacity: 140 000 tonnes of ingots, slabs, billets. Highly energy-intensive. Target date, 1984. (Deferred.)
5 Cement industry rationalisation	32	Coal conversion of cement plants. Need 1 613 million tonnes coal per year. (Conversion of 17 plants was completed by October 1984).
6 Petrochemical complex	920	Annual capacity: 50 000 tonnes polyethylene; 50 000 tonnes polypropylene. Target date, 1985–86. (Deferred.)
7 Diesel engine manufacturing	100	Set up of foundry shops; casting of piston rings, cylinder heads and blocks. Target date, 1982. (Deferred.)
8 Heavy engineering	100	To provide the machinery and equipment requirements of all the major projects. Target date, 1983. (Deferred.)
9 Coconut industry rationalisation (fatty alcohol project)	150	Intended to produce 100 000 tonnes/yr of fatty alcohol substitute for imported petro-chemical and alkyl benzene. Target date, 1985. (Operating as of 1985.)
10 Integrated pulp and paper	190	PICOP plant to expand from 450 tonnes/day to 900 tonnes/day of pulp capacity. Target date, 1985. (Deferred.)
11 Alcogas program	124	Capacity: 234 million litres of alcohol per year. (Deferred.)
12 ASEAN copper fabrication project	280	Annual capacity: 100 000 tonnes of fabricated copper per year. Approved by ASEAN ministers in Jan 1982. Target date, 1985. (Deferred.)

Note: a Status in November 1985
Source: Ministry of Industry (1981)

Appendix Table 8.1.6 Philippine crude mineral exports (quantity in thousand units; value in US$ million)

	Unit	1970		1975		1980		1983	
		Q	V	Q	V	Q	V	Q	V
Copper (ore and conc.)	Tonne	604.4	185.19	816.3	212.08	1 141.9	562.2	790.2	287.3
Gold (secondary)	Kg	–	(a)	–	73.07	0.42	238.9	11.4	160.5
Nickel	Tonne	500.0	0.21	0.2	32.21	608.8	176.8	7.1	30.3
Chromite									
Ref ore	DTonne	501.1	7.89	407.1	7.74	361.3	27.1	115.0	10.2
Met conc.	DTonne	94.9	2.41	94.2	2.69	122.8	13.9	24.8	1.5
Iron									
Ore and agglomerates	Tonne	1 171.9	13.22	16.7	–	4 383.0	118.2	–	–
Conc	Tonne	660.0	4.17	1 305.1	11.90	24.3	0.2	–	–
Lead conc.	Tonne	0.4	0.01	3.7	0.60	3.18	1.5	–	–
Zinc conc.	Tonne	7.7	0.63	14.8	3.13	14.9	2.8	4.2	0.8
Total			213.73		343.42		1 141.6		490.6
Percentage of total exports			18.7		15.0		19.7		9.8
Percentage of total imports			18.4		9.9		14.8		6.5

Notes: [a] From 1960 to 1971 all gold was purchased by the Central Bank of the Philippines. Gold was produced as a byproduct of copper production.
Sources: NCSO *Philippine Foreign Trade and Statistics* (for 1970–80); BMGS *Mineral News Service* (for 1983).

Appendix Table 8.1.7 Exports of Philippine processed mineral products (quantity in '000 tonnes; value in $US million)

	1970		1975		1980	
	Q	V	Q	V	Q	V
Metallic Manufactures						
Iron and steel	27.4	14.53	4.0	1.79	142.1	48.74
Copper	2.0	0.93	0.07	0.18	0.06	0.10
Nickel	nil	nil	7.0	32.92	22.9	137.95
Aluminum	0.5	0.56	1.8	2.86	3.0	2.80
Lead	nil	nil	nil	nil	0.9	0.85
Zinc	nil	nil	nil	nil	0.1	0.07
Subtotal		16.02		37.75		190.51
Non-metallics						
Cement	123.0	2.02	743.6	26.83	767.2	35.75
Fertilisers, manufactured	17.6	0.53	0.02	0.02	1.5	0.42
Subtotal		2.55		26.85		36.17
Energy						
Petroleum products	1 217.1	19.40	185.3	37.38	16.0	36.66
Gas, manufactured and natural	8.0	0.39	nil	nil	4.0	1.54
Subtotal		19.79		37.38		38.20
Total value of processed mineral exports		38.36		101.98		264.88
(percentage of total exports)		(2.6)		(4.4)		(4.6)

Source: NCSO *Philippine Foreign Trade and Statistics* (various issues)

Appendix Table 8.1.8 Summary of crude mineral imports (quantity in '000 tonnes; value in US$ million)

	1970		1975		1980	
	Q	V	Q	V	Q	V
Metallics						
Aluminum (bauxite ores)	0.013	0.001	7.6	0.171	8.2	0.2
Iron (ore and conc.)	0.011	0.001	nil	nil	4 650.6	74.4
Non-metallic						
Crude fertilisers	1.169	5.243	174.4	10.025	30.1	2.9
Energy						
Crude oil	8 080.7	101.2	9 138.1	709.8	12 036.8	646.9[a]
Total crude mineral imports		106.45		720.0		2 724.4
(percentage of total imports)		(9.2)		(20.8)		(35.3)

Note: a Assumed at US$30 per barrel
Source: NCSO *Philippine Foreign Trade and Statistics* (various issues)

Appendix Table 8.1.9 Imports of processed mineral products (quantity in '000 tonnes: value in US$ million)

	1970		1975		1980	
	Q	V	Q	V	Q	V
Metallic manufactures						
Iron and steel	912.46	118.08	628.09	175.62	933.0	398.87
Copper	6.07	9.40	4.23	8.60	7.35	25.42
Nickel	0.06	0.25	0.10	0.57	0.19	1.54
Aluminum	18.11	6.81	10.64	11.08	23.08	46.25
Lead	5.96	2.30	7.11	4.11	5.78	7.72
Zinc	17.79	5.47	9.51	7.77	16.23	13.06
Tin	0.86	3.21	0.30	2.15	0.55	7.82
Subtotal		145.52		209.90		500.68
Non-metallics						
Cement	24.3	0.55	14.60	0.71	11.8	4.00
Fertilisers, manufactured	195.1	8.70	241.1	42.77	872.25	139.40
Subtotal		9.25		43.48		143.40
Energy						
Coal, coke and briquettes	13.69	0.93	15.66	1.95	336.0	22.12
Petroleum products	191.28	15.70	578.18	56.40	1 856.19	350.95
Gas, manufactured and natural	2.43	0.11	13.35	1.73	54.54	18.17
Subtotal		16.74		60.08		391.24

Total value of processed mineral imports	171.51	313.46	1 035.32
(percentage of total imports)	(14.8)	(9.1)	(13.4)

Source: NCSO *Philippine Foreign Trade and Statistics* (various issues)

Appendix Table 8.1.10 Potential mineral trade, Philippine exports (1980 estimate) ('000 tonnes unless otherwise stated)

Product	Philippine exports	Other ASEAN imports	Australian imports
Crude minerals			
Copper concentrates	1 140	nil	nil
Lead concentrates	3.2	nil	nil
Zinc concentrates	14.86	nil	nil
Chromite			
— ores	356	nil	nil
— concentrates	123	nil	nil
Processed minerals			
Aluminum	3.0	nil	8.600
Copper	1.503	56.636	0.200
Nickel	27.348	12.508	2.036
Cement (US$ million)	35.00	nil	nil
Petroleum products (million tonnes)	0.038	11.79	nil

Note: Data from Singapore are too aggregated for inclusion.
Sources: Data from country papers.

Appendix Table 8.1.11 Potential mineral trade, Philippine imports (1980 estimate) (tonnes)

Product	Philippine imports	Other ASEAN exports	Australian exports
Crude minerals			
Crude mineral (million)	12.04	55.02	nil
Crude fertiliser (million)	0.03	0.34	nil
Iron ore and conc. (million)	4.2	0.02	80
Processed minerals			
Petroleum products (million)	4.7	26.3	nil
Fertiliser (million)	0.872	0.23	nil
Copper (thousand)	5.41[a]	6.51	96
Iron and steel (thousand)	441	893	821
Lead (thousand)	7.32[a]	22.21	nil
Zinc (thousand)	9.95[a]	1.30	nil
Aluminum (thousand)	17.25[a]	nil	46
Tin (thousand)	0.47[a]	148.16	1.59

Note: [a] Based on 1975 data. Survey series discontinued after 1975.
Sources: Country papers, Asean–Australia Joint Research Project

Appendix Table 8.1.12 Taxes and subsidies on oil (as percentage of price; price in US$ per litre)

Taxes and subsidies (per cent)	Jan 71	Oct 73	Jan 74	May 75	Apr 77	May 79
Premium gasoline						
special fund	–	–	–	3.7	6.6	13.1
special tax	22.5	28.9	26.5	24.0	31.7	23.6
industry cost	77.5	71.1	73.5	72.3	61.6	55.1
energy dev't impost	–	–	–	–	–	8.2
Total (per cent)	100	100	100	100	100	100
Wholesale posted price ($)	0.05	0.07	0.134	0.161	0.235	0.314
Regular gasoline						
special fund	–	–	–	3.9	5.4	6.7
special tax	26.7	32.6	28.1	25.2	31.5	25.3
industry cost	73.3	67.4	71.9	70.9	63.1	58.4
energy dev't impost	–	–	–	–	–	9.6
Total (per cent)	100	100	100	100	100	100
Wholesale posted price ($)	0.05	0.06	0.126	0.153	0.215	0.266
Diesel						
special fund	–	–	–	3.9	3.1	5.9
special tax	0.2	16.4	13.7	11.8	15.3	13.1
industry cost	99.8	83.6	86.3	84.3	81.6	81.0
energy dev't impost	–	–	–	–	–	–
Total (per cent)	100	100	100	100	100	100
Wholesale posted price ($)	0.04	0.05	0.10	0.119	0.155	0.179
Kerosene						
special fund	–	–	–	3.05	3.3	2.8
special tax	10.4	8.1	3.6	5.0	6.6	5.6
industry cost	89.6	91.9	96.4	91.95	90.1	91.6
energy dev't impost	–	–	–	–	–	–
Total (per cent)	100	100	100	100	100	100
Wholesale posted price ($)	0.04	0.05	0.007	0.135	0.144	0.168

Sources: Ministry of Energy

Appendix Table 8.1.13 Summary of production of Philippine mineral-based industries (value in US$ million)

Industry major group	1965	1970	1976	1980	Percentage share, 1980
Petroleum and coal	113.1	213.3	1 173.6	1 986.7	42.3
Non-metallics	64.9	72.7	221.0	953.7	13.5
Basic metal	42.8	139.9	380.6	1 148.7	16.3
Fabricated metal products	90.5	78.9	185.2	346.0	5.9
Machinery	7.4	17.7	71.9	70.3	0.9
Electrical machinery	59.2	80.3	191.4	472.8	6.7
Transport equipment	79.2	97.5	323.8	959.9	13.6
Total	457.4	700.3	2 547.5	7 051.3	100.00

Sources: NEDA (1983)

Notes

1 Equivalent to a substantial increase in gold prices.
2 The late 1940s and early 1950s.
3 As shown in Appendix Table 8.1.1.
4 See Appendix Table 8.1.1.
5 These industrial projects should be linked to the national infrastructure program, especially those concerned with the improvement of transportation, communications and electric power.
6 Based on United States Geological Survey Bureau of Mines Schemes.
7 Appendix Table 8.1.2. summarises the country's mineral resources.
8 Ministry of Energy (1982). More coal discoveries are still being reported.
9 Asian Development Bank (1982).
10 Philippines Chamber of Coal Mines (1982).
11 Appendix Table 8.1.4. presents the nonmetallic outputs for 1970, 1975 and 1980.
12 See Appendix Table 8.1.3 for details.
13 Ministry of Industry (1981).
14 The figures in parentheses are barrels of oil equivalent for 1982.
15 From the Philippine Sinter Corporation.
16 Until May 1983.
17 Dependence on imported energy was 78 per cent in 1981.
18 Bagasse, alcohol, coconut oil, and so on.
19 As shown in Appendix Table 8.1.12.
20 In the case of coal, its use is being made difficult because most of the equipment has been designed to use imported coal.
21 In 1982.

References

Asian Development Bank (1982) *Asian Energy Problems* New York: Praeger
Bacuta, G. Jr (1980) Development Prospects of Mineral Resources in the

Philippines, unpublished study conducted for the Bureau of Mines and Geo-Sciences

Bureau of Mines and Geo-Sciences (various years) *Mineral News Service*

—— (various years) *Annual Report*

—— (various years) *Philippine Metallic Ore Reserves*

De Jesus, C. ed. (1978) *Benquet Consolidated, Inc.: 1903–1978, A Brief History* Philippines: Benquet Consolidated

Hong, W. (1978) 'Natural Resource Problems in a Resource-Poor Developing Country: The Korean Case' in Lawrence B. Krause and Hugh Patrick (eds) *Mineral Resources in the Pacific Area* Papers and Proceedings of the Ninth Pacific Trade and Development Conference, San Francisco, 22–26 August 1977, Federal Reserve Bank of San Francisco

Kindleberger, C.P. (1973) *International Economics* Homewood, Illinois: Richard D. Irwin

McDevitt, J.F. and W.G. Jeffrey (1976) 'Minerals and Developing Economies' in W. Vogely (ed.) *Economics of the Mineral Industry* New York: AIMMPE

Media Systems (1977) *The Philippine Oil and Mineral Industry* Manila: Media Systems Inc.

MIRDC (Metals Industry Research and Development Center) (1976) *Non-ferrous Metals Industry of the Philippines* Manila: MIRDC

—— (1979) *Metal Working Industry of the Philippines* Manila: MIRDC

—— (1980) *Primary Iron and Steel Industry of the Philippines* Manila: MIRDC

Nurota, Y. (1978) 'Options for a Resource-Poor Developed Country — Japan' in Lawrence B. Krause and Hugh Patrick (eds) *Mineral Resources in the Pacific Area* Papers and Proceedings of the Ninth Pacific Trade and Development Conference, San Francisco, 22–26 August 1977, Federal Reserve Bank of San Francisco

Ministry of Energy (Philippines) (1980) *Five-Year Energy Program 1981–85* Manila: Ministry of Energy

—— (1981) *Philippine Industry in the 1980s* Manila: Ministry of Industry

—— (1982) *Philippine Energy Development, 1982–1987* Manila: Ministry of Energy

Ministry of Industry (Philippines) (1981) *Philippine Industry in the 1980s* Manila: Ministry of Industry

Ministry of Natural Resources, Resource Policy and Strategy Research Division (NRMC) (Philippines) (1982) *A Report on Copper Material Flow* Quezon City: Ministry of Natural Resources, Natural Resources Management Center

NCSO (National Census and Statistics Office) (Philippines) (various years) *Philippine Foreign Trade and Statistics*

—— (1975) 'Cement — the Industry and as a Non-Traditional Export Product' *Journal of Philippine Statistics* 26, 4

—— (1977) 'Petroleum and Other Alternative Sources of Energy' *Journal of Philippine Statistics* 28, 1

—— (1980) 'Development of Conventional and Non-Conventional Energy in the Philippines' *Journal of Philippine Statistics* 31, 3

NEDA (National Economic and Development Authority) (1982) *Philippine Development Report, 1981* Manila: NEDA

—— (1983) *Philippine Statistical Yearbook, 1982* Manila: NEDA

Philippine Cement Corporation (1977) Study of the Philippine Cement Industry, unpublished study by the Philippine Cement Corporation

Philippine Chamber of Coal Miners (1982) *Philippine Coal Production and Markets* Manila: Philippine Chamber of Coal Miners Inc.

Radke, D., C. Engell, A. Schlarb, D. Wessel and M. Zintl (1980) *Development Prospects of the Philippine Mining Sector* Berlin: German Development Institute

Roemer, M. (1979) 'Resource-Based Industrialization in the Developing Countries' *Journal of Development Economics*

Santos, T.M. (1980a) 'Philippine Energy Policy: A Developing Country's Response to the Energy Crisis' *Philippine Geographical Journal* 24, 1

Santos, T. and D. Salita (1980b) 'The Changing Legal Environment and Its Impact on the Development of the Philippine Mining Industry' *NRCP Research Bulletin* 35, 4

Santos, T.M. (1982a) 'Research on Minerals for Philippine Development Planning: A Survey' PIDS *Survey of Philippine Development Research II* Makati: Philippine Institute for Development Studies

Santos, T.M., M.S. de los Angeles and P. Abayan (1982b) *Deep Seabed Mining and the Philippine Mineral Economy* Pasig, Metro Manila: Development Academy of the Philippines, Secretariat to the Cabinet Committee on the Philippine Law of the Sea

Santos, T.M., V.G. Abiad and M.S. de los Angeles (1982c) *The Role of Mining in National Development: The Case of Benquet Province* Taguig, Metro Manila: National Research Council of the Philippines

Securities and Exchange Commission (various issues) *Business Day Top 100 Corporations of the Philippines* Manila: Securities and Exchange Commission

Siddayao, C.M. (1980) *The Supply of Petroleum Reserves in Southeast Asia* Kuala Lumpur: Oxford University Press

Smith, B. (1978) 'Australia's Mineral Production and Trade: Case Study of a Resource-Rich Developed Country' in L.B. Krause and H. Patrick (eds) *Mineral Resources in the Pacific Area* Papers and Proceedings of the Ninth Pacific Trade and Development Conference, San Francisco, 22–26 August 1977, Federal Research Bank of San Francisco

Toombs, R.B. and P.W. Andrews (1976) 'Minerals and Modern Industrial Economies' in W. Vogely (ed.) *Economics of the Mineral Industry* New York: AIMMPE

Figure 8.1

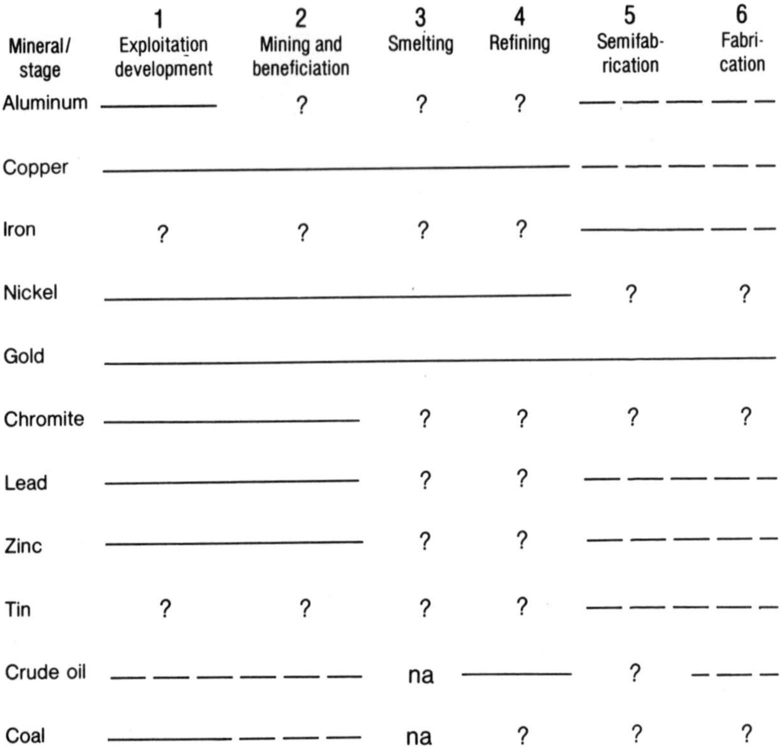

Note: Solid line indicates presence of the stage in the country; dashed line indicates presence of stage operating at low capacity; question mark indicates absence of stage; na—not applicable.

Copper smelting and refining came on stream in 1983.

9 The changing role of Singapore in the trading and processing of mineral resources

JOHN WONG AND HANK LIM

THE notion that geography and natural resources are major determinants of economic development has long been controversial, and remains so (Ginsberg, 1960). In the development experience of Asia, this argument is confirmed in Southeast Asian countries. However, at the same time it is clearly not persuasive in the case of East Asia, including Japan. The economist's contribution to the debate on the relationship between natural-resource endowment and the level and pattern of economic development lies in the theory of comparative advantage: a country tends to specialise in the production of commodities which it can produce relatively more cheaply than can other countries.

A simple static interpretation of the theory of comparative advantage yields the simple argument that if a country is rich in certain natural resources it can easily take advantage of its natural-resource base to foster the development of its economy. In the dynamic context, however, from a more realistic perspective, predictions of the theory of comparative advantage on the role of resource endowment in economic development are much more complex. It is not just the size of the resource base, but its diversity and accessibility, as well as a host of other vital economic factors, such as capital, labour, technology, price structure, institutional characteristics and government policy, that must be taken into account (Higgins, 1968: ch. 11).

Singapore's natural-resource base is virtually zero, but its growth and development have not been impeded by this. In fact, natural-resource-related activities—basically trading and processing—once occupied a significant place in the Singapore economy, although their importance has recently diminished as the economies of Singapore and other countries in the region have undergone rapid structural change.

We might first consider the nine countries on the western rim of the

Pacific Basin: Japan, South Korea, Taiwan, Hong Kong, Singapore, Indonesia, Malaysia, the Philippines, and Thailand, considered to be the world's most dynamic economies in terms of impressive growth performance for the past two decades (shown in Table 9.1). Japan clearly belongs in the league of the industrially advanced countries, and its achievement of a high level of industrialisation from a minute resource base is a well known success story. South Korea, Taiwan and Hong Kong are located in East Asia, and, together with Singapore, are often dubbed the Asian newly industrialising countries. These countries are noted for their dynamic economic growth, largely propelled by the rapid expansion of manufactured exports. Much like Japan, these four nations are primarily land-scarce and resource-poor countries. The city-states, Hong Kong and Singapore, possess virtually no natural resources, although South Korea and Taiwan are known to have developed a sound agricultural base for their economic development. These four Asian newly industrialising countries must depend on external sources of energy, and on other raw materials and mineral resources for their industrial growth. But, significantly, they have overcome their natural-resource constraints through intensifying their human-resource development and cultivating higher levels of skill, enterprise and industrial discipline.

By contrast, the four ASEAN countries (Indonesia, Malaysia, the Philippines and Thailand) are distinguished by their abundance of both renewable and non-renewable natural resources. Together they annually export the bulk of the world's natural rubber, tin, palm oil and coconut products, and a good proportion of other commodities and mineral resources, including petroleum and rice. The continuing export of these primary products and mineral resources has in fact been the main source of economic growth for these countries. In 1977, the proceeds from mineral exports alone, excluding other primary commodities, accounted for 70 per cent of total export earnings for Indonesia, 25 per cent for Malaysia, 15 per cent for the Philippines, and about 5 per cent for Thailand. In these resource-based economies, the exploitation of mineral resources for export, in either processed or semiprocessed form, has had a profound impact on the level and character of their economic growth, despite considerable efforts in recent years towards industrialisation as a means of diversification.

The impact of natural-resource endowment on the pattern of economic development for these two groups of Asian countries is reflected in their respective export structures, summarised in Table 9.2. For the four resource-based ASEAN countries, exports are clearly dominated by mineral resources and primary products, although varied progress has been made during the past two decades towards 'export substitution' with a higher proportion of non-traditional manufactured exports. In the resource-poor countries of East Asia, exports of

Table 9.1 Economic performance indicators

	GNP per capita (US$)	GDP annual growth rate (%)		Distribution of GDP (%)						Value added in manufacturing (US$m, 1975)		Gross manufacturing output per capita (US$m, 1975)	
				Agriculture		Manufacturing		Services					
	1980	1960–70	1970–80	1960	1980	1960	1980	1960	1980	1970	1979	1970	1978
ASEAN													
Indonesia	430	3.9	7.6	54	26	8	9	32	32	1547	4136	50	92
Malaysia	1620	6.5	7.8	37	24	9	23	45	39	946	2597	311	–
Philippines	690	5.1	6.3	26	23	20	26	46	40	2816	5339	193	–
Singapore	4430	8.8	8.5	4	1	12	28	78	62	827	2080	1628	3064
Thailand	670	8.4	7.2	40	25	13	20	41	46	1675	4154	210	–
East Asia													
Japan	9890	10.9	5.0	13	4	34	29	42	55	115 497	207 566	2867	4556
South Korea	1520	8.6	9.5	37	16	14	28	43	43	2 346	9 955	182	621
Hong Kong	4240	10.0	9.3	4	1	26	27	57	–	1 620	3 596	–	1920
Taiwan	2272	9.3	9.5 (1960–80)	29	8	22	42	–	–	–	–	180	935

Sources: World Bank *World Development Report 1982*; for Taiwan, computed from the Executive Yuan *Statistical Yearbook of the Republic of China 1981* Taipei.

Table 9.2 Structure of merchandise exports (percentage share)

	Fuels, Minerals and metals		Other primary commodities		Textiles and clothing		Machinery and transport equipment		Other manufactures	
	1960	1979	1960	1979	1960	1979	1960	1979	1960	1979
ASEAN										
Indonesia	33	69	67	28	0	1	b	1	b	1
Malaysia	20	29	74	53	b	2	b	11	6	5
Philippines	10	18	86	47	1	6	0	2	3	27
Singapore	1	27	73	22	5	5	7	26	14	20
Thailand	7	12	91	63	0	10	0	4	2	11
East Asia										
Japan	11	2	10	2	28	4	23	54	28	38
South Korea	30	1	56	10	8	31	b	20	6	38
Hong Kong	5	1	15	2	45	43	4	16	31	38
Taiwan	4.7	3.2	–	–	–	–	17	25	58	61

Note: a For Taiwan, 'Fuels, minerals and metals' includes 'Other primary commodities'; and 'Other manufactures' includes 'Textiles and clothing'.
b Negligible

Source: World Bank *World Development Report 1982*; for Taiwan, Executive Yuan *Statistical Yearbook of the Republic of China 1981*

primary products and minerals, which never figured prominently, have dwindled even further, and are dominated by manufactured goods exports.

The case of Singapore is most interesting. Geographically, Singapore is part and parcel of Southeast Asia; politically, it is a member of ASEAN. However, Singapore is more akin to the resource-poor East Asian newly industrialising countries in economic and social structure. This dual identity arises from both geography and history, and is clearly reflected in its development pattern. Singapore has developed a large export-oriented manufacturing base, as have other East Asian newly industrialising countries. However, it retains a sizeable natural-resource-based economic sector, as shown in Table 9.3. In other words, the Singapore economy has combined the structure of the resource-rich and the resource-poor economies of Asia. As an economy with no natural resources, Singapore has had to industrialise in order to sustain high economic growth and to provide more employment. At the same time, it is taking advantage of its strategic location in the resource-rich Southeast Asian region by developing resource-related activities, largely through trading and processing.

Singapore's role in minerals and energy of the region

Mineral industries

Historically, Singapore owes its early development and prosperity to its entrepôt activities. Situated at the crossroads of international sea routes and in the midst of richly endowed primary producing countries,

Table 9.3 Changing manufacturing structure in Singapore ($S thousand)

	1971		1981[a]		1979	1981
Industry	Output	% Total Manufacturing	Output	% Manufacturing	Value	added
Petroleum refinery and products	1 553.5	29.0	14 453.8	38.5	916	1 707
Nonmetallic mineral products	33.6	0.6	138.2	0.4	46	62
Iron and steel	57.0	1.1	343.1	0.9	112	138
Nonferrous metals	22.0	0.4	130.5	0.4	19	25
Fabricated metal products	222.7	4.2	1 492.7	4.0	326	538
	1 888.8	35.7	16 558.3	44.1	1 419	2 470
Total manufacturing	5 288.8	100.0	37 559.7	100.0	6 501	9 758

Note: [a] 1979 and 1981 figures are based on the three-digit ISIC (see Appendix 1.1).
Source: *Yearbook of Statistics, Singapore 1971; Report on the Census of Industrial Production 1981.*

Singapore has developed over the past century as the major port and trading centre of Southeast Asia, complete with strong and reliable ancillary services such as financing and warehousing. Thus agricultural and mineral products from neighbouring countries were sent to Singapore for transshipment or simple handling before shipment. In 1959, when the present People's Action Party government took power, the entrepôt trade sector was the single most important sector in the economy, constituting 19 per cent of total GDP, compared with 9 per cent for manufacturing. The importance of entrepôt activities is also directly reflected in Singapore's export structure. In 1960, domestic exports constituted only 6.2 per cent of Singapore's total exports. Even by 1973, when the Singapore economy was highly industrialised with full employment, domestic exports represented only 49.3 per cent of total exports.[1]

Yet Singapore's role in the mineral-product economy of the region involves more than transshipment and simple handling of these products. Its entrepôt trade role encompasses, in addition to buying and selling, a whole range of important economic services from storage, bulk-breaking, grading and financing to today's futures markets. In an overall sense, Singapore performs an essential economic service in facilitating and linking the extraction activities in the ASEAN region with the end-use activities in other industrial countries.

As well as trading, Singapore also processes mineral products. There have been two stages in this processing role. During the 1950s and earlier, when Singapore was an urban centre and Malaysia was the primary-producing hinterland of British Malaya, minerals from Malaysia were brought to Singapore for processing, particularly for example, tin smelting. Subsequently, Singapore gained greater economic autonomy under the People's Action Party government, and began to develop a technology-intensive and regionally-oriented processing industry, undertaking many industrial projects aimed at transforming the Singapore economy from one based on entrepôt trade to one based on manufacturing exports.

So began Singapore's oil refining industry, which has since grown to make Singapore the world's third largest oil refining centre after Rotterdam and Houston. Due to the overwhelming importance of energy minerals in the development needs of ASEAN, all member countries aim for oil self-sufficiency. The establishment of national oil companies, the refinery buildup in ASEAN, and the development of indigenous renewable energy resources, are all calculated to provide greater control over domestic and imported mineral resources.

Singapore's mineral industry involves the processing of imported raw materials, intermediate materials and domestic scrap into finished products. Production is highly capital-intensive, and demands complex technology and significant foreign investment. Singapore's minerals processing role goes back to the 1950s. In 1950, Singapore produced nearly 3000 tons of tin metal, nearly 45 per cent of the Malayan total. By the mid-1950s, Malaysia had built its own large smelter to handle tin ores from North Malaysia and Thailand. Singapore's tin smelting capacity declined over the 1960s and 1970s. In 1979 it smelted less than 5000 tons of tin-in-concentrate, compared with Indonesia's 33 000 tons, Malaysia's 130 000 tons, and Thailand's 38 000 tons, reflecting the shrinking share of tin smelting in the region (International Tin Council, 1980). Not only does Malaysia possess a large production capacity, but heavy Malaysian export duties render exports of tin-in-concentrate less profitable than that smelted in metal.

Singapore's iron and steel mills had an output of $S343.1 million in 1981, representing a sixfold increase since 1971 (see Table 9.3). However, their share in the output of the total manufacturing sector remains about 1 per cent. In terms of value added for 1979, iron and steel accounted for $S112 million, as compared with $S19 million for nonferrous metals and $S46 million for nonmetallic minerals, showing the relative importance of the iron and steel industry. By comparison, Singapore's petroleum industry earned $S916 million in value added from refining crude oil into petroleum products.

The iron and steel industry is of high priority to ASEAN partly because its output has high value added, and also because ASEAN has large recoverable iron ore deposits. Increasingly, ASEAN is interested in using domestic iron ore in forward linkage industries. For example, Malaysia's iron and steel mills consumed 50 per cent of the country's iron ore, while the Philippines has the largest iron and steel mills that utilise scrap, and is gearing itself for the export markets. Other ASEAN countries are also expected to expand their production of raw steel, making it increasingly difficult for Singapore to find export markets for her iron and steel in the ASEAN region. Table 9.4 shows raw steel production in ASEAN.

ASEAN was once also an important source of supply of ores and scrap to Singapore's iron and steel industry. In 1960, 80 per cent of metallic ores and scrap came from ASEAN, mainly Indonesia and Malaysia. However, this share had declined to 25 per cent by 1981 (Table 9.5). Singapore cannot meet the increasing domestic demand for its iron and steel, principally within the construction industry, leading to increased imports, mainly from Japan. But the bulk of Singapore's imported iron and steel products is for re-export to its neighbouring ASEAN countries, principally Malaysia, as summarised in Table 9.6.

Singapore is also an active trader in non-metal minerals manufactures. Most non-metal minerals products are imported from the industrially advanced countries. In 1981, two-thirds of the imported non-metal minerals manufactures were for domestic consumption, while the rest were for re-export, mainly to the ASEAN region (Table 9.7).

In nonferrous metals such as silver, copper, nickel, aluminium, zinc

Table 9.4 Raw steel: production by country (thousand short tons)

	1976	1980
Indonesia	153	660
Malaysia	200	240
Philippines	433	430
Singapore	224	330
Thailand	303	500

Source: US Department of the Interior *Mineral Yearbook* (Volume I, 1980) US Government Printing Office, Washington, DC, 1981

Table 9.5 Imports of metallic ores and scrap to Singapore ($S thousand)

	1960	%	1970	%	1981	%
Imports from						
Malaysia	3 172	29.2	14 176	58.0	23 594	13.5
Indonesia	4 885	45.0	na	–	na	–
Thailand	592	5.4	65.0	0.003	18 304	10.4
Total ASEAN (%)	8 745	80.4	14 905	61.0	44 528.7	25.4
Total imports	10 877	100.0	24 435	100.0	176 307	100.0
Exports to						
Japan	7 081	51.9	14 645	25.0	23 809	21.7
Malaysia	3 546	26.0	19 837	34.0	13 285	6.5
Total ASEAN[a] (%)	4 478.5	32.8	22 175.4	37.8	15 468	7.6
Total exports	13 654	100.0	58 665	100.0	203 529	100.0

Note: [a] Indonesian statistics excluded.
na—not available.
Source: *Singapore External Trade Statistics* December 1960, 1970 and 1981.

Table 9.6 Singapore's trade in iron and steel ($S thousand)

	1970		1981	
	Amount	%	Amount	%
Imports				
ASEAN total	6246.5	2	28 326.6	1.2
Japan	223 500	72	1 614 972	68
Total imports	312 324	100	2 360 552	100
Exports				
Malaysia	23 855	66.1	244 768	41.7
ASEAN total	26 927	68.8	272 043	46.4
Total exports	39 138	100	586 300	100

Source: *Singapore External Trade Statistics* (relevant years)

Table 9.7 Singapore's trade in non-metal mineral manufactures ($S thousand)

	1970		1981	
	Amount	%	Amount	%
Imports				
ASEAN	13 237	11.6	44 918	4.7
Japan	32 728	28.7	296 648	31.0
United Kingdom	10 270	9.0	33 526	3.5
United States	9 129	8.0	62 484	6.5
Total imports	114 113	100.0	955 693	100.0
Exports				
West Malaysia	15 033	34	104 463	31
East Malaysia	2 269	5.1	39 578	11.8
Thailand	117	0.3	8 770	2.6
Philippines	97	0.2	–	–
ASEAN Total	17 492	39.6	152 180	45.4
Hong Kong	5 546	12.6	71 357	21.3
Total exports	44 171	100.0	335 199	100.0

Source: *Singapore External Trade Statistics* (relevant years)

and tin, Singapore's most significant sources of supply are Australia, Japan and Malaysia, while its major export markets are the United States and Malaysia, as shown in Table 9.8.

Table 9.9 summarises the entrepôt trade role of Singapore in mineral products for the region. More than three-quarters of iron and steel and nonferrous metal exports are for re-export. About half the exports of crude minerals, metallic ores and scrap, non-metal minerals, and metal manufactures are for re-export. The lowest re-export ratio is found in petroleum and petroleum products, due to the existence of a large oil refining industry in Singapore. On the whole, the generally high re-export ratios for minerals suggest that Singapore to date remains an important port for the transshipment of minerals to the surrounding countries.

Singapore's role in the minerals economy of the region is not confined to trading. Singapore has also developed a significant sector for processing and manufacturing minerals products. In fact, Singapore's minerals-based industries remain important in terms of their contribution to employment, output, and value added, despite rapid structural change and diversification in the overall manufacturing sector over the years. Table 9.10 summarises the development and related changes of the four large minerals-based industries in Singapore during the past decade.

Petroleum refining is the single most important industry to Singapore with respect to total output, value added and capital expenditure. Petroleum industry output has increased fourteen times from 1972 to

Table 9.8 Singapore's trade in non-ferrous metals ($S thousand)

	1970		1981	
	Amount	%	Amount	%
Imports				
Malaysia	4 419	6.6	54 297	11.8
ASEAN total	4 558	6.7	65 296	14.1
Australia	13 102	19	57 074	12.3
Japan	16 768	24.7	138 873	36.0
Total imports	68 033	100	463 095	100
Exports				
Malaysia	5 952	51.7	80 748	12.4
ASEAN total	5 976	51.9	82 561	12.7
United States	4 145	36.0	160 596	40.0
Total exports	11 514	100	650 088	100

Source: *Singapore External Trade Statistics* (1970 and 1981)

Table 9.9 Singapore's entrepôt trade in mineral products, by commodity division, 1981 ($S thousand)

Commodity division	Re-export	Total export	Re-export × 100 / Total export (per cent)
Crude minerals	43 006	88 150	48.7
Metallic ore and scrap	111 276	203 529	54.7
Petrol/petroleum products	341 659	13 980 572	2.4
Non-metal mineral manufactures	190 398	335 199	56.8
Iron and steel	462 185	586 300	78.8
Nonferrous metal	551 294	650 088	84.8
Metal manufactures	258 305	578 451	44.7

Source: Department of Statistics, *Singapore Trade Statistics, Import and Export* (December 1981)

1981, but employment in this sector has grown only from 2659 to 3511. Petroleum refining is becoming more capital-intensive and higher in value added as older plants are gradually phased out. We can expect this industry to become even more capital-intensive in future.

Since 1979 the growth of the iron and steel industry has declined somewhat. This trend applies both to output and value added. In terms of absolute value, however, iron and steel still have a strategic position in the economy, despite their decline in relative significance.

The nonmetallic mineral products industry has maintained a fairly constant level of employment over the past ten years. However, its value added has increased substantially, reflecting the efforts of the industry to upgrade its products, rather than remaining exclusively in

Table 9.10 Development in mineral-based industries in Singapore, 1972–81

Industrial code		1972	1973	1974	1975	1976	1977	1978	1979	1980	1981
353, 354	**Petroleum refinery and products**										
	Number of establishments	9	10	11	11	11	12	11	10	10	11
	Employment (workers)	2659	3064	3103	3331	3167	3094	3082	3212	3342	3511
	Output ($S million)	1068	1968	5706	4753	6119	7022	7498	9308	11 520	14 454
	Value added ($S million)	281	362	865	605	699	774	787	916	1470	1707
	Capital expenditure ($S million)	261	341	65	99	98	154	161	381	529	376
371	**Iron and steel**										
	Number of establishments	13	16	16	16	15	15	15	16	17	17
	Employment (workers)	1306	1447	1692	1557	1522	1456	1518	1802	1878	1920
	Output ($S million)	73	105	177	148	152	145	193	272	338	343
	Value added ($S million)	27	52	100	51	54	66	92	112	132	138
	Capital expenditure ($S million)	7	13	24	14	3	5	5	5	18	13
369	**Non-metallic mineral products**										
	Number of establishments	19	20	21	20	21	22	23	23	28	24
	Employment (workers)	1281	1442	1745	1700	1735	1440	1427	1205	1315	1261
	Output ($S million)	31	58	79	88	85	97	89	103	115	138
	Value added ($S million)	14	29	36	37	39	43	40	46	53	62
	Capital expenditure ($S million)	2	5	9	5	3	6	3	2	3	7
372	**Nonferrous metals**										
	Number of establishments	4	6	6	6	5	7	8	10	12	11
	Employment (workers)	389	526	558	445	139	516	620	489	459	449
	Output ($S million)	21	33	49	41	34	53	56	135	190	131
	Value added ($S million)	5	11	13	11	7	13	13	19	22	25
	Capital expenditure ($S million)	8	1	0.4	0.7	0.8	1.0	0.9	5.2	4.7	10.6

Source: Department of Statistics, Singapore *Report on the Census of Industrial Production 1981* (December 1982)

crude processing, as was the case in the early period.

The same can be said of the nonferrous metal industry. With the exception of a few products, its growth has been rather slow. In fact, in 1980 output declined from 190 million to 131 million. With the Singapore economy facing full employment and labour shortage, value-added and capital expenditures provide more important criteria to assess the future prospects of the respective industries, in the context of Singapore's domestic economy as well as internationally.

Because of the paramountcy of the petroleum industry to the Singapore economy and its modern characteristics (contrasting with the more traditional minerals-based industries such as tin smelting) the petroleum industry merits special treatment.

Petroleum refining and products

In 1981 Singapore stood out as the world's third leading oil refining centre. The refining area began in 1961 when the Shell Company brought the first refinery output of 20 000 barrels a day on stream. By the end of 1981 Singapore's refining capacity had grown to 1.1 million barrels per day. The refining capacity of the various oil companies in Singapore is summarised in Table 9.11. Over 80 per cent of Singapore's refinery output comes from the three big oil companies, Shell, Esso and Mobil. Individual companies have their stable sources of crude supplies, most of which come from Kuwait and Saudi Arabia, as shown in Table 9.12. Over half of the refining capacity in Singapore produces fuel oils and aviation fuels as refined products (Table 9.13).

The reasons for Singapore's rapid emergence as a major refining centre in the late 1960s and early 1970s were several. First, the development of very large commodity carriers made it possible to transport at low cost massive amounts of oil direct to the markets for

Table 9.11 Refining capacity in Singapore, 1980

Company	Parent	Capacity (thousand bpd)		
Shell	Shell, Dutch/UK	460	460	500
Esso	Exxon, USA	231	230	230
Mobil	Mobil, USA	200	180	180
BP	BP, UK	27	27	27
SRC	BP/Caltex/SPC (Amoco, Oceanic Petroleum, C. Itoh, DBS)	170	170	170
Caltex				170
		1088[a]	1067[b]	1277[c]

Note: SRC—Singapore Refinery Corporation; DBS—Development Bank of Singapore

Sources: [a] Lim Siau Chua 'The Contribution of Oil to Singapore' Paper presented at AIESEC Seminar *The Role of Oil in our Economy*, National University of Singapore (31 August 1981)

[b] 'Trends in Singapore in the Oil Refining Trading and Offshore Oil Exploration Industries' American Embassy, Singapore (undated)

[c] Gill, R. and J. Curtis 'Singapore: Refining the Refining Process' *Resources Asia* (May 1981)

Table 9.12 Singapore refineries, 1980

Refinery	Capacity (bpd)	Actual output (bpd)	Main suppliers
Shell	460 000	325 000	Kuwait, Iraq, UAE, Indonesia, Malaysia
Esso	230 000	160 000	Saudi Arabia, Indonesia
Mobil	180 000	150 000	Saudi Arabia, Indonesia
SRC	170 000	60 000	Kuwait, Saudi Arabia
BP	27 000	25 000	Kuwait, Malaysia
Total	1 067 000	720 000	

Notes: Caltex imports 50 000 bpd of refined oil from its Bahrain refinery, but this will probably decrease with participation in the Singapore Refinery Corporation (SRC).

Singapore is the world's third largest oil refining centre after Houston and Rotterdam (and the world's third busiest port after Rotterdam and Yokohama). Its total refinery capacity rose to about 1.1 million bpd in December 1980, when the SRC refinery expansion began operations. Actual output in 1979 and 1980 was 720 000 bpd.

In 1979 Singapore consumed about 170 000 bpd, 48% marine fuel oil (bunker), 28% fuel oil for the Public Utilities Board, 9% aviation fuel, 6% automotive diesel, 6% other diesel oil (bunker), 3% gasoline, 1% LPG, 1% asphalt, 1% naphtha, and 1% high-speed diesel oil (bunker). Most of the remaining 550 000 bpd was exported to customers in Japan, Hong Kong, Indonesia, New Zealand, the Pacific Islands, Malaysia, Thailand, Australia, and the Philippines.

All Singapore refineries do contract third-party processing, primarily for Malaysia and Indonesia. This accounts for 33% of Shell's output, and accounted for 27% of Singapore's total crude run in 1979. Malaysia's Petronas processed 35 000 bpd in Singapore refineries in 1980 and expects to increase this to about 20% in 1981. Indonesia's Pertamina refined 160 000 bpd in 1980 and plans on a similar volume in 1981. Both countries are actively working on domestic refinery projects which when completed in the mid-1980s may reduce their shares of third-party processing in Singapore.

Source: See Table 9.11.

Table 9.13 Singapore: domestic production of refined products, 1979

	Thousand barrels	%	US$ million	%
Fuel oils[a]	148 009	56.1	2780.3	50.7
Aviation fuel	31 221	11.8	738.0	13.5
Kerosene	20 322	7.7	512.2	9.3
Naphtha & sulfur	18 714	7.1	405.3	7.4
Petrol (gasoline)	17 664	6.7	461.9	8.4
Petrochem feedstock	9 874	3.7	266.8	4.9
Lubricants	4 316	1.6	168.3	3.1
Asphalt and bitumen	2 542	1.0	47.0	0.8
Other[b]	11 149	4.2	103.9	1.9
Total	263 811	100.0	5483.7	100.0

Notes: [a] Fuel oils include:

gas oil	53 818	20.4	1228.7	22.4
diesel oil	8 486	3.2	181.4	3.3
residual	85 705	32.5	1370.2	25.0
total	148 009	56.1	2780.3	50.7

[b] Other includes refinery fuel and refinery losses or gains whose values are not available; thus total output value is slightly underestimated.

Source: American Embassy Singapore annual survey as reported in 80 Singapore # A-26, Petroleum Industry questionnaire enclosure.

refined products. With its excellent deep-water harbour, Singapore was strategically situated between the rapidly growing Arabian Gulf supplies and the expanding East Asian markets for oil products. Second, Singapore possessed a well developed infrastructure to serve the refining industry. Attractive economic incentives and the political stability of the city-state also reinforced Singapore's comparative advantage as the refinery centre for the region.

All refineries in Singapore undertake third-party processing, primarily for Malaysia and Indonesia. The refineries lease out capacity through crude-oil processing deals with neighbouring oil producing countries which have good supplies of crude but insufficient refining capacity. In 1981, third-party processing accounted for 40 per cent of Singapore's total crude run. Both Indonesia and Malaysia are actively working on domestic refinery projects which, when completed in the mid-1980s, will reduce Singapore's share of third-party processing contracts. Already, in 1983, Indonesia and Malaysia had cut back the amount of crude oil processed in Singapore because of their declining revenue from oil and commodities exports and excess supplies of refined oil products in their domestic markets.

Another important consideration is that, out of Singapore's capacity of 1.1 million barrels per day, only about 170 000 barrels per day are consumed locally, mostly in the form of fuel oil (66 per cent), diesel (15 per cent) and aviation fuel (11 per cent). The remaining 550 000 barrels per day of actual output are exported to Japan, Hong Kong, Australia and ASEAN. Some 30 per cent of capacity is at present unused.

Singapore's oil industry has undergone important structural changes since 1960. Before 1960, Singapore was mainly a supermarket for the trading and bunkering of oil. After that date, it moved into petroleum refining and became a base for oil exploration and offshore services. A third phase of diversification in the downstream sector began in the late 1970s. In the 1980s, Singapore faces the challenges posed by the changing matrix of petroleum refining. First, the refinery buildup in ASEAN could hurt Singapore, since exports to ASEAN comprise about a quarter of Singapore's total crude run. By 1984, Indonesia is expected to double its refineries' capacity from 483 000 barrels per day to some 1 052 000 barrels per day (*Petromin Asia*, November 1981). The Philippines, Thailand and Malaysia are all gearing up to increase their refining capacity in future. To overcome this problem Singapore is focusing on secondary processing to petrochemicals, or on the conversion of the heavy end of the barrel into higher grade refined products. The second challenge will stem from the decline in the third-party refining for Malaysia, Indonesia and Thailand, as a result of refinery buildup in ASEAN. However, despite the current glut in petrochemical products, it is the government's view that the prospect is viable in the long term because petrochemicals are important inputs for a large number of industries.

Petroleum and petroleum products have played a dominant role in Singapore's manufacturing, trade and service (bunkering) sectors. Before the 1973 oil crisis, mineral fuels imports constituted 14 per cent of total imports. After 1973, the percentage increased to 34 per cent of total imports because of the high cost of oil and the increasing importance of bunkering and refining in Singapore. As a result of Singapore's large refining capacity, it became a leading refinery and trading centre. In 1960 ASEAN was Singapore's largest importer and exporter of mineral fuels. Indonesia was the largest source of mineral fuels, supplying Singapore with 40 per cent of its total import requirements. Important buyers of petroleum products were Japan, Hong Kong and Australia. In 1981, two decades later, ASEAN's share of Singapore's export market had declined to 23 per cent as a result of its expanding refineries and oil production, while Japan and Hong Kong had emerged as important export markets for Singapore's petroleum products (Table 9.14).

Several current trends in Singapore's petroleum industry are worth noting. First, the refineries are planning reduced energy consumption in operation in order to remain competitive with the more energy-efficient facilities currently being installed. Second, Singapore refineries are redesigning or investing in new plant facilities to upgrade production into higher value products, for example, refining fuel oil into distillates or acquiring the capability to handle heavier crudes, a process often referred to as 'barrel whitening'. On average, Singapore refineries have been quicker than their European counterparts to invest in modern

Table 9.14 Singapore's trade in petroleum and petroleum products ($S thousand)

	1970		1981		1983	
	Amount	%	Amount	%	Amount	%
Imports						
Malaysia	184 586	1.8	2 478 366	12.6	3 543 630	19.1
Saudi Arabia	8 730 000	86.2	10 769 480	54.0	6 488 498	34.9
Kuwait	360 284	3.6	1 994 929	10.0	1 543 078	8.3
Other Middle East	854 128	8.4	1 922 320	9.8	4 101 111	22.1
ASEAN total	184 586	1.8	2 478 366	12.6	3 549 710	19.1
Total imports	10 130 000	100.0	19 819 179	100.0	18 601 400	100.0
Exports						
Malaysia	93 821	12.4	2 024 553	14.5	2 432 734	19.1
Philippines	1 231	0.2	147 229	1.1	498 674	3.9
Thailand	101 848	12.5	1 031 112	7.4	1 040 304	8.2
Hong Kong	65 327	8.0	2 612 857	18.7	1 860 776	14.6
Japan	180 314	22.0	3 134 960	22.3	2 556 692	20.0
South Vietnam	274 158	33.5	14 000	0.1	10 934	0.08
ASEAN total	197 230	25.1	3 215 532	23.0	3 971 712	31.1
Total exports	813 383	100.0	13 980 572	100.0	12 762 000	100.0

Source: *Singapore External Trade Statistics, 1970, 1981 and 1983*

secondary processing plants in order to stay ahead of refineries specialising in basic crude oil distillation. Singapore is also well on the way towards upgrading for producing more of the higher value middle distillates (diesel and kerosene) and top-end products (petrol). A third development is the attempt by Singapore-based refineries to reduce their dependence on the Middle East crudes and to strengthen ties with regional suppliers. Government-to-government trade in crudes is of increasing significance, and Singapore's initiatives could reduce the role of oil majors and their ability to shuffle crude supplies to their various customers. A fourth factor to be borne in mind is ASEAN's need for large injections of capital for exploration and production of mineral resources. Table 9.15 gives an indication of the financial requirements of the petroleum industry in the region for the period 1981–85. By far the largest portion of financial support envisaged for the region in the next few years will be required for oil and gas exploration and production.

Problems and prospects

Despite its lack of natural resources, Singapore has become a very important centre for the processing and trading of minerals and oil. This success was made possible by factors such as its strategic location, well developed infrastructure, effective government policy, and prompt government responses to the changing international environment. Accordingly, mineral and oil trading has experienced significant structural changes in the past two decades. In the past, Singapore had an important role in the processing and marketing of tin, iron and steel, but gradually other countries in the region have built up their own processing and production capacities. Consequently, the economics of minerals has changed dramatically for Singapore. Domestic

Table 9.15 Financing requirements of the petroleum industry in ASEAN, 1981–85 (US$ million)

Category	Indonesia	Malaysia	Philippines	Singapore	Thailand	TOTAL
1	19 832	831	641	–	–	21 304
2	–	335	56	168	792	1 351
3	2 004	1 175	–	–	–	3 179
4	2 730	1 134	40	478	450	4 832
5	1 165	–	56	–	–	1 221
6	820	300	52	54	–	1 226
Total	26 551	3 775	845	700	1 242	33 113

Notes: a Categories.
1 Exploration and production—oil and gas
2 Pipelines/storage/production facilities
3 Liquefied natural gas plants
4 Refining
5 Tankers
6 Marketing/distribution/others

Source: Saldivar-Sali, Arthur (ed.) *Proceedings of ASCOPE '81* Manila, Philippines, p. 385

production is now geared basically for internal use rather than for export.

In the economics of oil a different pattern of change is apparent. Singapore began as a centre for bunkering and trading of petroleum and its related products. Then, in the early 1960s, multinational oil companies took advantage of Singapore's effective economic measures and other incentive schemes and began to expand their refining capacities in Singapore, responding to growth in international trade, cheap energy and the development of an efficient and economic system of sea transport. The 1970s saw the advent of high energy prices, a general slowdown in world economic growth, and increasing competition from other ASEAN countries. In response, Singapore's oil industry began in earnest to increase efficiency and step up diversification into petrochemical activities.

The Singapore government has also been instrumental in effecting structural transformation within the industry. Government-owned companies have initiated changes in the industry. For example, the Singapore Petroleum Company, the Singapore Refining Corporation and, more recently, the Singapore National Oil Company. The petroleum industry generally sees the ideal role for government in complementing rather than directly competing with established oil companies. Nonetheless, in most countries governments are increasingly involved in all aspects of the operation of the petroleum industry, partly because of the vital importance of this industry to the overall economy, and partly for security considerations. In Singapore, the role of government is principally to secure a stable supply of crudes for the industry.

Another significant development in both minerals and oil economics is the enhanced role of Singapore in financing, shipping, storage and offshore services. In the field of energy financing Singapore offers a wide range of financial services, such as credit for rig-building, project lending, funds management, leverage leasing and fixed-rate credits under government programs. ASEAN countries require large injections of capital for exploration and production of mineral resources. Banks and other financial institutions in Singapore have demonstrated their ability to raise such large amounts with maximum flexibility. Singapore's financial role will continue to grow as long as national policies adopted by other ASEAN governments are conducive to private investment. As other ASEAN countries plan to increase their degrees of self-sufficiency in the production of mineral energy, they become increasingly dependent on Singapore for capital funding and for facilities such as storage and terminalling services.

For the rest of the decade, Singapore will continue to play an important role in financing minerals and energy development in ASEAN. This is particularly true for such projects as LNG plants and transportation, oil and gas pipelines and refining and distribution systems, where security of return on investment is most favourable. For example, a semi-government agency, the Export Credit Insurance Corporation of Singapore, underwrites loans to finance up to 80 per cent of the local component of equipment. Singapore is in a strong position to continue its role as financier for energy projects within the region. Despite reduced expenditure by the oil majors worldwide, Asia will demonstrate an increased need for large investments in energy development throughout the 1980s. As a result of reduced cash flow due to the oil glut, many companies are faced with capital constraints. Virtually all segments of the industry are seeking alternatives to equity and conventional corporate debt as a means of generating capital.

With high success ratios in exploratory drilling for both oil and gas, by the mid-1980s the Asian region will account for half of all capital spending by the major oil companies. Much of this spending is expected to go towards the continued development of Malaysia, Indonesia and Thailand, and on marketing in China. Because of the vast amounts of capital required the industry is expected to look increasingly to the external capital market to meet its financial needs. This development augurs well for Singapore. Also favourable to Singapore is another aspect of oil and gas financing. The world market for oil and gas, and activities related to exploration, production and development, are based on US dollar price fixing. Singapore possesses the third largest US dollar trade after New York and London, so that banks there are well placed to capitalise on projects requiring outside finance. This is particularly true for the capital-intensive development of natural gas.

In the area of energy minerals, terminalling is of growing importance to Singapore and has contributed to the changing role of Singapore in the oil business. As new refineries start up in the Middle and Far East, Singapore is preparing to play an important intermediary role as an oil trader. The growth of terminalling services caters for increased exports of refined products and smaller supplies of unrefined petroleum from the oil producing countries. The growing demand for terminalling services will expand Singapore's distributive and trading functions into the area of liquid products.

Apart from oil refining, storage and financial services, Singapore also has a role to play in the region of providing other support services and industries such as oil-rig construction, metal fabrication and offshore supply bases. The need for high-level technical and management skills in offshore services bodes well for Singapore's future.

However, Singapore's refining industry faces challenges from industrialisation in the ASEAN region. Traditional patterns will be altered markedly when planned capacity expansions come on stream in Malaysia and Indonesia. The next five years will be critical for refineries in Singapore as they seek to readjust their markets. Although excess capacity is likely in the short term, Singapore refineries take a longer view. For one thing, Malaysia can be counted on to continue its processing in Singapore, although at a reduced rate, until the Malacca refinery comes on stream in 1987. With Indonesia's current account deficits soaring due to the oil glut, and with its overcommitment to a variety of other large development projects, a new refinery may be postponed for several years. Moreover, Singapore refiners offer an attractive pricing package. Under the new third-party processing arrangement the charge is US$1 per barrel and a small percentage of the yield, a very difficult price for state-owned refineries to beat.

It is also indicative of a long-term view that Singapore is spending US$700 million to upgrade refinery facilities, taking it several steps ahead in the installation of secondary processing units. Large investments have also been made in facilities that will increase flexibility in handling a wider range of crudes. The spot market, which currently accounts for 100 000 bpd, is projected to grow over the remainder of the decade, and processing could go up as high as 150 000 barrels per day.

The bunker market is equally important in maintaining flexibility. With the slowdown in demand for fuel oil, Singapore as the second busiest port in the world possesses a huge market for the lower end of the barrel. The bunker market makes up 100 000 bpd. With this outlet, refiners are able to engage in profitable third-party agreements for the middle distillates because they can dispose of the more unsaleable fuel oil on the bunker market. The top end of the barrel does not represent a problem since all Singapore's refiners, with the exception of the Singapore Petroleum Company's share of the Singapore Refining Corporation, serve their own markets in other countries. In any case, Singapore's refineries will probably run their crude distillation units as necessary to feed the higher-value-added secondary plants in which they have recently invested. In the long run, if refining overcapacity develops in Southeast Asia, with products flowing in from the new Middle Eastern refineries, Singapore oil companies may need to adopt an intermediate role, processing semifinished products from the Gulf or trading in their finished products. There is no doubt that the oil industry in Singapore can look forward to major structural adjustment.

Increasingly the Singapore government is involved in the exploration, production, marketing and security aspects of mineral energy, as well as in seeking investment in joint-venture projects with multinational companies in the areas of terminalling, petrochemicals and refining. However, government intervention is not aimed at gaining control of the industry but rather at diversifying the sources of growth. It also provides exposure in refinery operation to Singaporeans, and establishes a yardstick in performance for other companies. In marketing, Singapore continues to maintain its position as a free trader of mineral products and as an important link in the global distribution network.

As Singapore's economies continue to expand and diversify in the 1980s, its vulnerability and dependence on imported oil will correspondingly increase. As a long-term measure for reducing its high dependence on oil, Singapore has moved to diversify energy resources by increasing the uses of natural gas and possibly coal. Efforts have also been made to diversify the sources of oil supply, and to economise on energy consumption. A feasibility study is under way for a project to pipe natural gas from Malaysia and Indonesia to Singapore.

It is clear that the Singapore government has throughout played a catalytic role in the development of Singapore's energy industry. The underlying motive has been to ensure security, infrastructure, incentives and forward planning. Such intervention was necessary to prepare this vital segment of the economy for fast adaptation to those rapidly changing international conditions which constantly threaten to alter the comparative advantage of countries that specialise in the production, processing and trading of minerals and energy. Despite the basic constraint of lack of natural resources, this general policy, coupled with appropriate responses from the private sector, will ensure the continued prominence of Singapore within the ASEAN region in the highly complex economics of mineral energy.

Notes

1 For a good discussion of Singapore's early development pattern, see Koh (1975).

References

Ginsburg, N. ed. (1960) *Essays on Geography and Economic Development* Chicago University Press

Higgins, B. (1968) *Economic Development: Principles, Problems and Policies* London: Constable

ITC (International Tin Council) (1980) *Tin Statistics 1969–79* London: ITC

Koh Foong Yin (1975) 'Singapore' in Shinichi Ichimura (ed.) *The Economic Development of East and Southeast Asia* Monographs of the Center for Southeast Asian Studies, Kyoto University, University Press of Hawaii, Honolulu

10 Minerals and energy in Thailand: production, consumption and trade

PRAIPOL KOOMSUP

IN terms of its contribution to national production, export earnings, government revenues and employment generation, the minerals sector in Thailand is of small importance in the economy relative to agriculture and manufacturing. However, since the first world oil crisis in 1973–74 and the commodity prices boom in 1974, high prices of oil and other minerals have caused several countries, including Thailand, to pay more attention to their own minerals sectors. International trade in minerals and energy has become increasingly important, and its impact on the economy, whether positive or negative, can be substantial.

As an oil importing country, Thailand now tends to regard its natural resources, especially mineral resources, as a critical factor in solving its external trade imbalance problem. While indigenous energy minerals are seen as substitutes for increasingly expensive imported oil, other non-energy minerals can be developed in order to earn more foreign exchange. Thailand regards more extensive processing of its own minerals as an effective method of import substitution. Several policy measures which have been taken by the government will affect the mineral production and trade patterns of the country in the future.

This chapter attempts to compile and analyse the data on production, consumption and trade for some important minerals and energy resources in Thailand during the period 1970–80. The most up to date data on the country's ore reserves and the policies of the government affecting production, consumption and trade in minerals and energy are also examined, with a view to identifying the key policy issues. Some remarks are also made upon possible local developments in minerals and energy activities over the next decade. The minerals and energy resources covered in this study are tin, tungsten, fluorite, iron, lead, copper, zinc, oil, and natural gas.

Some key characteristics of the minerals sector

The minerals sector in Thailand has been relatively small, with mining and quarrying activities contributing only about 2 per cent of the country's gross domestic product. This percentage dipped below 2 per cent in the early 1970s, but rose to a level above 2 per cent in the late 1970s (Table 10.1). The share of minerals in total exports averaged 9 per cent between 1971 and 1981, ranging from 6 per cent in 1977 to 13 per cent in 1971 (Table 10.1). During the same period, the government revenue collected from mineral exploration and production accounted for only 3 per cent of the total tax revenue. This share reached 4–5 per cent in some years (1974, 1979 and 1980) as the government received relatively large fees from the concessions for petroleum exploration. Being a small and capital-intensive sector, the minerals industry employs only a small number of workers. During the 1971–81 period, for instance, annual employment in mining was below 100 000, accounting for less than 1 per cent of the total labour force.

Between 1971 and 1981 the value of mineral production grew at an annual rate of 17 per cent, with large increases in 1974 and 1978. Changes in the total value of production were caused mainly by fluctuations in the export prices and domestic supply of tin, always Thailand's most important mineral. As shown in Table 10.2, the value of mineral exports in each year closely paralleled the value of mineral

Table 10.1 Contribution of minerals to GDP, export earnings and government revenue

Year	Value added of mining and quarrying as percentage of GDP	Exports of minerals as percentage of total export earnings	Government revenue from minerals as percentage of total tax revenue[a]
1970	2.03	na	na
1971	2.06	13.06	2.10
1972	1.78	10.23	2.79
1973	1.34	8.40	1.82
1974	1.77	8.40	5.22
1975	1.36	6.92	1.91
1976	1.53	6.65	1.75
1977	2.07	5.93	2.56
1978	2.22	10.62	3.60
1979	2.27	10.76	4.25
1980	2.12	11.21	4.65
1981	1.70	7.75	3.38
Average	1.85	9.08	3.09

Note: [a] Government revenue from minerals includes royalties, licences, fees, and other taxes on mineral exploration and production.
na—not available

Sources: Mineral export data are from Department of Mineral Resources *Mineral Statistics of Thailand* (various issues);
government revenues from minerals are compiled from mimeographed sheets produced by the Department of Mineral Resources;
other data appear in *Bank of Thailand Statistical Bulletin* (various issues)

Table 10.2 Production, exports and imports of minerals, 1971–81 (million baht)

Year	Value of mineral production	Value of mineral exports	Value of mineral imports[a]		Balance of trade for minerals	
			Excluding crude oil	Including crude oil	Excluding crude oil	Including crude oil
1971	2 462	2 257	2 768	4 709	−511	−2 452
1972	2 598	2 301	3 251	5 683	−950	−3 382
1973	2 747	2 708	5 083	8 655	−2 375	−5 947
1974	4 219	4 219	7 025	17 407	−2 806	−13 188
1975	3 414	3 117	5 787	17 863	−2 670	−14 746
1976	4 379	4 046	7 686	21 543	−3 640	−17 497
1977	6 849	6 413	10 861	27 309	−4 448	−20 896
1978	9 795	8 825	12 661	29 188	−3 836	−20 363
1979	12 407	11 646	17 392	40 817	−5 746	−29 171
1980	13 465	14 935	19 130	58 434	−4 195	−43 499
1981	11 901	11 856	19 712	66 953	−7 856	−55 097

Note: [a] The commodities included are crude minerals and base metals imported mainly for capital goods production, as classified by the Bank of Thailand.
Sources: Data on mineral production and mineral exports are from Department of Mineral Resources *Mineral Statistics of Thailand* (various issues); import figures are from *Bank of Thailand Statistical Bulletin* (various issues)

production, implying that the output of minerals has been directed mainly to foreign markets.

The role of foreign investment in Thailand's mining industry, particularly tin mining, has changed considerably since the early 1900s. Before the second world war, a large part of the industry was controlled by Chinese immigrants, while British and Australian capital made possible large-scale and capital-intensive mining, most significantly the offshore dredging of tin. The postwar government policy of restricting foreign ownership, and earlier wartime disruptions in foreigners' mining operations, resulted in greater participation in mining by Thai private investors. When, in 1975, Thailand nationalised the offshore concessions previously awarded to a foreign company, a government organisation was set up to operate the concessions. Subsequently, most of the country's minerals have been mined by small-scale Thai operators. The role of foreign companies in the non-energy minerals sector has been limited to tin smelting, and joint ventures with local companies in offshore tin mining, lead mining, and zinc mining and smelting. However, in petroleum exploration and production, because overseas technology and capital are essential, the government has allowed foreign oil companies to operate without local partners.

Thailand's mineral production and exports have been dominated by tin, tungsten and fluorite, these three minerals accounting for over 90 per cent of total mineral production and exports during the 1971–81 period (Table 10.3). Tin alone has contributed over 80 per cent of total mineral production and exports in recent years, and this share continues to rise. Tungsten and fluorite constituted on average 10 per cent and 6 per cent respectively, these shares declining over the period. Other locally extracted minerals of substantial value are lignite, lead, gypsum, limestone, columbite and tantalite. The recently discovered deposits of natural gas and crude oil are both expected to increase in importance in the near future.

As far as foreign trade is concerned, Thailand has remained a net importer of minerals since the beginning of its postwar economic growth. While the value of minerals exports grew between 1971 and 1981 by 18 per cent annually, the minerals import bill (excluding crude oil) increased at an annual rate of 22 per cent (Table 10.2). As a result, the trade deficit in minerals increased from 500 million baht in 1971 to almost 7900 million baht in 1981. If crude oil imports are included, then minerals imports are even more significant, growing at a higher annual rate of 30 per cent, and widening the trade gap from 2450 million baht in 1971 to 55 100 million baht in 1981. Minerals imports grew rapidly, not only because of heightened minerals demand associated with domestic industrialisation, but also because of the sharp rise in the price of crude oil imports after the 1973–74 oil crisis.

Thailand's mineral exports have depended largely on tin metal, which accounted for 69 per cent of its total mineral products exports

Table 10.3 Percentages of tin, tungsten, fluorite and other minerals in Thailand's mineral production and exports

Year	Tin		Tungsten		Fluorite		Other minerals	
	Production	Exports	Production	Exports	Production	Exports	Production	Exports
1971	62.52	69.17	14.37	14.06	15.46	13.73	7.65	3.04
1972	64.87	72.34	15.92	16.25	12.12	9.44	7.09	1.97
1973	67.78	76.73	11.35	10.97	10.04	7.44	10.83	4.86
1974	74.90	74.89	11.23	11.15	7.69	7.04	6.18	6.92
1975	65.03	74.54	12.82	12.07	6.20	6.48	15.95	6.91
1976	70.03	75.09	13.78	12.61	3.39	6.76	12.80	5.54
1977	74.10	79.50	13.41	12.65	3.69	3.58	8.80	4.27
1978	77.16	83.10	11.75	11.54	2.54	2.36	8.55	3.00
1979	82.78	87.85	5.31	5.12	2.23	2.14	9.68	4.89
1980	82.68	88.78	4.38	4.27	2.18	2.01	10.76	4.94
1981	81.13	87.81	4.29	3.24	2.80	2.75	11.78	6.20
Average	73.00	79.07	10.78	10.36	6.21	5.79	10.01	4.78

Source: Department of Mineral Resources *Mineral Statistics of Thailand* (various issues)

during the 1970s (Table 10.4). Next most important are ores, concentrates and scrap of ferrous and nonferrous metals, contributing 14 per cent of the total. The remaining minerals, namely crude fertilisers and other crude minerals, fuel minerals, semiprocessed iron and steel products, and nonferrous metals (excluding tin), each constituted less than 10 per cent of the total.

As in other developing countries, most of Thailand's mineral exports have been destined for industrialised countries. Between 1970 and 1979 about 85 per cent of its mineral exports went to the EEC (36 per cent), the United States (29 per cent) and Japan (20 per cent). This concentration in exports is even greater in the case of tin metal. Ninety-eight per cent of tin metal exports went to these countries, 41 per cent to the EEC, 36 per cent to the United States, and 21 per cent to Japan. Although less concentrated, the market for most other mineral exports shows a similar pattern, the two exceptions being fuel minerals and nonferrous metals (excluding tin), which are exported to other developing countries.

During the 1970s other ASEAN countries and Australia imported only 4 per cent and 0.5 per cent, respectively, of Thailand's total mineral exports. The major items, accounting for more than half of this trade, were semiprocessed iron and steel products and some crude minerals. By contrast, tin metal accounted for three-quarters of mineral imports from Thailand by Japan, the EEC and the United States.

Of mineral imports to Thailand, over 80 per cent was contributed by fuel minerals (57 per cent)—the major item being crude oil—and semiprocessed iron and steel products (27 per cent) (Table 10.5). Two of the most important sources of imports were the Middle East countries, supplying over 60 per cent of Thailand's fuel minerals, mainly crude oil, and Japan, supplying 24 per cent of Thailand's mineral imports. Over 80 per cent of mineral imports from Japan were semiprocessed iron and steel products, for which Japan was Thailand's dominant supplier.

Supplies of mineral products from other ASEAN countries and Australia constituted 16 per cent and 3 per cent respectively of Thailand's total mineral imports. From the ASEAN countries, Thailand imported mainly crude oil from Indonesia and Malaysia, and petroleum products from Singapore. Over 60 per cent of Thailand's mineral imports from Australia were nonferrous metals, mainly zinc, lead and aluminium, Australia being Thailand's most important supplier of zinc and lead.

Tin

Historical background

Tin mining is Thailand's oldest and most important mining industry, dating from the Sukothai period (AD 1238–1350), when tin was used

Table 10.4 Percentage distribution of Thailand's exports of major mineral products, 1970–79

	Other ASEAN countries	Australia	Japan	United States	EEC	Rest of World	Total
Crude fertiliser and other crude minerals	12.59 25.15	0.97 14.91	37.92 14.89	22.53 6.02	6.48 1.40	19.51 13.95	100 7.75 (201 799)
Ores, concentrates and scrap of ferrous and nonferrous metal	3.16 11.54	0.10 2.67	11.98 8.58	17.46 8.52	50.28 19.80	17.02 22.23	100 14.16 (368 584)
Fuel minerals	20.36 14.74	0.97 5.35	3.30 0.47	0.59 0.06	0 0	74.78 19.38	100 2.81 (73 122)
Semiprocessed iron and steel products	40.76 36.30	11.22 76.57	1.42 0.25	0.72 0.08	0.30 0.03	45.58 14.54	100 3.46 (89 976)
Tin metal	0.39 6.92	0 0	21.51 75.50	35.69 85.31	40.81 78.71	1.60 10.25	100 69.39 (1 805 552)
Other nonferrous metals	8.55 5.35	0.10 0.50	2.65 0.33	0.11 0.01	0.84 0.06	87.75 19.65	100 2.43 (63 186)
Total	3.88 100 (101 021)	0.51 100 (13 188)	19.77 100 (514 502)	29.02 100 (755 289)	35.97 100 (936 097)	10.84 100 (282 122)	100 100 (2 602 219) (2 602 219)

Note: Figures in parentheses are export values in thousand US dollars for the period between 1970 and 1979. In each box, a figure above a diagonal is the value of a commodity (or group of commodities) exported to a country (or group of countries) expressed as a percentage of Thailand's total export value of that commodity. A figure below a diagonal is the value of a commodity (or group of commodities) exported to a country (or group of countries) expressed as a percentage of Thailand's total mineral export value to that country.
Source: Compiled from data supplied by the International Economic Data Bank, Australian National University

Table 10.5 Percentage distribution of Thailand's imports of major mineral products, 1970–79

	Other ASEAN countries	Australia	Japan	United States	EEC	Rest of World	Total
Crude fertiliser and other crude minerals	1.67 / 0.21	8.51 / 5.70	6.40 / 0.56	5.62 / 2.60	4.03 / 1.79	73.77 / 3.21	100 (184 253) / 2.07
Ores, concentrates and scrap of ferrous and nonferrous metals	1.47 / 0.51	8.40 / 15.31	5.16 / 1.22	35.80 / 45.06	25.34 / 30.66	23.83 / 2.82	100 (500 818) / 5.63
Fuel minerals	27.94 / 97.22	0.37 / 6.90	1.49 / 3.56	1.58 / 20.15	1.32 / 16.13	67.30 / 80.55	100 (5 070 426) / 56.99
Semiprocessed iron and steel products	0.93 / 1.54	1.10 / 9.64	72.73 / 82.86	3.00 / 18.21	5.96 / 34.70	16.28 / 9.26	100 (2 410 630) / 27.10
Nonferrous metals	1.04 / 0.52	23.50 / 62.45	34.19 / 11.80	7.62 / 13.98	9.48 / 16.72	24.17 / 4.16	100 (730 093) / 8.21
Total	16.38 / 100 (1 457 347)	3.09 / 100 (274 775)	23.78 / 100 (2 115 812)	4.47 / 100 (397 949)	4.66 / 100 (413 977)	47.62 / 100 (4 236 360)	100 (8 896 220) / 100 (8 896 220)

Note: Figures in parentheses are import values in thousand US dollars for the period between 1970 and 1979. In each box, a figure above a diagonal is the value of a commodity (or group of commodities) imported from a country (or group of countries) expressed as a percentage of Thailand's total import value of that commodity. A figure below a diagonal is the value of a commodity (or group of commodities) imported from a country (or group of countries) expressed as a percentage of Thailand's total mineral import value from that country.

Source: Compiled from data supplied by the International Economic Data Bank, Australian National University

locally for weapons, utensils, Buddha images and various equipment. Tin exportation began some time during the Ayudhaya period (AD 1350–1767), when foreigners from Portugal, France and Holland were allowed to operate tin mines in the country (Sivarong, 1980). Later, Chinese immigrants dominated the tin industry. They came to southern Thailand to work the tin mines many centuries ago, and until the first decade of the twentieth century, when the Europeans reappeared, Thai tin production was almost entirely in Chinese hands (Ingram, 1978: 98–9).

In the early 1900s, foreign investment in Thailand's tin mining grew rapidly. With British and Australian capital came the more efficient and capital-intensive technology of dredging. The world's first tin dredge was put into operation by an Australian company at Tongkah Harbour, off Phuket Island in southern Thailand. More dredges financed by foreign investors were later introduced, and by the outbreak of the second world war over 60 per cent of the country's total tin output was produced by dredging (ESCAP, 1979). During the war, all foreign-owned mines were abandoned, leaving a vacuum in the industry which was filled later by Thai entrepreneurs.

From the early 1950s, the role of the Thai private sector has been strengthened further by the government policy of restricting foreign ownership. Until this postwar period, the government did not intervene directly in tin production and trade, apart from issuing licences and collecting taxes. A major step towards increased state involvement took place in 1975. Faced with intense political pressure against foreign investment in southern Thailand, the government nationalised offshore concessions obtained by foreign companies and set up the government-run Offshore Mining Organization to operate them. However, due to lack of expertise in offshore mining, the Organization had to employ foreigners and several large local companies to operate in the area for the first five years.

Tin has remained one of Thailand's top five export earners since the 1870s. Its export value has always constituted at least 90 per cent of the combined value of all mineral exports. Thailand is at present the free world's largest tin producer and exporter, after Malaysia and Indonesia. Because of low domestic consumption, almost all the tin produced has been exported. After 1965, when the first local tin smelter began operation, all tin has been exported in the form of tin metal rather than as concentrates.

Deposits and reserves

The most important sources of tin ore in Thailand are alluvial, colluvial and eluvial placers located onshore in nine southern provinces, and offshore from Phang-nga and Phuket provinces, as well as in the south (Figure 10.1). Other less important sources are found to the north and in some areas of the central region west of Bangkok. The tin ore mined in the country is predominantly cassiterite. The common associated minerals are ilmenite, zircon, monazite, garnet, tourmalite and quartz. Other minerals, mined as byproducts of tin from several locations, include xenotime, columbium and tantalum. In the northern provinces of Chiang Mai and Chiang Rai, wolframite and sheelite also occur with tin. Most of the tin deposits are less than 50 metres thick and contain sand, gravel, silt and clay, together with cassiterite (Scholla, 1981: 134).

It has been estimated that between 1907 and 1976 some 1.089 million tons of tin in concentrates were recovered in Thailand. At the end of 1976, tin ore reserves were estimated at 883 450 tons. At the present rate of exploitation—around 40 000 tons a year—these reserves should last until 1998 (Trangkochasarn, 1982). The measured, indicated and inferred reserves of tin in Thailand were estimated at 1.5 million tons in 1975 and 1.2 million tons in 1980 (US Bureau of Mines, 1975).[1] These figures represent about 15 per cent of total world reserves of tin, putting Thailand in third place after China and Indonesia. Additional resources of 4.36 million tonnes have been estimated to lie in submarginal, hypothetical and speculative deposits.[2] This figure accounts for 16 per cent of the world's total additional tin resources.

It is clear that Thailand has reserves sufficient for mining on a profitable scale for many years to come. Large deposits in the south could probably accommodate at least five inland dredges (Scholla 1981: 168). However, this potential has not been realised because of legal and security problems.

Thailand also possesses fairly large offshore reserves of tin at depths of more than 160 feet. Once new mining techniques are perfected for mining at these depths, it is anticipated that offshore deposits will become economically feasible. In similar vein, much of the area of the north has not been subjected to thorough geological investigation. However, the region offers potential for hard-rock tin mining. To date, in short, most of Thailand's tin is produced from deposits which are relatively cheap to explore, finance and operate. Consequently, given its reserves, Thailand should remain competitive in the international tin market for the foreseeable future.

Mining technology

In Thailand, tin has usually been prospected using the time-honoured methods of panning, 'Banka' drilling and pitting. These techniques have been preferred to other, more sophisticated methods, because they do not need large investment and high technology.

Most tin mines are open-pit, operating in alluvial deposits. They are simpler, less expensive and less risky than shaft mining, which is very rarely used in Thailand. Mining methods employed onshore include gravel pumping, dredging, ground sluicing, dulang washing, hydraulicking and gophering.

To break up the tin-bearing material, gravel-pump and hydraulic mining both use water jets, power shovels, bulldozers, or explosives. However, methods differ for conveying the material to the launders. In gravel-pump mining a pump is used, while hydraulicking employs the pressure developed by a natural head of water. In each method, the material is washed through a riffled launder (a 'palong' in Malaysia), where the tin ore is collected by the riffles and the waste is washed free.

In other open-pit operations, the material is excavated and transported to the washing plant by various other methods. In ground sluicing, for instance, the material is washed off by a rush of water from natural or artificial sources. Although the method is simple and may not require large investment, it is limited by the need for a large quantity of water, sometimes as much as fifty times the volume of the excavated earth.

Dredging, which can be employed in large placer mines both offshore and onshore, uses a bucket-line dredge to excavate and transport the ore to the interior of the dredge, where it is washed on screens and concentrated on jigs and shaking tables. The waste is then discharged at the rear of the dredge to fill in the mined pond, and the rougher tin concentrate is sent to the dressing sheds for final concentration. Dredging is usually a large-scale operation needing heavy capital investment. The Offshore Mining Organization offshore dredger took over two years and 360 million baht to build.

After 1975, offshore mining operations in Thailand also involved an armada of between 1500 and 2000 suction boats. These are small fishing boats converted to local dredges, equipped with a suction pump, a four-inch suction hose and collar, an air compressor trommel and a palong. Generally unlicensed and illegal, these dredges operate in water mostly at depths of less than 60 feet. The operation is technically inefficient, the tin recovery rate being about 30 per cent, compared with the normal 90 per cent for gravel-pump mines and 95 per cent for dredging operations (US Bureau of Mines, 1980: 951). Moreover, the rich, shallow water deposits become so contaminated by the suction boat operations that it is not economic to recover the remaining ore by subsequent operations within the same area. Further, it is a dangerous venture; divers must risk their lives working under water without adequate preventative measures and equipment. Nevertheless, the problem should diminish, since it is believed that tin mining by the suction boats will gradually cease as the shallow-water deposits become depleted.

Most of the active tin mines in Thailand are gravel-pump mines. This type of mining has the advantages of low costs for exploration, capital investment and excavation. However, with the increases in oil prices since 1974, the cost of energy has become the largest component of total operational costs, and the major consideration. In representative gravel-pump mines, expenditure on fuel and electricity accounts for 35–40 per cent of the total, while wages, the second biggest item, represent only 12–14 per cent. Other onshore mining methods, except for inland dredging, show operational costs comparable to, and in some cases less than, gravel-pump operations, and their suitability depends on topographical and geological conditions.

Production

TIN CONCENTRATES

Figures for tin concentrates production between 1970 and 1981 are given in Table 10.6. Total production ranged from over 22 000 tons in 1975 to over 46 000 tons in 1979, while production value increased from 1.5 billion baht to 1.8 billion baht in 1970–73, to a peak of 11.5 billion baht in 1980. Thailand's tin supply accounted for 13 per cent of total world production during this period.

About 85 per cent of Thailand's tin came from twelve provinces in the south, mainly Phang-nga Province, which alone has produced at least half of the country's total output in the last four years. Of the remaining 15 per cent, 13 per cent was mined in the central region and 2 per cent in the north. Since 1970, most of the country's tin output has been produced by Thai operators working in small onshore mines, supplemented more recently by the offshore operations of small suction boats. In 1981 about 85 per cent of total tin production came from nearly 600 onshore mines, each producing less than 30 tonnes of tin, and from hundreds of small suction boats. The remaining 15 per cent came from inland and offshore dredging, usually a large-scale operation.

Until 1978, gravel pumping was the most common mining method in Thailand and produced the largest share of total output. However, this share declined from about 55 per cent in 1970 to 34 per cent in 1978, subsequently stabilising at about 30 per cent. Meanwhile, in the early 1970s, offshore dredging increased its share of production, growing from 12 per cent in 1970 to 22 per cent in 1974. Offshore production dropped suddenly in 1975/76, when the Thailand Exploration and Mining Company (TEMCO), a foreign mining company operating offshore, stopped production once its concessions were revoked by the government. This action was of some significance, since the company had produced about 10 per cent of Thailand's tin in 1974. Its stoppage not only reduced dredging's share of output, but also diminished the total tin output for 1975 by 20 per cent. Also contributing to the production drop in 1975 were floods in the south, which temporarily interrupted onshore mining, and the imposition of International Tin Council (ITC) export quotas to curb excess supply.

The revoking of the TEMCO offshore concessions allowed several hundred small locally owned suction boats to mine illegally within the TEMCO lease.[3] Tin production from suction boats increased sharply from over 6000 tonnes in 1976 to nearly 20 000 tonnes in 1980. In the late 1970s, their share of production was the largest of the various

Table 10.6 Production of tin concentrates, primary tin metal and tinplate

	1970	1971	1972	1973	1974	1975	1976	1977	1978	1979	1980	1981
Tin concentrates												
Volume (tonnes)	29 731	29 609	30 132	28 561	27 767	22 397	27 921	33 044	41 210	46 364	45 986	42 968
Value (million baht)	1 627	1 539	1 685	1 862	3 160	2 220	3 066	5 076	7 558	10 271	11 546	9 654.9
Average price (baht per tonne)	54 724	51 977	55 921	65 194	113 804	99 120	109 810	153 613	183 402	221 530	251 076	224 702
Primary tin metal												
Volume (tonnes)	22 097	21 771	22 281	22 927	19 827	16 630	20 337	23 102	28 945	33 058	34 689	32 636
Value (million baht)	1 634.2	1 565	1 714	2 159	3 033	2 243	3 047	4 892	7 290	9 825	11 549	9 910
Tinplate												
Volume (thousand tonnes)	10.9	12.5	14.6	20.0	26.8	20.9	26.1	36.2	43.9	64.8	69.9	na

Note: na—not available

Sources: Data for tin concentrates and primary tin metal are from Department of Mineral Resources *Mineral Statistics of Thailand* (various issues); data for tinplate are from International Tin Council *Tin Statistics* (various issues)

mining methods, exceeding the output from gravel-pumping mines in 1979 (Scholla, 1981). Spurred on by higher tin prices and the abolition of ITC quotas in 1976, total tin production grew significantly between 1976 and 1979, and much of this increase can be attributed to the spectacular growth of small boat mining.

Also offshore, the Offshore Mining Organization, which had been awarded TEMCO's former mining leases, contracted with a foreign company (Billiton) and ten other firms to mine by dredging in its concessions. Production from these offshore dredges averages 2770 tonnes of concentrate annually. In 1981 Offshore Mining began operating its own dredge in one of its concession areas, the dredge being the largest in Thailand and the second largest in the world. It is specially designed, too, for monsoon conditions, when wind and waves are very high, and its annual production capacity is over 1200 tonnes of concentrate.

In 1981 tin production again dropped as a result of the economic recession, which softened the demand for tin and pushed down prices. Mining costs also increased significantly with higher oil prices after the 1979–80 oil crisis. The decline in tin production continued into 1982, and ITC export quotas were reintroduced in that year.[4]

Production levels also fluctuate with movements in domestic tin prices, which in their turn are largely influenced by export prices. Between 1970 and 1981, tin prices first showed a sharp increase in 1974 with the so-called 'commodity boom'. This pushed up the prices of several minerals and agricultural commodities. Subsequently, tin prices rose significantly almost every year, apart from 1975 and 1981 when they declined slightly, peaking in 1980 at about 250 000 baht per ton, almost five times the average price for the period preceding 1974. It is now widely agreed that movements in tin prices depend more on the level of demand from major industrialised countries than on supply conditions, which remain relatively stable.

With the increased dependence of the tin industry on offshore mining (offshore production accounted for over 40 per cent of total output in the late 1970s and early 1980s), production has exhibited seasonal fluctuations. The monsoon season (April to August) makes offshore mining difficult. Moreover, heavy rains during this period can cause excessive soil erosion and reduce the productivity of onshore mines.

Official statistics on tin production must be interpreted with caution. It is generally recognised that some of the tin produced in these years was smuggled out to Singapore and Malaysia to avoid paying royalties to Thailand. Since the Department of Mineral Resources usually estimates tin production on the basis of royalty-paid tin, actual tin production tends to be underestimated by official sources. Tin smuggling is believed to be closely associated with suction boat mining, which has not been effectively regulated by the government (ESCAP, 1979: 23). The Offshore Mining Organization estimated that in 1975 alone some 4800 tons of tin concentrates were mined and smuggled out by suction boats in its own offshore lease.[5] This is equivalent to 20 per cent of the official total production for that year. If the Organization's estimate is reliable, the actual decline in 1975 production would be much less than that implied by the official figures. Tin smuggling seemed to decline in later years as the government began to buy tin directly from the suction boats through a provincial organisation.

Tin metal and tinplate

Production data for primary tin metal and tinplate in Thailand are shown in Table 10.6. Normally, a tonne of tin concentrates in Thailand yields about 0.72 tonnes of tin metal. Over 90 per cent of tin metal is produced by the Thailand Smelting and Refining Company (Thaisarco), Thailand's first and biggest smelter, owned by the Billiton Company of the Royal Dutch Shell Group. The remaining output comes from a few small smelters owned by local investors. Because of the government's export ban on tin ore or concentrates, every kilogram of tin concentrates must be smelted locally before exportation or domestic consumption. It is therefore not surprising that changes in tin metal production paralleled those in output of tin concentrates for most of the 1970–81 period.

As for tinplate production, before 1982 Thailand boasted only one tinplate factory, that of the Thai Tinplate Manufacturing Company. Its output had increased steadily from 10 000 tonnes in 1970 to 70 000 tonnes in 1980, and in 1982 its capacity was expanded to 120 000 tonnes per year. A second tinplate plant (The Cha-Um Pineapple Tinplate Company) was scheduled to start operation in 1983 with an initial capacity of 60 000 tonnes per year. The rapid growth of tinplate production in Thailand has been stimulated by the demand for tinplate by the canned food industry, particularly the canned pineapple industry, which has expanded very rapidly during recent years.

Smelting

In 1907 over half of Thailand's total tin output was smelted locally by the Chinese. By the early 1920s virtually all output was going to Malaysian smelters, which outstripped the primitive, small-scale smelters of the Chinese. Local smelting revived during the second world war, but died out again at the end of the war (Ingram 1978: 100–1).

In 1965 the Thailand Smelting and Refining Company (Thaisarco), the first modern tin smelting plant in the country, started operations at Phuket in southern Thailand. Since then, exports of tin concentrates have been prohibited by government regulation and most tin production has been sold to Thaisarco for processing. The company was granted tax privileges by the Board of Investment, including exemptions from income and import taxes for the first five years of its operation. These privileges were based on the stipulation that Thaisarco must develop the capacity to produce a minimum of 15 000 tonnes of refined tin metal per year, and that its sales and purchase policies be in line with those of Malaysia's established tin smelting company—that is, the prices paid by Thaisarco must be based on Penang tin market prices.

Thaisarco was initially jointly owned by the Union Carbide Corporation and the Billiton Company. However, after the TEMCO incident in 1975, Union Carbide sold its interest to Billiton, which is now the sole owner. The plant is operated entirely by Thai personnel, with the exception of the chief chemist, who is a Dutch national. Output is marketed by a Billiton operating company, Billiton Metals, and Ores Nederland BV of the Hague (ESCAP, 1979: 34).

The Thaisarco facility consists of four reverberatory furnaces, one electric furnace and one roasting furnace. Its initial capacity was 30 000 tonnes of tin metal per year, and this has since been expanded to 38 000 tonnes. From the beginning, the plant has relied solely on local tin for its raw material.

Thailand's second tin smelting plant, Thai Pioneer Enterprise, located 40 kilometres north of Bangkok, was opened in 1981. The plant has an initial capacity of 3600 tonnes of tin metal per year, which can be increased to a maximum of 5200 tonnes per year. The location of the plant means considerable savings on transport costs for mining companies in the northern and central regions. The company is wholly Thai-owned, and has signed an agreement establishing a West German company, Metallgesellschaft AG, as its sole marketing agent, mainly for exports (Scholla, 1981: 155).

Thailand also has a number of very small smelters, but their combined smelting capacity is not significant and their products are sold exclusively for domestic consumption.

At present, Thailand's total smelting capacity is over 40 000 tonnes of tin metal a year, about 10 per cent of the world's total tin smelting capacity.

Markets and uses

Marketing channels

The marketing channels for tin in Thailand are illustrated in Figure 10.2.

Tin concentrates are sold by miners through two main channels. The first involves direct sales to large smelters, mainly Thaisarco, which determine the tin metal content and pay a price based on the Penang tin price for that day. The second channel, which seems to be more important than the first, involves selling to local dealers, most of whom are themselves large miners. These dealers then bring the ore to the smelters. The price received by the miners through this channel is lower than that obtained through direct sales because of the marketing costs which are borne by the dealers. Most of the tin metal from the large smelters is exported through their marketing agents. Some of the tin concentrates are smuggled out of the country, mainly to Singapore, by both the miners and the local dealers.

Domestic consumption

A small amount of tin metal is supplied by small and large smelters to local tin using industries. However, only 1–2 per cent of tin metal production is consumed locally, despite a rapid increase in consumption from 270 tonnes in 1975 to over 800 tonnes in 1980. Most of this increase was due to the expansion of local canning industries, which use tinplate in the manufacture of metal containers for food and beverages. The use of tin metal in tinplate production is the most important application of tin in Thailand, accounting for some 75 per cent of total tin metal consumption. Other, less important, uses of tin metal are the manufacture of solder, bronze, brass, babbitt, bar tin and powder, and miscellaneous alloys for pipes and tubes.

Exports

Thailand is the world's third largest tin exporting country after Malaysia and Indonesia, accounting for over 10 per cent of total world tin exports. Nearly all tin output is exported, so that changes in exports parallel movements in production. During the period 1970–81 most exports were in the form of tin metal, although small quantities of other tin products were also included—that is, final slag, tin–lead alloys, tin slag and metallic residue.

As shown in Table 10.7, the volume of tin metal exports from Thailand declined steadily from 1970 to 1975. However, the value of these exports rose in some years, when strong demand for tin in industrialised countries pushed up export prices. After the commodity boom in 1974, the world tin price was maintained at a very high level. However, there was a two-year lag before export volumes reflected a response to these price increases, the first growth in volume being apparent in 1976. Exports continued to grow until 1981, when world demand for tin was weakened by the worldwide economic recession. Increases in export prices between 1970 and 1981 were so large that Thailand increased its export earnings from tin by over 500 per cent, although export volume grew by only 50 per cent.

The three largest importers of Thailand's tin in this period were the Netherlands, the United States and Japan. Their combined imports accounted for over 90 per cent of Thailand's total tin exports. Until 1973 the main importer of Thai tin was the United States, which took more than half of the total tin exports. Since 1978 the Netherlands has become Thailand's most important customer, accounting for half of its total tin exports. This may be due partly to the location in the Netherlands of the common parent company of Thaisarco and its marketing agent. It is also possible that some of the tin exported to the Netherlands was re-exported to other European countries.

There was no diversification in Thailand's tin export markets

Table 10.7 Exports of tin metal and imports of tinplate

	1970	1971	1972	1973	1974	1975	1976	1977	1978	1979	1980	1981
Exports of tin metal												
Volume (tonnes)	22 246	21 705	21 351	20 529	19 468	16 552	22 579	27 904	37 194	36 864	36 796	32 008
Value (million baht)	1 618	1 561	1 643	1 912	2 996	2 240	3 038	5 098	7 334	10 232	12 702	9 653
Average fob price of tin metal (baht per tonne)	72 732	71 919	76 952	93 136	158 893	135 331	149 368	211 145	251 656	296 642	333 951	301 590
Imports of tinplate												
Volume (tonnes)	38 509	38 306	37 749	46 504	30 969	24 613	44 746	50 434	43 882	65 667	11 408	na
Value (million baht)	153.0	176.4	145.1	208.0	205.8	161.7	284.1	283.1	315.4	441.7	119.5	na
Average cif price (baht per tonne)	3 973	4 605	3 844	4 473	6 645	6 570	6 349	5 613	7 187	6 726	10 475	na

Note: na—not available
Sources: Department of Mineral Resources *Mineral Statistics of Thailand* (various issues); International Tin Council *Tin Statistics* (various issues)

between 1970 and 1981. Rather, Thailand relied heavily on the three industrialised countries as markets, despite some sporadic and insignificant exports to other developing and developed countries.

IMPORTS

Thailand's only tin import item of significant value is tinplate, which is used in the manufacture of metal containers, mainly for pineapples and seafood. Thailand's tinplate consumption increased steadily from 29 000 tonnes in 1970 to 130 000 tonnes in 1979, resulting in excess demand which could be satisfied only by imports. Table 10.7 shows Thailand's imports of tinplate during the period 1970–80, classified by import sources. The important suppliers of tinplate were all industrialised countries, the two most important being Japan and the United States which together provided over 70 per cent of Thailand's imports of tinplate. Other sources were South Africa, West Germany, the United Kingdom, Australia and Canada, but imports from these other sources show erratic and, in some cases, declining trends. With an increase in Thailand's tinplate production capacity to 180 000 tonnes per year, a drastic reduction in imports was anticipated after 1982.

Tungsten

Thailand's first tungsten mine was opened during the second world war in a northern province near the Burmese border. Production in later years fluctuated markedly in response to variations in price and availability. The discovery of two major tungsten deposits in 1970 and 1971 led to significant production increases, and since 1972 tungsten has been second to tin as Thailand's most important mineral commodity.

Tungsten in Thailand is found in two ore minerals, scheelite and wolframite. A significant proportion of tungsten is also mined from combined tin–tungsten operations.

Reserves

Several quite extensive tungsten bearing deposits are located in the north, the south and the central west of Thailand (Scholla, 1981: 196; and see also Figure 10.3). Thailand's total tungsten reserves are estimated by the US Bureau of Mines at about 25 000 tonnes of 'reserve base'. The Department of Mineral Resources estimates 17 000 tonnes of indicated reserves and 9 500 tonnes of inferred reserves, while ESCAP estimates reserves at 34 000 tonnes. Geological data suggest that Thailand has the potential to become the free world's largest tungsten producer. Significant deposits at Khao Soon, Doi Mok and Doi Ngoem show extensive mineralisation at widely separated sites, and are good indicators of the existence of other relatively large and commercially viable reserves.

Production

The four main tungsten producing areas are located in Khao Soon (Nakhon Si Thammarat Province), Doi Mok (Chiang Rai Province), Doi Ngoem (Phrae Province), and Kanchana Buri Province. Between 1970 and 1980 over 80 per cent of Thailand's tungsten came from these areas, and changes in total production during the period were greatly influenced by fluctuations in their output, particularly from the first three locations.

As shown in Table 10.8, production of tungsten ore increased dramatically from 1300 tonnes in 1970 to over 4800 tonnes in 1971. This was made possible by the 1970 and 1971 discoveries of two major tungsten ore bodies, one at the Doi Mok scheelite deposit in northern Thailand, and one at the Khao Soon wolframite deposit in the south. Production at Doi Ngoem started in 1977 with the discovery of another deposit. Total output remained at a high level until 1978, Thailand ranking fifth in the world after China, the USSR, Bolivia and the United States.

However, these mines are located in remote areas, where communist influence is strong and government authority either not recognised or nonexistent. This led to the problem of repeated invasions by illegal, generally nomadic, pick-and-shovel miners (Scholla, 1981: 172). High rates of crime and violence were common, as gangsters and adventurers became involved. In 1978, for intance, as many as 30 000 illegal miners were working at Khoa Soon, and 5000 at Doi Ngoem.

The most common mining methods are ground sluicing, gophering, and open pitting. These inefficient and labour intensive techniques have tended to exhaust surface deposits and leave behind significant amounts of low-grade ores. To exploit these underground deposits requires technical expertise and financial backing, as well as security. To date these needs have not been met, although most concession holders in these areas appear ready to resume operations once the situation is under control.

In 1979, many mines in Doi Mok and Khao Soon with substantial low-grade reserves were closed as uneconomic to operate, or for security reasons. Some mining continued at Doi Ngoem, but a police presence was required to prevent invasions. As a result in 1979 and 1980 production of tungsten ores dropped drastically to just over half the 1978 output.

Processing

The concentrating processes for tungsten and tin are similar. However, all tungsten minerals are friable, and the processes are hindered by excessive fines and slimes. Tungsten ore is recovered by stage crushing and grinding in jaw crushers. Flotation, chemical treatment, and magnetic separation procedures are sometimes used in large mines to separate wolframite, scheelite and other minerals from tungsten. Most

Table 10.8 Production and exports of tungsten ores and imports of tungsten alloys

	1971	1972	1973	1974	1975	1976	1977	1978	1979	1980
Production of wolframite and scheelite										
Volume (tonnes)	4 867	6 485	5 049	4 286	3 447	3 986	4 276	6 182	3 543	3 143
Value (million baht)	353.9	413.6	311.8	473.5	437.7	603.6	919.0	1 150.6	659.0	611.3
Average local price (baht per tonne)	72 714	63 778	61 755	110 476	126 980	151 430	214 920	186 121	186 000	195 054
Exports of wolframite and scheelite										
Volume (tonnes)	4 384	5 843	4 756	4 942	3 354	4 023	4 199	5 774	3 517	3 594
Value (million baht)	317.4	373.7	297.2	470.5	376.1	510.2	811.8	1 018.6	596.4	638.3
Average fob price (baht per tonne)	72 400	63 957	62 489	95 204	112 135	126 821	193 332	176 411	169 576	177 601
Imports of tungsten alloys										
Volume (tonnes)	na	0.70	10.80	2.10	3.70	3.80	1.24	0.87	3.13	na
Value (million baht)	na	2.60	5.30	6.20	5.10	3.30	8.36	8.45	1.81	na

Note: na—not available

Sources: Production and export data are from Department of Mineral Resources *Mineral Statistics of Thailand* (various issues); import data are from Department of Customs *Foreign Trade Statistics of Thailand* (various issues)

tungsten processing in Thailand involves small-scale, labour-intensive operations located near tungsten mines.

The most processed tungsten product in Thailand is in the form of tungsten concentrate. In 1981 the Board of Investment approved a proposal by United Tungsten Industries Ltd for the construction near Bangkok of Thailand's first plant to produce an intermediate product of tungsten, called 'ammonium para-tungstate'. The plant will have a capacity of 3000 tonnes per year.

Markets and uses

Virtually the entire output of tungsten concentrates is exported to industrialised countries, major markets being West Germany, the Netherlands, the United States, Sweden, the United Kingdom and Japan. The volume exported between 1970 and 1980 fluctuated with production (Table 10.8) rather than demand, production itself being affected by deposit discoveries and ore exhaustion.

In industrialised countries tungsten is used in cutting and wear-resistant materials, particularly in the form of tungsten carbide. Other end uses are mill products, specialty steel, hard-facing rods and materials, superalloys and chemicals. Thus demand for tungsten is heavily influenced by industrialisation and military development.

In parallel with other minerals, the export price of tungsten increased sharply in 1974 as a result of the commodity boom. It has remained high since then, and has had a positive influence on the domestic price of tungsten. (See the price data in Table 10.8.)

Thailand imports small amounts of tungsten in the form of alloys. As shown in Table 10.8, the volume and value of these imports have fluctuated widely from year to year.

Fluorite

Reserves

Thailand's probable fluorite reserves are estimated at 21 million tonnes. Under the present conditions of production, markets and security, the industry has a theoretical life of 40 years (Scholla, 1981: 69). However, if recoverable reserves are assumed to be 50 per cent of total reserves, apparent industry life is reduced to twenty years. Most fluorite reserves are in the north, with the biggest deposit at Bang Hong District in Lamphun Province, which contains estimated reserves of some 3.5 million tonnes. Other sizeable deposits are located in the central west and in Surat Thani Province in the south (Figure 10.4). Most fluorite deposits in Thailand have been discovered without the aid of modern exploration techniques. With more systematic and modern exploration techniques, new ore bodies may be identified.

Production

Thailand's fluorspar mining industry is relatively young, having started in 1960 with a production of 3500 tonnes of fluorspar. The country's largest ore deposit was discovered in 1959 at Lamphun Province, 500 kilometres north of Bangkok. This discovery, together with increasing Japanese demand for the mineral for use in its steel and aluminium plants, led to rapid growth of the industry.

Fluorite production in Thailand peaked in 1971, when over 400 000 tonnes were mined, Thailand then being the second largest fluorite producer in the free world. Thereafter, output declined slowly to over 300 000 tonnes in 1980, due to decreasing demand and the entry of Chinese fluorspar into the international market, providing substantial supplies to Japan (Table 10.9).

Most mining is done by open pitting, and there are very few small underground mines. The four main mining companies in the industry are Universal Mining, Thepnithi, Ba Tha Mining, and United Fluorite Mining. The first three operated in Lamphun and the fourth in Kanchana Buri Province in the west. Their combined production accounted for a substantial portion of the country's total fluorite production between 1970 and 1980, although the bulk of the output came from relatively small mines in the north and the central west of Thailand.

Processing

Upgrading of natural fluorspar is necessary before fluorite can be marketed. There are three methods of fluorite beneficiation:

1 *Hand forting*: the crude ore is crushed before screening by hand to select the high-grade ore. One limitation of this labour-intensive method is that excessive low-grade ore is left out.
2 *Sink and float, or heavy medium*: fluorite is separated from other materials by using a suspension of finely ground ferrosilicon in water as a separating medium. This process is more effective in controlling the grade and minimising losses. Because it requires high investment and skilled labour, it can be adopted only by large mines with substantial ore reserves. In fact, at the moment, Universal Mining and S.K. Mineral are the only two companies in Thailand using this technique.
3 *Flotation*: ceramic and acid grades of fluorite are produced by froth flotation, a process which ensures a high level of metallurgical efficiency. The Thai Fluorite Processing Company is the only establishment with flotation facilities. With a feed capacity of 15 000 tonnes per month, it is also the largest plant of its kind in Asia.

A study commissioned by the Board of Investment suggested the establishment of a fluorite briquette plant, a hydrofluoric acid facility, and a synthetic cryolite plant. However, no action has been taken, as yet, to set up these fluorite processing facilities.

Table 10.9 Production, consumption and exports of fluorite

	1970	1971	1972	1973	1974	1975	1976	1977	1978	1979	1980
Production											
Volume (tonnes)	318 222	427 398	400 717	398 070	479 206	338 650	253 686	294 451	314 786	316 526	366 439
Value (million baht)	204.8	380.5	314.9	275.9	419.1	298.2	237.4	273.3	286.7	319.1	383.4
Consumption											
Volume (tonnes)	290	1 025	–	–	750	385	76	26	71	292	220
Value (million baht)	0.2	0.9	–	–	0.6	0.3	0.07	0.02	0.06	0.3	0.3
Exports											
Volume (tonnes)	340 599	346 782	261 208	268 401	314 291	210 496	292 208	241 700	208 459	204 714	214 286
Value (million baht)	219.2	301.0	217.1	201.4	296.9	200.9	274.0	229.5	208.4	249.8	299.9

Source: Department of Mineral Resources *Mineral Statistics of Thailand* (various issues)

Uses and markets

Different grades of fluorite are needed for various purposes. In general, fluorite can be classified into four grades:

1 Low-grade fluorite, of little use in industry. Most has to be upgraded before application.
2 Metallurgical-grade fluorite, used almost exclusively by the iron and steel industries for desulphurisation and dephosphorisation during steelmaking. Some fluorite is also used in the refining of aluminium, copper, lead, nickel, tin, zinc and other minerals.
3 Ceramic-grade fluorite, used by the ceramics industry for the production of glass and enamel.
4 Chemical or acid-grade fluorite, used as a feedstock for the manufacture of hydrofluoric acid, a key ingredient in the aluminium and fluorochemical industries.

The use of fluorite in Thailand is limited to the iron and steel, glass, and ceramics industries. Since these industries are still relatively small, domestic fluorite consumption accounts for a small share of total fluorite production. As shown in Table 10.9, domestic consumption ranged from 1000 tonnes in 1971 to nil in 1972 and 1973, and generally seemed rather erratic.

Most of Thailand's fluorite output has been exported in the form of metallurgical and acid grades. Large fluorite mining companies sell their fluorite concentrates directly to importers, while small miners normally sell their products to dealers and processing companies before they are exported. Since 1970 the main markets for Thai fluorite have been Japan, the USSR, the Netherlands, Australia, South Korea, Poland and Taiwan. Table 10.9 shows a decline in export volume during the 1970s. A sharp drop in 1975 was caused by a slump in the Japanese steel industry. Despite the subsequent recovery of steel production to earlier levels, Japan has not increased its purchase of Thai fluorite, but has turned instead to fluorite supplies from China.

Since limited amounts of fluorite are found in Asia, with the exception of China, the market outlook for Thailand seems promising, given the plans for new steel and aluminium plants in the region.

Iron

Before 1960 Thailand's requirements for iron and steel products were satisfied by imports. Since then, government promotion of industrial investment has encouraged the establishment of iron and steel factories within the country. However, the range of products manufactured locally is still limited, the largest group being bars and wire rods.

Reserves

Thailand's iron ore deposits are scattered, none can be considered large, and the ore types discovered are of poor quality. Thailand has four main types of iron ore: magnetite, hematite, limonite and laterite. Ore reserves with a content of more than 40 per cent iron are estimated at about 50 million tonnes, most of which are located in the northeast and the central region. Additional low-quality reserves of 27 million tonnes are located in the central region and the south. Compared with the world's total iron-ore reserves of 103 000 million tonnes (US Bureau of Mines, 1980), Thailand's reserves are very small indeed.

Production

Thailand had only twelve iron mines in 1981, some of which were inactive. All mining is open-pit, using a combination of labour, machines and explosives. The mines are Thai-owned, and three are operated by two iron and steel mills.

The data on iron ore production between 1970 and 1980 appear in Table 10.10. Annual production fluctuated between 1970 and 1976. In 1977 there was a sudden spurt when iron ore production increased from 25 000 tonnes in 1976 to 63 000 tonnes. Thereafter it fluctuated between 85 000 tonnes and 103 000 tonnes.

The bulk of Thailand's iron ore is smelted in three wholly Thai-owned smelters. The production of smelted iron or pig iron was relatively stable in the 1970s (Table 10.10), ranging between 11 000 tonnes and 20 000 tonnes annually, except for the 1979 peak of almost 30 000 tonnes. The shrinking domestic market and higher energy costs affected the smelters badly, and in 1980–81 their total output dropped, one smelter being forced to close down.

The most important sector of Thailand's iron and steel industry is the manufacture of steel bars and wire rods, mainly from imported non-flat steel and scrap. Altogether there are in the country between 70 and 80 factories producing these iron and steel products. Of these, the nine largest, with electric-arc furnaces, have a total capacity of 606 000 tonnes a year. These companies are either wholly Thai-owned or joint ventures between Thai investors and Japanese, Taiwanese or Indians. The remaining 60–70 factories are relatively small, and use rerolling techniques to produce low-grade products. These rerolling plants are capitalised mainly by Thai investors, with secondary foreign investment principally from Japan and Taiwan. Total rerolling capacity is estimated at about 360 000 tonnes a year. Production of bars and wire rods grew steadily in the 1970s (Table 10.10). However, the recent worldwide slump in the steel industry has meant capacity under-utilisation in most of Thailand's bar and wire rod producing plants.

The manufacture of flat steel products remains limited to two main product groups, tinplate and galvanised sheet and pipes. Thailand's one tinplate factory has a capacity of 70 000 tonnes per year.[6] Three companies produce galvanised sheet, their total capacity being some 250 000 tonnes per year. These three companies are joint ventures between Thai and Japanese investors. Production figures for both product groups are shown in Table 10.10.

Table 10.10 Production of iron ore and iron and steel products (tonnes)

	1970	1971	1972	1973	1974	1975	1976	1977	1978	1979	1980
Iron ore											
–volume	22 523	41 591	27 818	36 309	36 303	32 476	25 000	63 470	88 121	103 101	84 966
–value (million baht)	3.6	6.7	4.4	5.8	5.8	5.2	4.0	10.2	14.1	16.5	13.6
Pig iron	11 172	14 042	12 678	15 069	17 011	12 626	18 214	19 557	20 812	29 224	17 738
Steel bars	na	na	na	na	na	na	329 977	347 949	411 671	494 281	780 644
Wire rods	na	na	na	na	56 129	57 861	69 960	76 699	83 732	58 674	56 291
Tinplate	10 900	12 500	14 600	20 000	26 804	20 917	26 149	36 170	43 939	64 761	69 852
Galvanised sheet	na	na	na	na	92 652	102 669	115 136	129 618	126 936	127 880	124 015

Note: na—not available
Source: Department of Mineral Resources *Mineral Statistics of Thailand* (various issues)

Processing

Most iron and steel plants in Thailand are located in the central region, near Bangkok, due mainly to the importance of proximity to Bangkok, as the biggest market for iron and steel products; the savings on transport costs for imported raw materials made possible by closeness to Bangkok; and the lack of major ore deposits in the country.

Iron smelting employs two technologies, the blast furnace and the electric-arc furnace. The ore smelted is mined locally, and both imported and local scraps are used. Other raw materials, for example, silica sand, limestone, dolomite and quartzile come predominantly from local sources. The main energy source is charcoal, also of domestic origin.

Steel bars are produced either by electric-arc furnace or by rerolling. The first method yields bars of better quality than the second. While 70 per cent of iron and steel scrap, the main raw material of production, comes from domestic sources in electric-arc furnace plants, in rerolling mills all the scrap is imported. The electric-arc furnace method consumes a large amount of electricity, its cost accounting for over 10 per cent of the total and representing the second biggest item after iron and steel scrap, at 40 per cent. On the other hand, in the rerolling method the cost of scrap represents over 70 per cent of the total, but electricity costs are lower than for the furnace technique.

For galvanised sheet production, most raw materials are imported. Cold coiled iron sheet is imported from Japan, and zinc metal from Australia. Both materials account for nearly 80 per cent of total production costs.

Recently a Netherlands-based company was commissioned by the government to conduct a feasibility study of the construction of a large integrated steel plant to be located in South Thailand.[7] The plant is expected to produce initially 1.6 million tonnes of flat steel a year. Most of the raw materials will need to be imported (see *Bangkok Post*, February 1983).

Uses and markets

The major end uses of iron and steel products (that is, mainly bars and wire rods) are in construction. Thus the demand for these products depends substantially on construction activities, in both the public and the private sectors. In the 1960s and 1970s, the Thai economy grew rather rapidly, accelerating growth in the construction sector. Consequently consumption of iron and steel products grew steadily, apart from 1974 and 1975 when construction activities were retarded by the oil crisis. Estimates of Thailand's consumption of pig iron, iron and steel bars, and apparent steel consumption are given in Table 10.11.

Thailand is a net importer of iron and steel manufactures, importing large quantities of these products in various forms, either semiprocessed or processed. The imported products fall into five main groups:

Table 10.11 Consumption, imports and exports of iron and steel products

	1970	1971	1972	1973	1974	1975	1976	1977	1978	1979
Consumption (apparent steel consumption)										
Volume (thousand tonnes)	792	754	936	1 160	1 025	916	1 346	1 544	1 697	1 802
Imports										
Volume (thousand tonnes)	1 023	836	1 446	1 692	1 548	1 366	2 051	2 514	2 566	2 317
Value (million baht)	1 899	1 425	1 647	3 984	5 436	3 769	5 671	4 620	9 546	9 856
Flat steel products (thousand tonnes)	284	230	419	490	445	403	698	769	644	608
Non-flat steel products (thousand tonnes)	286	262	208	217	244	180	217	300	238	252
Exports										
Volume (thousand tonnes)	2.634	1.565	16.482	18.002	10.339	1.992	4.149	0.065	1.215	1.813
Value (million baht)	594	34	390	134	264	105	127	97	247	449

Sources: Office of Basic Industry Development, Ministry of Industry *The Iron and Steel Statistics in 1971–1980* Industrial Finance Corporation of Thailand *Iron and Steel Industries in Thailand* (1981); Department of Customs *Foreign Trade Statistics of Thailand* (various issues)

pig iron and scraps, bars and sections, sheets and plates, tubes and pipes, and others.

Thailand's imports of iron and steel products increased annually until 1974 and 1975, when volume dropped with a lull in construction (Table 10.11). Imports recovered between 1976 and 1979, but 1980 and 1981 saw another decline. The table shows imports of flat and non-flat steel products separately. It is clear that while imports of non-flat steel products remained relatively stable during the 1970s flat steel products showed a significant rise. This is explained by the fact that Thailand was able to manufacture substitutes for non-flat steel products, particularly bars and wire rods, while it relied heavily on foreign sources for flat steel products.

Thailand imports iron and steel products mainly from Japan, the United States and West Germany. Between 1977 and 1981 imports from Japan accounted for 50–60 per cent of the total, and those from the United States and West Germany constituted 6–10 per cent and 3–5 per cent respectively. The remainder were imported from Australia, the United Kingdom, India, Taiwan, and other industrialised countries.

Thailand exported bars, wire rods, pig iron, scraps, plates and sheets in small quantities, mainly to Laos and Vietnam (Table 10.11).

Copper, lead and zinc

Reserves

In the mid-1960s, a survey conducted by the Department of Mineral Resources together with the UN and US Geological Survey, reported two significant copper deposits, one in Phu Hin Lek Fai and the other in Phu Thong Daeng, both in Loei Province of northeastern Thailand. The proven reserves at these two sites total 120 000 tonnes, while proven and probable reserves are estimated at 16 million tonnes of ore, averaging between 0.4 per cent and 0.7 per cent copper. However, future development of the deposit is unlikely unless there are improvements in infrastructure in the area.

A high-grade zinc deposit was discovered near the Thai–Burma border at Mae Sod, Tak Province. The estimated proven reserves are 3 million tonnes of ore with up to 35 per cent zinc content. The life of the mine is estimated at seven to nine years. However, zinc mining may be possible beyond this time, a further countrywide geological investigation being expected to locate additional zinc deposits.

Some of Thailand's lead deposits are located in Kanchana Buri Province, with the estimated proven reserves of 1.26 million tonnes of ore grading between 5 per cent and 18 per cent lead. Small deposits have also been found in the north and the south, although they are too small for viable operations (Figure 10.5).

Production

Thailand's copper–lead–zinc deposits have been known and sporadically mined for many years. However, few have proved large enough for long-term mining operations.

Lead production is confined to one mine in Kanchana Buri Province near the Thai–Burma border. The mine has been operated since 1971 by the Kanchana Buri Exploration and Mining Company, a Thai–German joint venture with Metallgesellschaft AG as the German partner. The mine is the largest underground operation in the country, and expanded its output significantly in the late 1970s (Table 10.12). Production in 1981 was over 40 000 tonnes, more than ten times higher than the levels achieved before 1979. Very small amounts of lead are mined in the north and the south, mostly as a byproduct of tin mining.

Copper was mined for only three years, between 1972 and 1974. The mine was located in Uttaradit Province and yielded only 38 tonnes of copper, closing down because of ore exhaustion.

In 1973 and 1975, zinc was mined from a sizeable high-grade zinc deposit in Tak Province to the north. However, the mining company, New Jersey Zinc, closed its operation in 1975 after it had failed to construct a zinc smelter within three years of operation, as required by the government. Total zinc output during the three years was over 150 000 tonnes.

Zinc mining rights are presently held by the Padaeng Industry Company, which was set up in 1980 to implement the zinc mining and smelting project. The company is a joint venture between the government (with a 20 per cent shareholding), a group of Thai commercial banks (40 per cent equity), a group of Belgian investors (30 per cent equity), and several local and Japanese companies. Financial backing and smelting technology were provided by the Belgian partners, Méchim SA, Société des Mines, and Foundaries de Zinc de la Vieille-Montagne SA. The estimated cost of the project is US$150 million, and the smelter, located 120 kilometres east of the mine, will have a capacity of 60 000 tonnes of zinc metal per year. The smelter was scheduled for completion in 1984.

So far, the only processing associated with the three ores is preliminary ore beneficiation. All the ores mined are smelted abroad.

Uses and markets

Lacking domestic smelting facilities, Thailand has had to export most of the ores mined from local deposits (Table 10.13). Lead was exported mainly to the Netherlands, the remainder going to Japan, Belgium, West Germany and Italy. Thai zinc ore was shipped to the United States, Australia and West Germany. Japan was the sole buyer of Thai copper in 1973, the only year of exports.

To satisfy domestic demand, Thailand imported lead, zinc and copper in processed forms. Lead was imported mostly unwrought and

Table 10.12 Production of lead

	1971	1972	1973	1974	1975	1976	1977	1978	1979	1980	1981
Total volume (tonnes)	3 465	4 279	8 716	3 631	3 608	2 127	1 190	3 945	20 515	24 847	40 667
Total value (million baht)	8.1	9.0	21.3	17.1	13.1	6.4	7.7	26.4	261.3	228.6	298.5

Source: Department of Mineral Resources *Mineral Statistics of Thailand* (various issues)

then cast locally. Two casting plants gave Thailand a total capacity of 950 tonnes per month. Thailand's major supplier of lead was Australia, which provided over 80 per cent of Thailand's lead imports in the 1970s. The main application of lead in Thailand came in the manufacture of car batteries.

A high percentage of Thailand's zinc was imported in the form of zinc smelter and unwrought zinc. As in the case of lead, Australia was the dominant supplier, accounting for 60–70 per cent of total zinc imports between 1970 and 1980 (Table 10.13). The main application of zinc was in the manufacture of galvanised sheets, used mainly in construction.

Data on copper imports are also shown in Table 10.13. Copper was imported in two forms, unwrought copper, and copper telephone and telegraph wire. Between 1970 and 1980, over half of Thailand's copper imports came from Japan. Other important suppliers were the United States, Australia, and several European countries. The bulk of copper was consumed locally as refined metal in the production of electrical wire. Other important uses were in the manufacture of brass and bronze products, for example, utensils, jewellery and furniture.

Petroleum

Petroleum exploration in Thailand dates from 1921, when the government hired an American geologist to conduct an oil search survey, the oil to be used as a substitute for wood in steam train engines. Subsequent surveys by various government agencies indicated a small oil deposit in northern Thailand. However, not until the late 1950s did the Ministry of Defence start oil production, at a rate of less than 500 barrels per day (bpd).

After 1961 the government started to promote private investment in petroleum exploration. A few multinational corporations were awarded concessions, but their explorations failed to discover significant petroleum reserves. In 1971 new petroleum legislation replaced the former petroleum exploration permits granted under Thai mining laws. This new legislation covered ownership and the work obligations of concession holders, and made exploration in Thailand more attractive to large international oil companies. Encouraged also by the surge in oil prices after 1973, over ten oil companies accelerated their exploration activities in Thailand, and several have been fortunate to discover oil and natural gas deposits large enough to justify production.

Reserves

By December 1982 the following reserves of oil and natural gas, at the locations shown in Figure 10.6, had been identified:

Crude oil: in TSEPCO's Sirikit Field in Block S1, the maximum reserves of 'stock tank oil initially in place' were estimated at

Table 10.13 Exports and imports of lead, zinc and copper

	1971	1972	1973	1974	1975	1976	1977	1978	1979	1980	1981
Exports of lead											
Volume (tonnes)	5 180	–	800	885	1 490	1 225	2 220	2 500	13 097	20 971	37 580
Value (million baht)	12.2	–	1.8	1.2	1.7	2.2	6.5	8.7	103.3	165.0	238.5
Imports of lead and lead alloys											
Volume (tonnes)	na	na	8 269	5 777	8 345	8 051	7 931	9 882	11 024	17 101	12 803
Value (million baht)	na	na	68	86	87	83	105	159	290	365	264
Imports of zinc and zinc alloys											
Volume (tonnes)	20 214	26 578	23 975	24 816	18 795	21 004	29 617	33 848	34 492	34 780	40 098
Value (million baht)	146	196	246	419	341	356	471	456	582	668	813
Imports of copper and copper alloys											
Volume (tonnes)	na	na	na	8 205	10 149	8 933	18 289	17 788	19 593	17 752	15 191
Value (million baht)	na	na	na	409	389	367	635	662	984	983	974

Note: na—not available
Sources: Department of Mineral Resources *Mineral Statistics of Thailand* (various issues); and some figures from the Metallurgy Division of the Department of Mineral Resources

196.2 million and 17.4 million barrels respectively. From these sources, the recoverable reserves range from 30 million to 60 million barrels (see *Economic Outlook*, February 1983).

Natural gas: the total natural gas reserves in the Gulf of Thailand are estimated at 16.1 trillion cubic feet. The breakdown is shown below.

Gas field	*Company*	*Block number*	*Estimated total reserve (trillion cubic feet, proven and probable)*
Erawan	Union Oil	B12	1.8
B	Texas–Pacific	B15, B16	7.2
Kaphong-Plathong	Union Oil	B10, B11	1.3
Baanpot	Union Oil	B13	0.8
Satun	Union Oil	B11, B12	3.2
E	Texas–Pacific	na	0.2
Pla-Dang	Union Oil	na	0.6
Jakrawan	Union Oil	na	1.0
		Total	16.1

(*Source*: Department of Mineral Resources, Mineral Fuels Division)

These estimates may be revised as new data from current and future drillings and seismic surveys become available. It is also possible that some natural gas will be produced in the future from the northeast concession areas (E1–E5) awarded to Esso Exploration and Production Khorat Inc.

Before 1983, the only oil produced onshore came from Fang District, Chiengmai Province, in the north. The operation was managed by the Ministry of Defence since the late 1950s, and production levels ranged from 50 bpd up to 300 bpd. No further increase in Fang output is likely in the future.

In 1981 Thai Shell Exploration and Production Co. (TSEPCO) discovered a substantial amount of crude oil at Lan Krabue Subdistrict, Kumphangphet Province. Oil production from the field, officially named 'Sirikit Oilfield', began in January 1983 at the rate of 5000 bpd, this level being expected to rise to 17 000 bpd by 1985. TSEPCO also produced associated natural gas from the production wells at the rate of 450–720 cubic feet per barrel of crude oil.

So far, natural gas discoveries in the Gulf of Thailand have been more significant than oil finds. In 1981 Union Oil, an American oil company which started exploring offshore in 1972, began to produce natural gas from its Erawan Field in Block 12. In 1982 the production levels ranged between 120 million and 150 million standard cubic feet (scf) per day, below the original target levels of 200–250 million scf per day. Union Oil's operation also yielded condensate of 4800 bpd. 1982 predictions were that with more investment natural gas production from the Erawan field would be boosted to 200 million scf per day by the end of 1983.

As for production in other areas, Union Oil was expected to supply natural gas from additional fields at the rate of 150 million scf per day by early 1985, increasing to 300–400 million scf per day from late 1986. Texas Pacific, another American oil company, discovered Thailand's largest natural gas field to date in its four concession areas (Blocks 15–16). Although the production plan is not firm, it is likely that production will start in 1987 at the rate of 150 million scf per day, increasing to 250 million scf per day from 1989.[8] Thailand's annual supply of natural gas is expected to increase from 120 million scf per day in 1982 to 980 million scf per day by 1990 (Table 10.14).

At the 1982 flow rate of 120 million scf per day of natural gas, some 5000 bpd of condensate were also produced. The value of natural gas and condensate amounted to over 3000 million baht for 1982. By 1986, natural gas and condensate production was expected to more than treble, increasing production value to a level close to that of tin in the early 1980s. In fact, from 1986 onwards natural gas and condensate are likely to become Thailand's most important minerals if planned production increases can be implemented.

Production

Three oil refineries have a total refining capacity of 175 000 bpd. Thailand's first refinery, the Thai Oil Refinery, is 48 per cent government-owned, the remaining equity being held by Shell, Caltex and other Thai and foreign investors. The refinery is managed by personnel from the Royal Dutch Shell group. It had a capacity of 65 000 bpd in 1982, which is expected to increase to 120 000 bpd in 1987. The refinery is located in Sri Racha District, Chonburi Province, on the eastern coast.

Table 10.14 Estimated production and uses of natural gas (million cubic feet per day)

		Uses		
Year	Production	Substitute for fuel oil	Raw materials	Total
1982	120	200	–	200
1983	171	250	–	250
1984	223	238	39	277
1985	380	354	60	414
1986	530	460	128	588
1987	780	599	295	894
1988	830	642	295	937
1989	880	620	334	954
1990	980	608	334	942
1991	980	731	334	1 065

Sources: The National Energy Administration; the Petroleum Authority of Thailand.

The second refinery is wholly government-owned, and is operated by the Petroleum Authority of Thailand, a government enterprise. Its refining capacity in 1982 was 65 000 bpd, scheduled to increase to 70 000 bpd by 1985. The refinery is situated in Bang Chak, an eastern area of Bangkok.

The third refinery, which is wholly owned by Esso, is located in Sri Racha District, near the Thai Oil Refinery. It has a 45 000 bpd capacity, and the government gave approval to expand capacity to 63 000 bpd in 1985 by 'de-bottlenecking' operations.

The products of these three refineries are gasoline, kerosene, jet fuel, diesel oil, fuel oil and LPG. Between 1970 and 1980, the distribution of production was in approximately the following proportions: fuel oil, 34 per cent; diesel oil, 29 per cent; gasoline, 21 per cent; jet fuel, 10 per cent; kerosene, 3 per cent; and LPG, 3 per cent.

Most of the crude oil refined locally is imported, mainly from the Middle East. Crude oil produced from the Sirikit Oilfield is refined at the government-owned Bang Chak refinery.

As of late 1982, natural gas from the Gulf of Thailand was not processed, but was used as fuel in two electricity generating plants. It is planned to separate the gas, and to use the extracted gas components (methane, ethane, propane and buthane) as feedstocks for a petrochemicals complex, a urea–ammonia fertiliser plant, an LPG plant, a soda-ash plant, and other gas-related industries. A gas separation plant is under construction on the eastern coast of Thailand, where natural gas is delivered onshore. The project, costing US$200 million, was expected to have an initial capacity of 350 million scf per day of gas in 1985. While some gas will be used as feedstocks, most of it will continue to be burned as a fuel in power generation plants and heavy industries.

Uses and markets

Oil products

Because of its limited production, Thailand relies almost exclusively on foreign sources for its oil. Both crude oil and finished oil products are imported, the locally refined output being insufficient for domestic consumption. Consumption of oil, Thailand's most important energy source over the past three decades, doubled from about 100 000 bpd in 1971 to over 200 000 bpd in 1980 (Table 10.15). During that period, diesel oil consumption accounted for 33–36 per cent of total energy consumption, fuel oil for 30–34 per cent, gasoline for 18–20 per cent, jet fuel for 7–10 per cent, kerosene for 2–3 per cent, and LPG for 2–4 per cent.

The main user of oil products was the transportation and communications sector, which accounted for over 40 per cent of all consumption during the period. The oil products taken up by this sector were mainly diesel oil and gasoline. Manufacturing and the electricity and water supply sector each accounted for over 15 per cent of the total oil

Table 10.15 Consumption and imports of crude oil and petroleum products

	1970	1971	1972	1973	1974	1975	1976	1977	1978	1979	1980
Consumption of petroleum products (barrels per day)	88 159	109 201	128 170	144 943	141 503	146 690	165 695	183 638	198 833	209 091	211 862
Imports of crude oil											
Volume (million barrels)	27.80	39.59	47.96	55.90	48.21	52.34	54.85	60.94	59.99	65.01	61.36
Value (million baht)	1 198	1 941	2 432	3 572	10 382	12 076	13 857	16 448	16 527	23 425	39 304
Imports of petroleum products											
Volume (million barrels)	9.59	6.71	4.59	6.89	6.55	5.10	7.69	10.71	16.00	15.80	22.17
Value (million baht)	718	509	373	737	1 589	1 650	2 241	3 360	5 253	7 698	17 040

Note: Petroleum products include gasoline, kerosene, diesel oil, LPG, and some special fuels.
Sources: National Energy Administration *Thailand Energy Situation* (various issues); *Bank of Thailand Monthly Bulletin* (various issues)

demand, the main product consumed being fuel oil. Oil consumption by the agricultural sector, mostly diesel oil, accounted for another 10 per cent of the total.

Oil importing, wholesaling and retailing is handled by three international oil companies,[9] Shell, Esso and Caltex, and by a government enterprise, the Petroleum Authority of Thailand.

Since all domestically consumed oil was imported during the 1970s, the volume of imports increased in line with oil consumption. Accordingly, imports of both crude oil and petroleum products doubled from 30 million and 10 million barrels respectively in the early 1970s to 60 million and 20 million barrels respectively in the late 1970s. However, because of sharp oil price increases after 1973, the value of oil imports rose spectacularly from around 2500 million baht a year just before 1973 to 56 000 million baht in 1980 (see Table 10.15). Since 1974, oil has been Thailand's single most important import item, accounting for 20–30 per cent of the total import bill, compared with 10 per cent before the first oil crisis.

Crude oil was once imported mainly from the Middle East, particularly Saudi Arabia. However, in the late 1970s and early 1980s relatively more crude oil was purchased from other sources, such as China, Indonesia and Malaysia. Finished oil products were always imported from Singapore, the largest oil refining centre in the region.

Before 1977, a small volume of oil products was exported to Laos and other neighbouring countries. However, this activity stopped after 1977 when domestic oil consumption began to exceed refined output.

NATURAL GAS

During the 1970s natural gas from the Gulf of Thailand was purchased by the Petroleum Authority of Thailand, which in turn delivered it to another government enterprise, the Electricity Generation Authority of Thailand, for use in two of its power plants. The gas was transported from the gasfield through a pipeline owned by the Petroleum Authority of Thailand, 590 kilometres long and in two sections, 425 kilometres under water and 165 kilometres on land.

The Petroleum Authority contracted to supply 30 million scf per day of gas to a cement company from 1983. Plans were also laid for developing other uses for offshore natural gas, including the construction of a gas separation plant. As a feedstock, the gas will be used mainly in an ethylene cracking plant, scheduled to produce 300 000 tonnes of ethylene for year. Gas will also be important raw material for downstream petrochemical industries, established in later stages. Figures on Thailand's natural gas prospects are shown in Table 10.14.

National policies on mining

Exploration and production

A policy of reform in the mining industry was initiated in 1960, when the first national economic and social development plan was drawn up.

The government began to encourage mineral development with the aim of increasing export earnings from mineral resources. After the first oil crisis of 1973–74, another policy goal was set, that of developing indigenous energy resources as substitutes for imported oil.

In general the government is not directly engaged in mineral production, apart from a few tin and tungsten mines operated by the Mines Organization and the Offshore Mining Organization. Some support services are also provided by the government through the Department of Mineral Resources, such as geological mapping, economic and geological surveys, and technical advice. In most cases, the detailed exploration and development of accessible mineral resources are left to the private sector. However, there are some exceptions where government enterprises themselves carry out mineral development. Examples include the roles of the Electricity Generation Authority of Thailand and the National Energy Administration in lignite development.

The principal legislation affecting the mineral industry is the 1967 Minerals Act, and its amendments of 1973 and 1979. The law governs onshore and offshore exploration, mining, selling, dressing, transporting and exporting of all minerals except petroleum. The main tenet of the law is that mineral resources are vested in the Crown, and thus the owner of the land is not entitled to ownership of the minerals beneath. No one can explore or mine mineral ores without a prospecting licence or a mining lease from the government. This also applies to petroleum exploration, but there are some differences between non-petroleum and petroleum minerals in the terms and conditions of concessions. A separate law, the Petroleum Act of 1971, amended in 1973, now forms the basic legislation governing petroleum exploration and production.

The Minerals Act is administered by the Ministry of Industry through the Department of Mineral Resources. It provides for the issue of three types of prospecting licences (ordinary, exclusive and special), two types of mining leases (provisional and ordinary), and other relevant licences, for example, mineral purchasing licences, mineral storage licences, mineral transporting licences, and mineral importing and exporting licences. Prospecting licences are granted for a period of one to three years, while a mining lease is valid for a maximum of 25 years.

In practice, the issue of a prospecting licence and a mining lease has become increasingly complicated and time consuming. The government's concern over the adverse impacts of mining on the environment, which includes forests, natural water sources and soil surface, has resulted in the involvement of more government agencies before a licence or lease application can be approved. All four departments concerned, the Department of Mineral Resources, the Department of Land Development, the Department of Forestry and the Department of Rural Irrigation, must have no objection to the proposed mining before a lease is granted. Recently, approval by military authorities has also been required if mining will be in a so-called 'sensitive' area where communist influence is strong. As a result, it can take years before a lease is approved, and this has led to widespread illegal mining.

The long delay in obtaining a mining lease is not the only factor causing illegal mining. Even when mining leases are legally obtained, the concession holders' rights are often violated by illegal miners. This is well illustrated by the invasion of the small suction boat operators mining tin in the Offshore Mining Organization offshore concessions to the south, and the operation of pick-and-shovel miners in wolfram deposits legally awarded to mining companies. The political pressure exerted on the government by local interests and the absence of law enforcement agencies are two important factors allowing these violations to continue.

Despite the problem of illegal mining, government policy is to open up new areas of sizeable magnitude for further minerals exploration and development. Minerals with good potential include lead, zinc, copper, tin, tungsten and potash. The government also aims to achieve increased domestic processing of mineral exports. Consequently, the setting up of processing facilities is strongly encouraged, and investments in this area are eligible for promotional privileges, including exemption from import duty and business tax on machinery and equipment, as well as an income tax holiday for a period of three to eight years.

Since Thailand is a signatory to the International Tin Agreement, the government is empowered to impose production quotas on tin miners during periods when export quotas are enforced by the International Tin Council. In practice, as in 1982 and 1983, the government usually distributes these production quotas on the basis of the miners' past production. Currently at issue is the question of whether this method of quota distribution is efficient and equitable.

Petroleum operations are covered by the 1971 Petroleum Act, amended in 1973, which provides a concessionaire with the right to explore, produce, store, transport and sell petroleum, but not the right to refine it. The duration of an exploration concession shall not exceed eight years and is eligible for only one renewal, this not to exceed four years. The concession holder is entitled to relinquish the whole or any part of his exploration area at any time. At the end of the fifth year of the exploration period, 50 per cent of each exploration block must be relinquished. At the end of the eight-year period another 25 per cent of the remaining area must be relinquished. At the termination of the exploration period, the production period may not exceed 30 years, although it may be extended once only, for a period not exceeding ten years. The law prohibits the government from nationalising the concessionaire's properties and rights to conduct petroleum operations, and from restricting the export of petroleum except in cases of national emergency.

To date, although the government has played no active role in petroleum exploration and production, it has been instrumental in making petroleum production possible by oil companies which find commercial deposits of oil and natural gas. Through the Petroleum Authority, the government is the sole buyer of local oil and gas, and has also invested a large sum in the construction of gas pipelines, control stations and other installations, all of which have facilitated the transportation of the natural gas from the Gulf. Indeed, without gas consumption by government-owned power plants, natural gas production would not be economic, there being no other large private consumers of natural gas in Thailand. Government-initiated feasibility studies on gas utilisation and future infrastructures and facilities required will further encourage natural gas exploration and production. The role of government in the development of crude oil is currently somewhat less important because of the strong domestic market and established local processing operations.

Foreign investment requirements

Until 1966, private sector mining was restricted to the isthmus of Thailand, the rest of the country above the 11th parallel being closed by Royal Decree to reserve the land for agricultural purposes. The law was subsequently modified, and currently stipulates that the mine owner above the 11th parallel must be of Thai nationality, and that at least 50 per cent of the shares in the mining company must be held by Thai nationals. However, in a case where foreign technology and capital are necessary, this policy can be relaxed provided cabinet approves.

The government first announced a policy on offshore tin mining in 1978, amending it in 1980. This industry had long been dominated by foreign companies. The objective of the policy was to allow greater participation by Thailand in existing and future offshore tin mining. Restrictions on foreign investments fell into three categories.

1 Foreign companies engaged in offshore tin mining for 25 years or more must establish a new company incorporated and registered in Thailand within two years, and must increase the shareholding of Thai nationals in that company to at least 70 per cent.
2 New offshore tin mining investments can be carried out by foreign companies, provided that non-Thai shareholders hold no more than 49 per cent of the issued shares at the outset, this proportion decreasing within five years of operations to no more than 30 per cent.
3 For offshore tin mining at depths greater than 200 feet, a wholly foreign-owned company may operate, provided that no less than 60 per cent of the issued shares are sold to Thai nationals within ten

years of operation. This provision recognises that foreign technology and capital are still essential for deepwater mining.

It should be noted that this policy of restricting foreign investment applies only to offshore tin mining. The mining of other non-petroleum minerals is subject to the less restrictive policies described earlier. Petroleum operations are even less restricted, there being no stated minimum level of Thai participation. However, Thai majority ownership in any mining venture is strongly advocated by the government.

Taxation and royalties

In the main, government revenues from minerals are based on royalties imposed under the Royalty Rates Act 1966 as amended in 1977 and 1979. In the case of tin, the royalty rate is levied on the actual tin content of concentrates. The royalty rates are applied progressively to six intervals of a posted price based on the Penang Market tin price.

In October 1983, to help the tin industry (which was adversely affected by price slumps in the world tin market), the government adjusted the royalty structure by making the rates less progressive when the posted price was below 14 000 baht per 60 kg, but more progressive at higher prices. The adjusted royalty rates are shown below.

Posted price (baht per 60 kg)	*Royalty rate (% of posted price)*
1–3 000	nil
3 000–7 000	5
7 000–9 000	10
9 000–11 000	20
11 000–14 000	30
14 000–17 000	40
17 000–20 000	50
above 20 000	60

In any case, the maximum effective royalty rate on tin shall not be over 30 per cent of the posted price.

For other non-petroleum minerals, royalties are levied as a flat percentage of posted prices. The royalty rates on the minerals under discussion are shown below.

Mineral	*Rates on ore*	*Rates on metal*
Tungsten	–	20
Metallurgical-grade fluorite	7	–
Chemical-grade fluorite	4	–
Iron	4.5	2
Lead	10	2.5
Copper	10	2.5
Zinc	10	2.5

Mining companies are also subject to corporate income tax, the rate being 30 per cent for companies registered with the Security Exchange of Thailand, and 40 per cent for others. Business and municipal taxes are also imposed at a rate of 4.4 per cent on mineral sales and exports.

Petroleum taxation is governed by another law, the Petroleum Income Tax Act of 1971. Royalties are collected in cash or kind at 12.5 per cent of petroleum sales, profits from production being subject to income tax at 50 per cent. For exploration in water depths exceeding 200 feet, involving sophisticated technology, large investment and higher risks, the royalty rate is 8.75 per cent of the petroleum produced, with the income tax rate unchanged. In addition to these taxes, the government usually requires and receives a 'signature bonus', a lump-sum payment made when a concession is granted.

With the discovery of more hydrocarbons, in 1982 the government added a 'production-sharing' feature to the existing 'royalties and tax' system. The new 'additional petroleum benefit' (APB) condition requires a concessionaire to pay, in addition to royalty and profit tax, a bonus to the government at a rate ranging from 27.5 per cent to 43.5 per cent of sales value, when oil production exceeds 10 000 bpd. Moreover, deductible expenses in each year must not exceed 20 per cent of the value of petroleum sold in that year. Under the new system, the government share would increase from 46.25 per cent of oil sales when production is below 10 000 bpd, to 68 per cent when production exceeds 30 000 bpd, as shown in Table 10.16. An American Company, the MCF Oil Corporation, was the first to obtain a concession area under this new condition. The area awarded was just west of the Sirikit Oilfield, and was believed to have good potential.

Since the government, through the Petroleum Authority, is to date the sole buyer of Thailand's natural gas and oil, it gains an additional benefit if the wellhead prices are low. The government's attempts to maximise this benefit have led to lengthy price negotiations between the Authority and oil and gas producers. The wellhead price of natural gas from the Erawan Gasfield was agreed upon on a cost-plus basis, taking into account also the market price of fuel oil, the product substituted by natural gas. To speed up production from the Sirikit Oilfield the government and TSEPCO agree to adopt a provisional price of US$28.84 a barrel, the average of the company's and the government's offered prices. Prices are limited by a legal requirement that the price obtained by a concessionaire cannot exceed the import and export prices for crude oil or natural gas.

Except for crude oil and natural gas, the government imposes no price controls on other mineral ores and metals. However, it does control the retail prices of two groups of finished mineral products, the steel products, bars and wire rods, and all oil products. This measure has been adopted because, in the government's view, these products are

Table 10.16 Government revenue from petroleum (per cent)

	Oil production			
	Below 10 000 bpd	10 000–20 000 bpd	20 000–30 000 bpd	Above 30 000 bpd
Sale value	100.0	100.0	100.0	100.0
Royalty	12.5	12.5	12.5	12.5
Deductible expense	20.0	20.0	20.0	20.0
Bonus	0.0	27.5	37.5	43.5
Profit tax	33.75	20.0	15.0	12.0
Net profit	33.75	20.0	15.0	12.0
Government share (royalty plus bonus plus profit tax)	46.25	60.0	65.0	68.0

vital to the livelihood of the population. Thus for oil products prices are controlled both at refinery and retail levels. Different taxes and subsidies apply to various oil products according to their classification as 'essential' products. Thus gasoline is heavily taxed, being consumed mostly by the wealthy, while, for distributional and political reasons, kerosene, LPG and fuel oil attract very low taxes or subsidies.

Institutions and support services

In an attempt to improve the local mining sector, and because it believed that the existing mining associations were ineffective in dealing with the industry's several problems, the government proposed the setting up of two independent institutions, the Chamber of Mines and the Mineral Resources Development Corporation. The Chamber of Mines was intended to exercise control over mine operators in areas where laws and regulations were frequently violated. Additionally, the Chamber was to act as an important link between industry and relevant government agencies.

The second institution, the Mineral Resources Development Corporation, was designed to mobilise technical and financial resources in Thailand's mineral development. The Corporation was expected to play an important role in helping small and medium-sized mining companies to obtain funds from financial institutions, which had been very cautious in lending to the miners. Joint ventures between local and foreign investors were also to be facilitated by the Corporation.

The Asian Development Bank recently provided a technical assistance grant to the Thai government for implementing the Accelerated Mineral Resources Development Project. The project is designed to strengthen and improve the institutional structure of the Department of Mineral Resources, and also involves an airborne geophysical survey. The area surveyed includes approximately 395 000 square kilometres in four separate blocks, accounting for 77 per cent of Thailand's total area. The information obtained is to be used in planning the further mineral development of Thailand.

Concluding remarks

Although the minerals sector has played a relatively small role in the Thai economy, its future growth is expected to be strong. Oil price increases since 1973–74 have meant high trade deficits, with petroleum the largest item among imported commodities. Efforts have been made to accelerate the exploration and production of petroleum and other minerals in order to reduce this growing trade gap. Increased production of natural gas, crude oil, zinc and lignite in the 1980s will probably increase the minerals sector's share of GDP and government revenues. Furthermore, because these minerals will substitute for imports, their production should also contribute to foreign exchange savings.

Despite the mineral sector's good growth potential, the above discussion highlights a number of key policy problems. These issues are summarised below.

First, in recent years mining companies have often complained that mining, particularly tin mining, is no longer highly profitable. This may have been due to oil price increases which pushed up mining costs considerably beyond the point where they were offset by increases in the export prices of tin. Moreover, the tin royalty, an ad valorem tax based on world prices, has not been adequately adjusted to reflect cost changes. The final result has been illegal mining and smuggling, leading to inefficient mining practices and loss of government revenues. Appropriate policy formulation will require answers to the following questions.

1 How do tin producing countries, individually or collectively, maintain the world price of tin at levels which are remunerative for producers and fair to consumers?
2 How can taxation, particularly royalties, on tin and other minerals be structured to ensure an efficient or optimal rate of exploitation, to minimise illegal activities—for example smuggling—and administrative costs, and to take into account cost and price changes?
3 Where export quotas are imposed by the International Tin Council, how can the government distribute quotas among miners to ensure the effectiveness of the quota system and the best tradeoff between efficiency and equity in the minerals sector?

Second, there are large areas in the country which remain generally unexplored. This seems to be attributable to long delays in issuing mining leases, and violations of concession rights by illegal miners. In some cases, the lack of a clear or consistent policy on tradeoffs between mining and conservation leads to uncertainties in the industry, and thus a lag in activity.

Third, in the case of petroleum exploration and production, the suitability of the present royalty–income tax system may be questioned. Even though a 'production-sharing' feature has been introduced, government insistence on being the sole buyer can make the process of price negotiations between the government and oil companies time-consuming and detrimental to the national energy program.

Fourth, it is widely accepted that energy pricing is the most effective instrument in energy demand management. As an oil importing country, it is very important that Thailand conserve energy. The government's policy on energy pricing must be formulated in such a way that energy prices encourage efficient energy use, subject to certain constraints related to government tax revenue, equity, and other political considerations. To date, energy pricing, particularly petroleum products, has been ad hoc and haphazard. As a result, the prices paid by consumers have been distorted, leading to unintended and sometimes undesirable interfuel substitution.

Finally, several minerals have been, and will continue to be, processed in Thailand, with the aim of creating added value and promoting industrial development. Large minerals processing projects, for example the zinc smelting plant and the petrochemicals complex utilising natural gas, will be launched into production in the near future. Since most of these projects are import-substituting, it is important to consider their impact on foreign exchange savings, employment, technology transfer, government revenue, and export prospects.

Notes

The section on mining technology relies extensively on the US Bureau of Mines (1980) and Scholla (1981). The section on national policies on mining relies extensively on the Department of Mineral Resources (1981).

1 Measured reserves are computed from dimensions revealed in outcrops, trenches, workings or drill holes. The quality is computed from the results of detailed sampling. The sites for inspection are spaced so closely, and the geological character is so well defined, that the mineral content of the resource is well established.

2 Indicated reserves are computed from information similar to that used for measured reserves, but the sites for inspection are farther apart. However, the degree of assurance is high enough to assume continuity beyond measured and indicated reserves.

Submarginal deposits or resources are the component of identified resources that does not meet the economic criteria of reserves and marginal reserves.

Hypothetical deposits or resources are undiscovered resources that are similar to known mineral bodies, and that may reasonably be expected to exist in the same producing region under analogous geological conditions.

Speculative deposits or resources are undiscovered resources that may occur either in known types of deposits in favourable geological settings where mineral discoveries have not been made, or in types of deposits as yet unrecognised for their economic potential.

3 Illegal mining in the TEMCO lease actually started before 1975, and was one of the important factors underlying the revoking of the concession.
4 Hundreds of small mines were reported to have ceased operation in 1981/82.
5 The quantity of tin concentrates is converted from 3500 tonnes of tin metal reported by the Offshore Mining Organization.
6 A more detailed discussion of tinplate production in Thailand is given above.
7 The study was to be completed by 1984.
8 The expected total production of natural gas in Thailand is shown in Table 10.14.
9 Until 1982, Summit was another foreign oil company operating in oil trade and leasing the Bang Chak refinery.

References

Department of Mineral Resources (1981) *Policy Guidelines for the Mining Sector* Bangkok (July)

ESCAP (1979) 'Transnational Corporations and the Distribution of Gains in the Tin Industry of South-East Asia' Working Paper No. 7 (August)

Ingram, J.C. (1978) *Economic Change in Thailand, 1850–1970* Stanford University Press

Scholla, P.F. and Associates (1981) *Mines and Mineral Deposits in Thailand* Bangkok

Sivavong, V. (1980) *Study Report on Tin* Siam Commercial Bank

Trangkochasarn, Y. (1982) *Economic Minerals in Thailand* Bangkok: Department of Mineral Resources

US Bureau of Mines (1975) *United States Mineral Resources*

——(1980) *Minerals Facts and Problems*

Table 10.17 Baht–US dollar exchange rates, 1970–81

Year	Baht per US dollar
1970	20.928
1971	20.928
1972	20.928
1973	20.375
1974	20.375
1975	20.400
1976	20.400
1977	20.400
1978	20.390
1979	20.425
1980	20.415
1981	21.795

Note: The figures are averages of selling and buying rates.
Source: *Bank of Thailand Monthly Bulletin* (various issues)

Figure 10.1

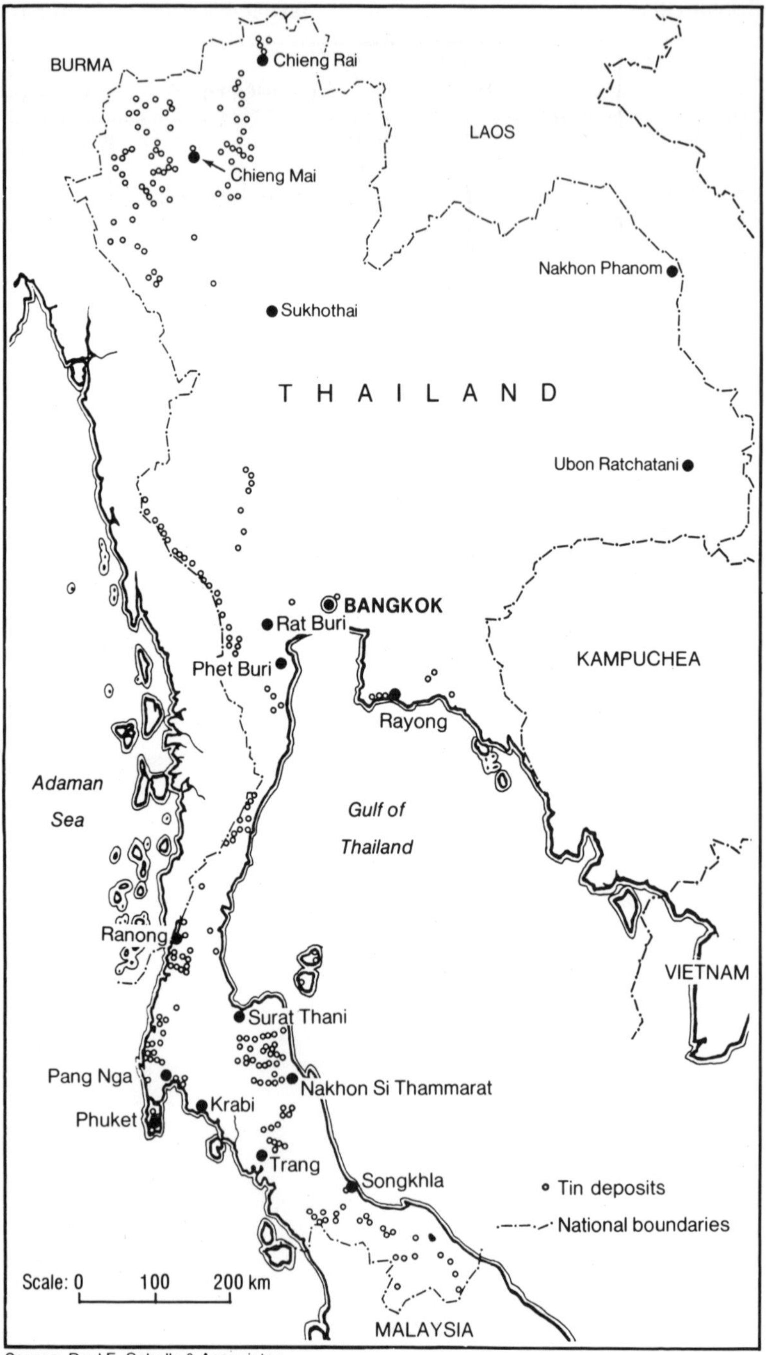

Source: Paul F. Scholla & Associates

Figure 10.2

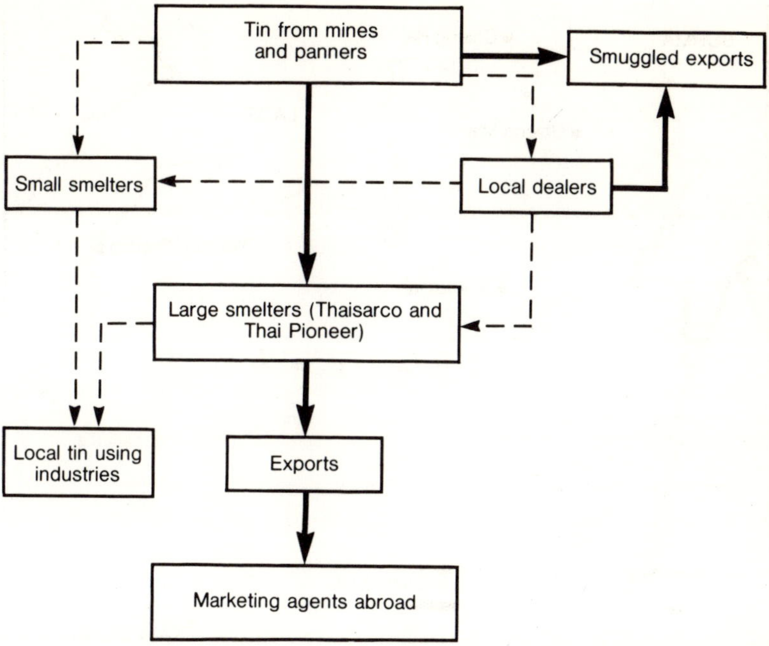
Tin from mines and panners
Smuggled exports
Small smelters
Local dealers
Large smelters (Thaisarco and Thai Pioneer)
Local tin using industries
Exports
Marketing agents abroad

Figure 10.3

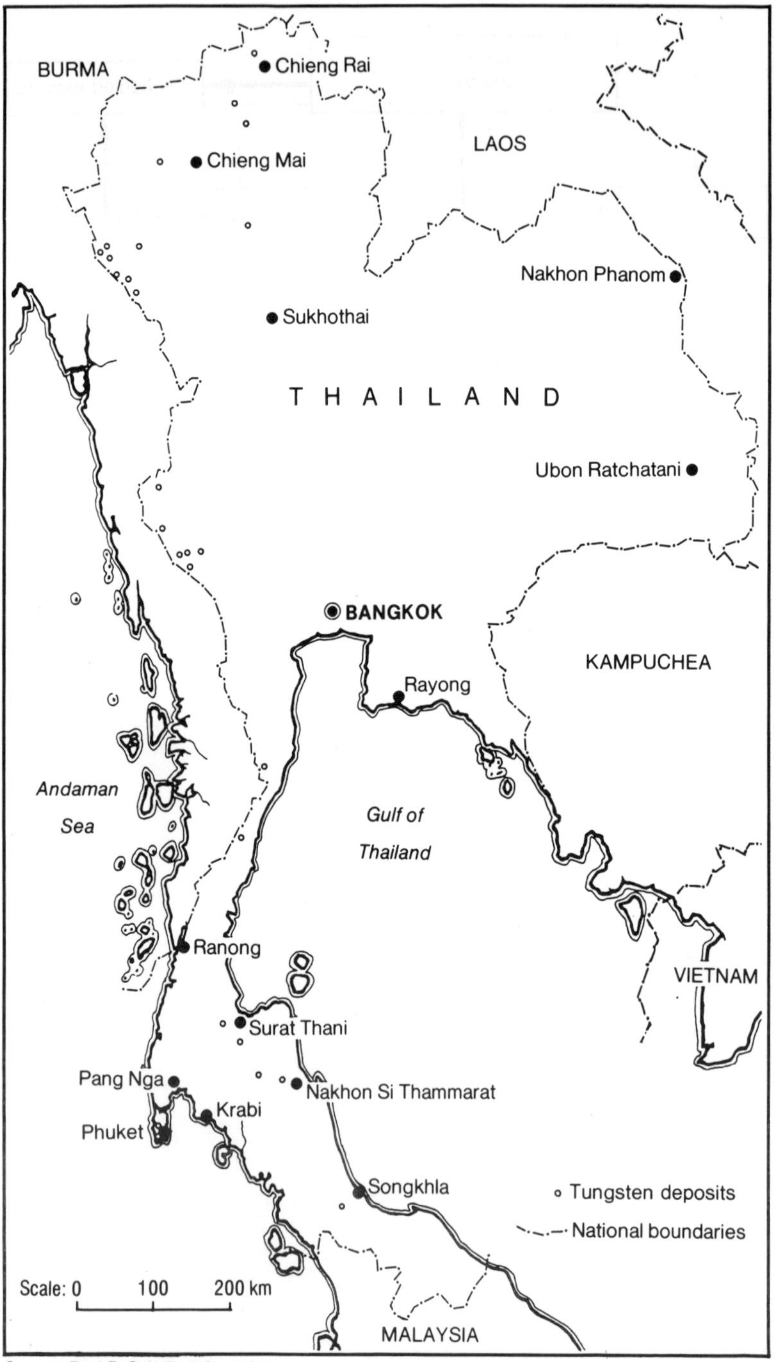

Source: Paul F. Scholla & Associates

Figure 10.4

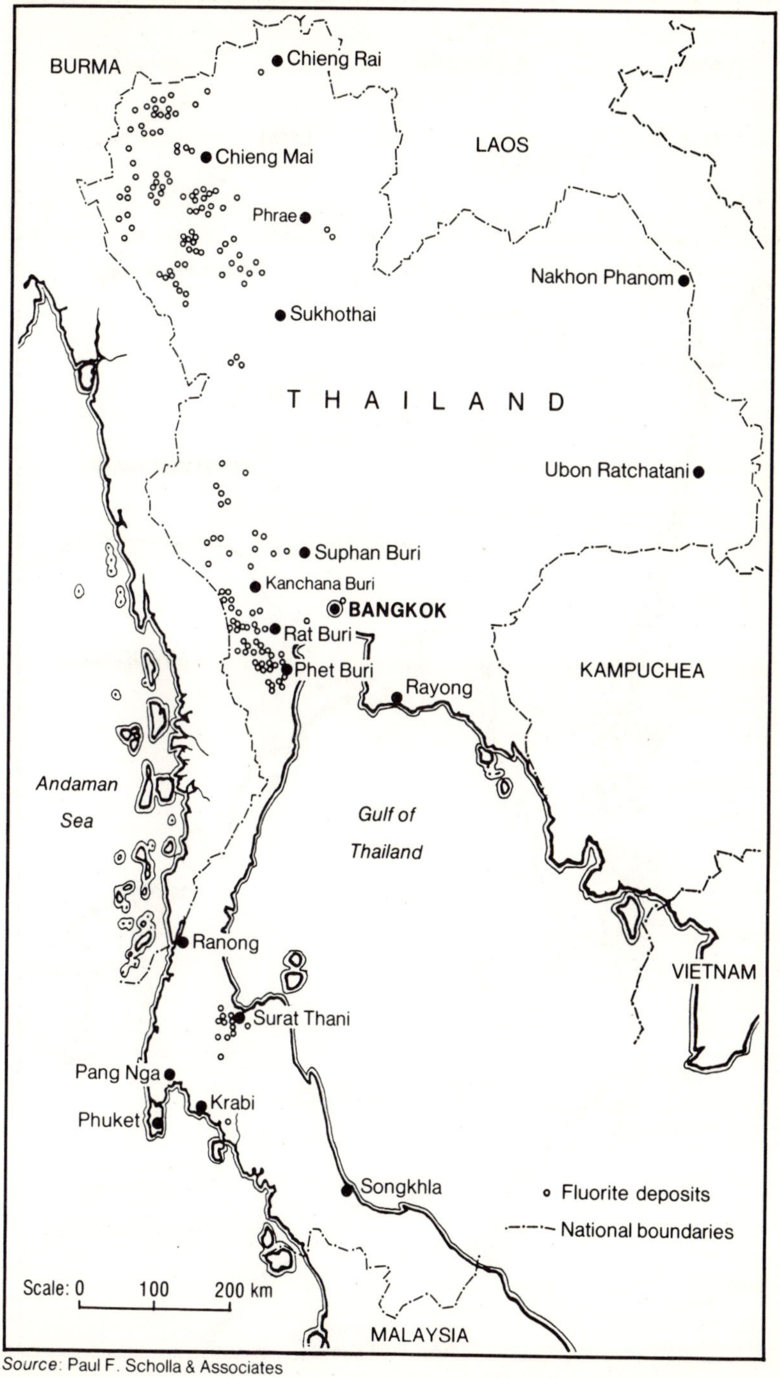

Source: Paul F. Scholla & Associates

Figure 10.5

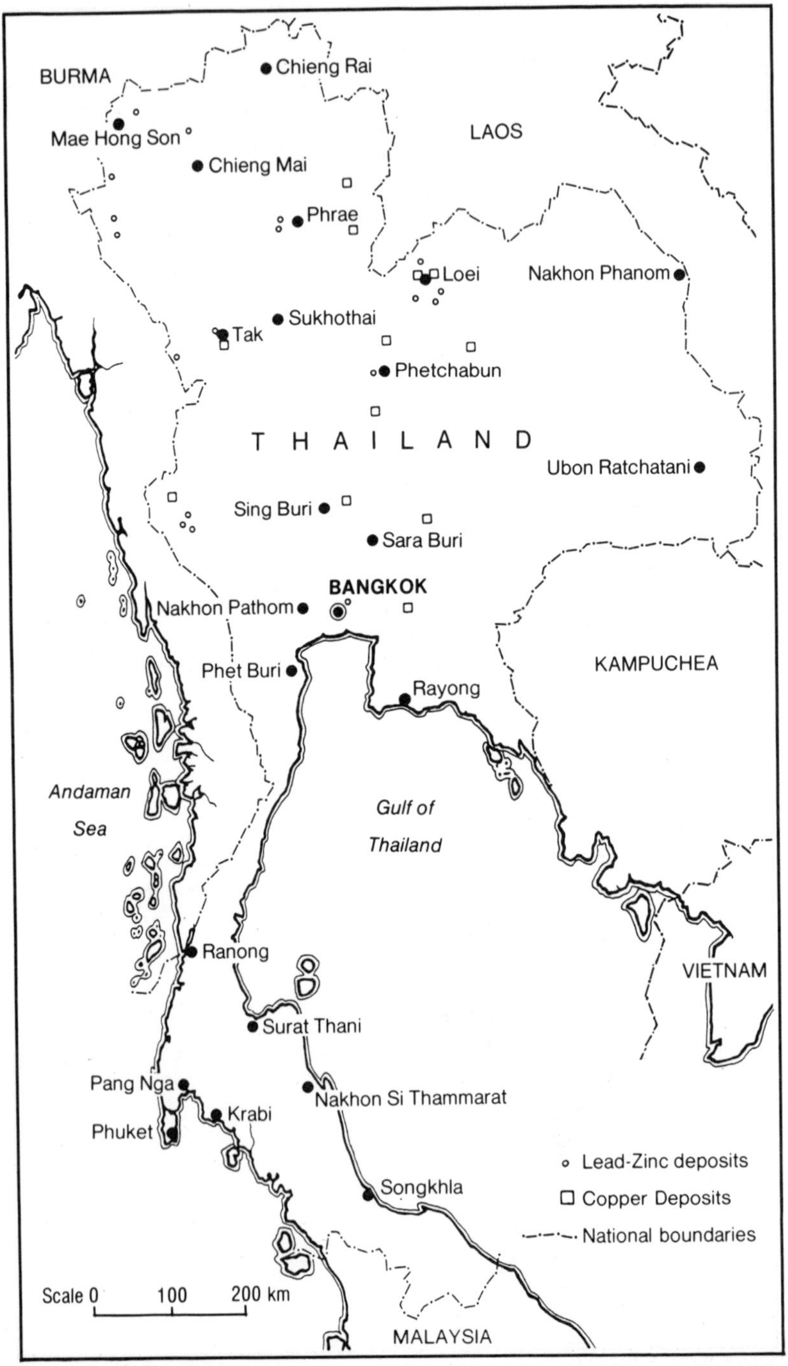

Source: Paul F. Scholla & Associates

Figure 10.6

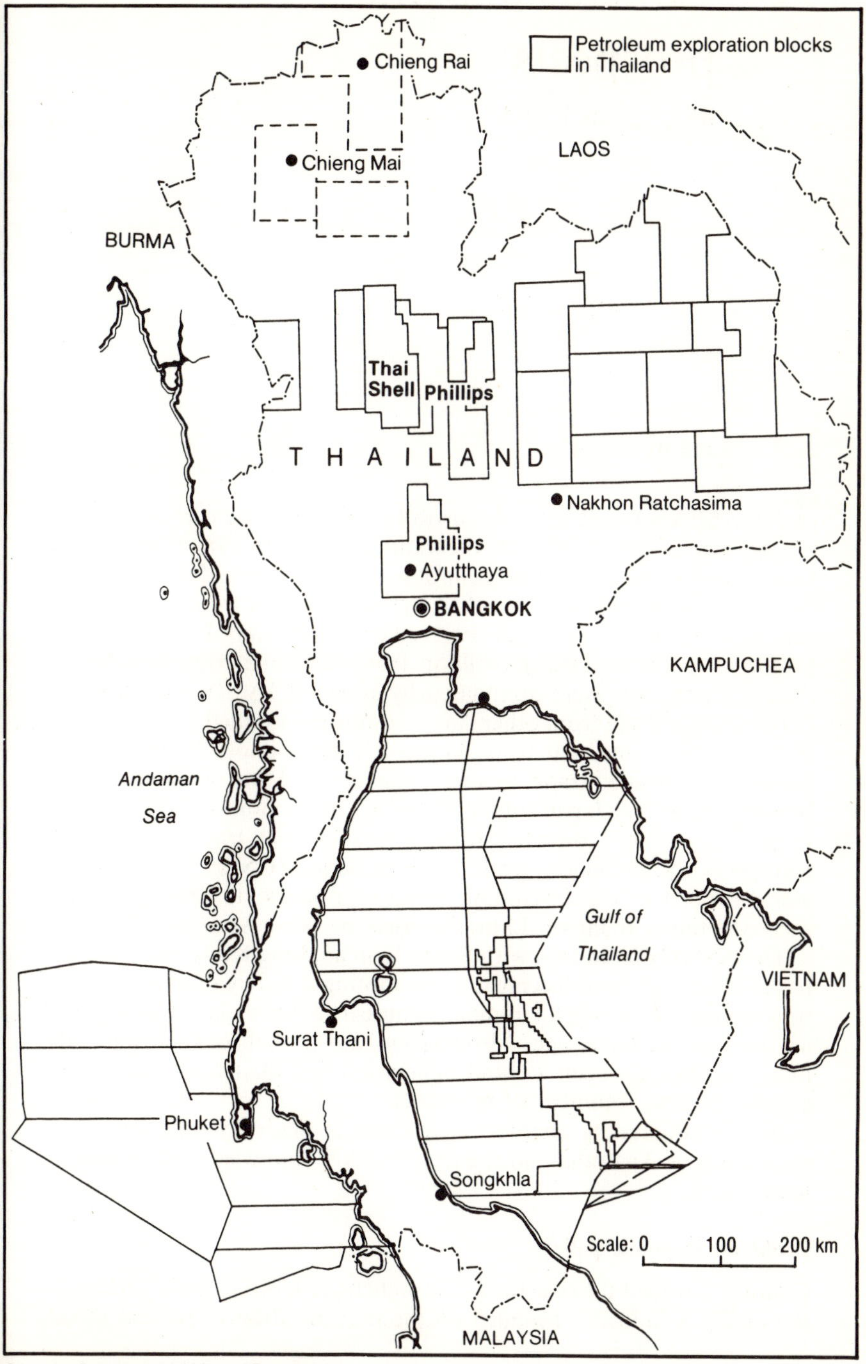

Scource: Department of Mineral Resources

11 Conclusions and prospects

BRUCE MCKERN

ASEAN and world trade

THE ASEAN countries and Australia today find themselves situated on the edge of the world's largest trading region. In 1984 trans-Pacific trade, accounting for 11.3 per cent of world trade, exceeded for the first time trans-Atlantic trade. North America's imports from countries of the Western Pacific totalled US$120.6 billion, and Latin America imported a further US$21.1 billion from the same countries. North America's imports were dominated by Japan (US$61.5 billion, or 51 per cent), but trade from other Asian countries[1] totalled US$54 billion (GATT, 1985), and that from Australia US$3 billion. In addition to trans-Pacific trade, trade amongst the other Asian economies and between them and Japan totalled US$39.3 billion and US$30.9 billion respectively. While dwarfed by the total export trade of Western Europe (US$729 billion) and North America (US$295 billion), the world exports of the developing Asian economies, at US$181 billion, exceeded those of Japan, Latin America, or West Asia[2] in 1984.

The ASEAN countries, with South Korea, Hong Kong and Taiwan, have, of course, been foremost in promoting a policy of export-based development, in which their minerals industries have played an important role, as we have seen. Nevertheless, the growth of their minerals industries has not been without difficulties, and their future prospects are clouded by a number of problems.

In this concluding chapter an attempt is made to summarise the experience of Australia and the five ASEAN countries studied and to look ahead.

Problems in public policy

Chapter 3 argued the case for a comprehensive national policy towards the minerals industry, founded on the maximisation of present social value. Rationally appealing as that argument may be, it is clear that the political complexities of policy formulation in each of the six countries studied has produced minerals policies which differ widely in their emphases. Mineral pricing is one such example, particularly where energy pricing is concerned. It is clear that, apart from OPEC and the International Tin Agreements, cooperative action amongst producer countries on minerals production and pricing has met with little success. Within individual countries of the region, pricing policies differ. Australia has, broadly speaking, followed a market-determined pricing policy for minerals products. In the case of petroleum, a Federal levy has been used to establish prices for domestic crude equivalent to the landed price of imported crude (which is not, of course, a free price, but a price reflecting OPEC-administered production controls). The selling price for refined products suffers from some price control and a retail sales tax, which favours diesel fuel, but there is no other cross-product intervention. As a consequence, there has been a shift in energy demand towards conservation and alternative sources.

In the case of Indonesia, domestic consumption of petroleum has been effectively subsidised, despite the aim of the Fourth Five-Year Plan to reduce dependence on petroleum. Domestic oil consumption growth has limited the surplus available for export, and there has often been a reluctance to let domestic prices rise, particularly for products used in households. In Malaysia, minerals are more heavily taxed than, for example, food, and household kerosene and diesel fuel have been subsidised. In the Philippines, price controls have also shifted fuel consumption towards diesel fuel with resultant shortages. Government ownership of the largest refiner (PNOC) has had a negative effect on the entry of private enterprise into the use of coal-based and geothermal energy. In Thailand, also, energy pricing policies have been seen by some as lacking consistency, resulting in distortions and interfuel substitution.

State ownership of the minerals industry in the region has increased during the period. In Australia private ownership has been the rule, with a significant degree of foreign ownership. Australian foreign ownership policy has shifted from laissez-faire to highly controlled and, more recently, to a more neutral stance. There is very little government ownership, except in coal. Foreign ownership in uranium mining is limited to 25 per cent, whereas in other minerals a general limit of 50 per cent can be relaxed if it is considered to be in the national interest. Indonesia has had a mixed approach: a high level of government ownership in the petroleum sector and considerable private and foreign ownership in hard minerals. It, too, has passed through several phases in its approach to foreign investment, recognising the need for foreign capital, technology and management skills in the development of new resources. In Thailand, government participation in mining is minimal, private enterprise being the rule. Offshore tin mining is the only sector in which foreign ownership is controlled. There is encouragement of Thai ownership in other sectors, particularly in projects situated above the 11th parallel, but majority foreign participation is permitted when foreign technology and capital are needed.

Malaysia has increased its domestic ownership of the tin industry, while State-owned and Federally-owned enterprises have become active in other sectors. Petroleum production is government-owned, with foreign enterprises contracted to undertake exploration and production under production-sharing contracts, as in Indonesia. In other minerals there is pressure for local ownership, and there is growing state ownership, in some cases with the assistance of foreign loans. The Philippines has for many years restricted foreign ownership of minerals to a minority position, sometimes employing service contracts to obtain foreign participation. Government ownership is not great, except in the energy sector, where PNOC dominates petroleum refining and distribution. The National Power Corporation controls power generation, and will influence the sourcing of uranium oxide for the country's first nuclear reactor. In the hard minerals sector, private ownership is the rule.

The general pattern in the region regarding ownership is to use government agencies to control petroleum production, either through state ownership or by regulation, often using foreign contractors. In non-fuel minerals most of the ownership is private, and foreign ownership is permitted, and in some cases strongly encouraged. Given the scarcity of capital and technology in the past, the encouragement of foreign investment has been of great importance in the development of the industry throughout the region, particularly where large-scale projects were concerned. Although at times there have been phases of greater restriction on foreign ownership, and in some countries a deterioration in relationships with foreign companies, the atmosphere in more recent years has been one of mutual cooperation under workable arrangements. Subject to the stability of the regime in each country, there is no reason why this accommodation should not continue.

It is interesting to note that there has been very little international direct investment in the minerals industry between Australia and ASEAN. There has been minor Australian ownership of tin mining in Indonesia and investment in offshore petroleum exploration, but otherwise little cross-investment. A number of reasons suggest themselves for this phenomenon. All of the ASEAN countries, as well as Australia, have been short of capital, tending to concentrate on domestic rather than foreign investment. Yet Australian firms have invested quite actively in ASEAN in other sectors, and there have been inflows of portfolio investment, as well as direct investment, to Australia from ASEAN. The most important reason probably lies with the nature of domestic minerals investment in each area. Australian minerals developments of the 1960s and 1970s were large-scale, concentrated in iron ore, coal and bauxite (minerals in which ASEAN is not rich) and dominated by foreign ownership. Australia-based entrepreneurs were not very active in the minerals industry during this period of great development, and foreign firms brought capital, and more importantly technical and managerial resources, to the development of the Australian industry. Their focus was Australia and the Japanese market, and they either did not consider ASEAN investments or did not see their Australian operations as a relevant investment vehicle. In petroleum, foreign-based firms predominated in both Australia and the ASEAN region, and the parent companies made their ASEAN investments directly, rather than via a subsidiary located in one of the other countries. In copper, zinc and tin, the Australian companies were preoccupied with their domestic operations. ASEAN-based firms, likewise, tended to be domestically preoccupied or lacking the large-scale experience to consider foreign investment. Australia's steel producer was also focused on the Australian market, seeing ASEAN as an export opportunity but not to be served by foreign investment. In recent years the expertise to invest abroad has grown in both ASEAN and Australia, but the poor state of the market for minerals commodities has been a restraining influence. While there has been further investment downstream in the building products industry, cross-investment in basic minerals has remained low. As markets improve, it can be expected that there will be more interest in foreign direct investment.

One of the key areas of confrontation between governments and private investors (domestic and foreign) has been the issue of fiscal policy towards the minerals industry. The country studies in this volume make it clear that each government sees the industry as a source of taxation revenue, and taxation policies have been erected in a rather haphazard way to extract revenue. In federal systems, such as Malaysia and Australia, there has been competition between State and Federal authorities for taxation benefits. All countries in the region use a combination of royalties (related to production volume or value) and income taxation to extract revenue. None has a system as comprehensive as the resource rent tax proposed in chapter 3 (although Australia has recently introduced such a tax for new petroleum discoveries, and a near neighbour, Papua New Guinea, employs such a system). Instead, a range of taxes, royalties and levies affect the industry. There are export taxes (for example, Malaysian tin, Australian coal, Filipino copper, unprocessed minerals in Indonesia), and production-sharing arrangements (Indonesia, Malaysia, Thailand), State government taxes disguised as freight charges (Australia), and a range of local and municipal taxes. The Philippines has the lowest nominal rate of tax on mining, yet appears to have gained considerable domestic benefit from its copper industry. Although government royalties, taxes and so on absorbed only 36 per cent of before-tax profit (about 7 per cent of copper industry revenue) in 1978, domestic purchases, local salaries and owners' dividends resulted in a high proportion of copper income accruing to the nation rather than overseas.[3] Local ownership was as important in retaining these benefits as was the level of taxation, but the most important factor in the Philippine case appears to have been the domestic availability of other inputs to the industry, that is, upstream linkages.

From the country studies presented here it is not easy to make definitive judgments about the linkage effects, both upstream and downstream, of the minerals industry. In the Malaysian case, it is concluded that the industry has induced backward linkages in the light engineering and construction sectors, and forward linkages through the smelting of tin. The proportion of value retained was high, and externalities associated with infrastructure development, commerce and settlement were very important. Similar benefits flowed from the oil and gas discoveries, although the backward linkages were fewer owing to the sophisticated technologies required. In the Australian case, the industry has also induced backward linkages in the development of a minerals construction and equipment industry. Because of the remoteness of mine sites from population centres, the externalities have probably been less important. The backward linkages in Thailand and Indonesia have probably not been substantial, although in Indonesia's case government revenues from petroleum have helped to initiate a diversification of downstream operations. The problems of integrating the minerals industry more fully into a modern manufacturing sector are considered in the concluding section. An important issue in this area is that of downstream processing of minerals, and the effect of reliance on minerals exports on the country's terms of trade.

Minerals terms of trade and industrialisation

The exportation of mineral commodities has played an important role in the growth of the six ASEAN countries and Australia over the past twenty years. As we have seen for Indonesia, exports of petroleum have contributed a quarter of GDP, a large part of government revenue and over 80 per cent of the country's exports. For Brunei, likewise, petroleum is the key source of revenue and foreign exchange, and for Singapore the processing of petroleum has been a highly important industry. While the exploitation of these resources has provided substantial growth to each economy, the benefits of this growth have been unevenly absorbed in each case. Indonesia has not yet succeeded in making the transition from an exporter of crude petroleum to a producer of a wide range of manufactured goods and processed products at an economic scale. This transition is a difficult one, given the previous state of development, the difficulties of implementing major projects, and the competing claims on fiscal resources and administrative skills. Government policy has been to encourage the growth of import-replacing industries, suited to the domestic market rather than at export scale.

The Indonesian petroleum industry has provided the government with funds to initiate a diversification of the industrial base and, together with other mineral resources, has stimulated a number of downstream linkages. Indonesia benefited greatly from the great leap in the price of crude petroleum, but the sharp decline in prices in early 1986, following upon falling demand in 1984 and 1985, precipitated severe difficulties. Whereas government spending reached US$20.5 billion in 1981, by 1984 spending had fallen to US$16.2 billion, with a further reduction expected during 1986.[4] The revenues from oil permitted an import-competing industry to develop without strong pressure to reach world standards of efficiency. The current restraints may force rationalisation and, perhaps, increased competitive strength in the manufacturing sector. However, many of the projects planned to take advantage of petroleum wealth, such as petrochemicals and energy-intensive processing projects, have had to be shelved. In the case of petrochemicals this has not been so serious a result, given the overcapacity, and fierce competition in parts of that industry (particularly in Europe and Japan). It is unfortunate, however, that the Krakatau steel plant, which might have provided a competitive base to the manufacturing sector, has not proved as efficient and low-cost as expected.

In Singapore, twin problems have beset the petroleum refining industry. The growth of refinery capacity in other ASEAN countries (Indonesia, Malaysia and Thailand) stimulated an effort to increase efficiency and shift output towards higher value products, particularly petrochemicals. The diversification strategy to deal with the first problem—the growth in refining capacity elsewhere—led to heavy investment in petrochemicals at a time when the market was oversupplied, as noted earlier. While an improvement in demand in the industrialised countries, coupled with reductions in European and Japanese capacity, have brightened the outlook for Singapore's petrochemicals industry, it will face strong competition from the petrochemicals operations of the oil-rich countries of the Middle East, as well as Malaysia, Brunei and Indonesia, which may well be tempted to sell at prices reflecting low opportunity costs for gas feedstocks. Singapore is responding to this problem in part by emphasising entrepôt activities, such as storage, shipping, finance and bulk-breaking of petroleum, and support services for offshore petroleum exploration. The lower price of petroleum, if sustained, will no doubt severely reduce the revenue from these activities, particularly as exploration diminishes. On the other hand, cheaper petroleum will provide a stimulus to the developed economies and should increase the demand for petrochemicals and refined petroleum products. In the past Singapore's ability to counter the obvious implications of comparative advantage has rested on the purchase of crude oil at world prices, world-scale efficiency in production, and the proximity of a regional market for a large part of its output, as a result of the shortfall in the processing capacity of its neighbours. The impending shift in its environment was clearly recognised and the strategy changed. On balance, it seems likely that Singapore will have to accept a shift away from petroleum processing as a major industry, although its entrepôt activities will provide breathing space.

Malaysia, while less dependent on the minerals industry than Indonesia or Singapore, has suffered the misfortune in recent years of declining terms of trade. Like all the ASEAN countries Malaysia relies heavily on raw materials for its export trade. Although it has a diversified commodity base, including crude petroleum, palm oil, tin, rubber and timber, the prices of most of these products have fallen while import prices have risen, so that Malaysia's terms of trade have fallen by 19 per cent since 1980 (*The Economist*, 30 November 1985). The position of Thailand and the Philippines is worse, largely because of their lack of petroleum exports, and will be more difficult for Malaysia following the 1986 fall in the price of crude oil. For Malaysia, Thailand and Indonesia, the near collapse of the tin market in 1985 and the soft prices which followed were a further aggravation. Thailand's plans for a gas-based petrochemicals complex are also affected.

The minerals industry in Australia, while affected by poor prices and reduced demand in the early 1980s, benefited to some extent by the strengthening demand in the United States, Japan and the EEC in 1984 and 1985. Coal and iron ore exports increased considerably, albeit at lower prices, while nonferrous metals remained depressed.

Although the North West Shelf commenced production of gas, no other major resource projects were undertaken, despite several discoveries of good potential. Australia, like its ASEAN partners, also suffered from a decline in its terms of trade, reflecting a similar emphasis on raw materials in its export mix. The fall in the price of crude petroleum in early 1986 was partly passed on to consumers, with adverse impact on Federal taxation revenue, and some US$1.0 billion dollars of exploration investment was curtailed by the Esso–BHP partnership.

The preceding review of the minerals developments of the region in recent years demonstrates the importance of the industry to most of these economies and their resultant vulnerability to shifts in world demand. In most cases minerals are of great importance to the trade balance, with exports of hard minerals offsetting petroleum imports. While, in the more diversified Australian economy, the softness of the minerals industry in recent years has been less calamitous than in the case of Indonesia for growth, employment and tax revenue, in both cases the raw material base has not resulted in downstream industries as highly developed as comparative advantage might suggest. In Australia, the degree of linkage varies with the mineral: very low for iron ore and coal; moderate for bauxite–aluminium; and higher for lead, zinc and copper.

It is now more difficult to recover lost ground and establish world-scale, export-oriented industries on a raw material base. The world steel industry has been in serious difficulty for more than five years, petrochemicals are generally in oversupply (with some exceptions) and more capacity is coming on stream. Yet, if ASEAN and Australia are to capitalise on their resource endowments and the comparative advantage flowing therefrom, more entrepreneurship will be needed in finding markets for minerals-based products with higher value added. A sustained economic recovery, built on the foundations of lower energy costs, could perhaps provide the markets for such a development. However, refined metals and commodity petrochemicals will not provide a sufficiently long-term solution, suffering as they do from the vagaries of commodity markets and inevitable deterioration in terms of trade. A long-term solution requires deeper industrialisation and specialisation in manufactured goods with lower elasticities. The resource wealth of Indonesia and Australia could provide some of the base for such industries, but in Malaysia, Thailand and the Philippines there will be limitations to what can be erected on the minerals base. Although tin can be used to produce tinplate, which can be used for domestic canning operations, a cheap source of sheet steel is also critical. Malaysia and Thailand have steel making operations based on scrap, but these may not be competitive in the long run, being dependent on the scrap market. In any case, domestic tinplate usage will not absorb a high percentage of the tin produced in ASEAN. Other users of tin will usually wish to buy metallic tin for their own operations, rather than buy a further processed form. Malaysia has the potential to develop an efficient steel industry, having built a sponge iron plant in Sabah utilising offshore gas and iron ore imported from Brazil. An unexploited opportunity exists here for Australia to provide the iron ore requirements for this industry.

Rather similar conditions apply to the Philippines in the case of copper, even if the proposed copper refinery goes ahead. While there may be some benefits from processing minerals further, in most cases the final product which fully exploits the country's comparative advantage will probably be itself a commodity product, facing a competitive world market. This is not to argue against the development of downstream minerals-based industries, where a satisfactory rate of return can be demonstrated, but rather to say that further processing of minerals will not permit escape from declining terms of trade unless the new industries provide a base for a highly developed and technology-intensive manufacturing sector. Low-cost energy and steel, in particular, could help to provide this base, as would other minerals to a lesser degree, but they will not of themselves solve the problem.

To the extent that each of the ASEAN economies strengthens its domestic iron and steel industry, there should be growing opportunities for them to export to other developing countries in the region and beyond. There should also be opportunities for Australia to export processed iron ore in the form of pre-reduced agglomerates, depending on the market for the alternative feedstock, scrap steel. This prospect is, however, overshadowed by the existing Korean steel industry and the likely growth of a substantial industry in China. We can expect a continuing market in Australia for Indonesian crude oil and refined products, unless there are unexpected large finds of new petroleum deposits. Unless new fields are discovered, Australia faces a growing demand for imported petroleum. Both countries, as well as Brunei and Malaysia and perhaps Thailand, will find themselves in competition for markets for their LNG exports, much of which will be for Japan.

If Australia eventually developed an export-oriented steel industry (a most unlikely development in the present state of the world industry), there could also be competition for markets. In other minerals there is a limited degree of competition for markets amongst the six countries. Indonesia, Thailand and Malaysia compete in the tin market, although the restoration of the International Tin Council to full authority would place greater emphasis on cooperation. Australia may seek to establish a new tin smelter and compete more directly in the tin market. Thailand's new zinc smelter will also compete more directly with Australia in the metallic zinc market. In copper, likewise, there is some competition in the markets for concentrates between Indonesia, Australia and the Philippines, and between Australia and the Philippines for blister copper. These are the normal conditions of selling in commodity markets and have not been a cause of conflict in the past, nor should they be in the future. Minerals trade between the countries of the region has been marked by a high degree of complementarity, and competition in third markets by a relatively high degree of differentiation, with little potential for conflict. A more aggressive steel or petrochemicals industry in ASEAN could widen the field for conflict, while also increasing the possibility of trade in some of the raw materials.

This discussion suggests that, to the extent that comparative advantages in mineral endowment can be exploited by each country, there is scope for a continuing relationship of complementarity. Were Australia to proceed successfully to develop the promise of a large-scale export-oriented steel industry on its rich iron and coal resource base, there could be conflict with the aspirations of ASEAN countries to develop domestic steel industries of efficient scale. The possession of a steel industry is probably not critical to industrialisation, despite the compelling examples of Japan and South Korea. It would probably be more effective to develop domestic steel fabrication and machinery construction, based on imported steel, as a precursor to deeper industrialisation. Given the likely state of the steel market in the next decade, ASEAN countries may well conclude that their domestic steel industries should be encouraged to remain small but efficient, a goal achievable with modern steel technologies,[5] provided energy costs are competitive. For the time being, Australia also seems to be set on this course.

There seems to be a need for the steps already taken towards diversification by the ASEAN members to proceed further. The nature of each country's strategy in developing a deeper and more diverse industrial sector will depend in varying degrees on its minerals base. Minerals will continue to provide income to support these shifts to some degree, but will also remain subject to the fluctuations of the world economy. As the studies of each country suggest, minerals exploitation provides valuable resources for economic growth. There is a risk, however, in placing too much reliance on their contribution and thus deferring longer term development of the industrial structure. Government policy has an all-important role in providing the environment in which such growth can be stimulated.

Notes

1 Including the Indian sub-continent, ASEAN, Sri Lanka, Hong Kong, Taiwan, and South Korea.
2 Saudi Arabia, Kuwait, the United Arab Emirates, Israel, Iran, Bahrain, Jordan, Lebanon, Syria, Yemen.
3 These conclusions are from a study (McKern, 1981) which employed the retained-value approach to compare the contributions of the copper industries in the Philippines, Indonesia and Papua New Guinea to national economic welfare. While retained value is not a satisfactory measure of the desirability of a prospective minerals project (see chapter 3), it can be useful in comparing outcomes between projects in the same minerals sector.
4 These comments on Indonesia's problems in 1986 are derived from the *Australian Financial Review*, 27 March 1986.
5 For a thorough discussion of alternative technologies of steel making and their economics, as well as of those for copper and aluminium, see Brown and McKern (1986).

References

Brown, M. and R.B. McKern (1986) *Minerals Processing, Technology and Economics for Developing Countries* Paris: OECD (in press)

GATT (General Agreement on Tariffs and Trade) (1985) *International Trade, 1984/85* Geneva: GATT

McKern, R.B. (1981) 'Transnational Corporations in the Copper Industry of Asia and the Pacific: A Comparison of the Experience of Indonesia, Papua New Guinea and the Philippines' in *Transnational Corporations and Primary Commodity Exports from Asia and the Pacific* Bangkok: United Nations

Index